edited by Michael Balfour

The Health Food Guide

foreword by Len Deighton

Pan Original
Pan Books London and Sydney

The information given in *The Health Food Guide* was believed to be correct when it went to press, and the editor and the publisher regret that they cannot be held responsible for errors or changes of fact that may have occured. The information is also subject to the laws of copyright, and no part of this book may be reproduced without written permission from the publishers. No payment or benefit of any kind has been requested or received from any individual or organization included in this book. The editor would be most grateful to receive from readers and users of this book their experiences and opinions of organizations and establishments included or omitted; kindly write to:

The Editor, *The Health Food Guide*, c/o Pan Books Ltd, Cavaye Place, London SW10 9PG.
Thanks.

First published 1981 by Pan Books Ltd,
Cavaye Place, London SW10 9PG
© Michael Balfour 1981
ISBN 0 330 25988 1
Printed and bound in Great Britain by
Cox & Wyman Ltd, Reading

Contents

A definition of health food

Health food is defined as having been produced from or reared on soil that is unpolluted by chemical fertilizers, is free of chemical sprays, artificial stimulants to growth and additives, has not had the goodness refined out of it, and is prepared for the table with the least possible delay and loss of nutrients.

I'm afraid to say I love tinned fruit – and I hate green vegetables.

– Richard Seifert, who is said to have changed London's skyline more than any architect since Christopher Wren

Foreword by Len Deighton

Faced with the offer of a well-stocked garden or a well-stocked cellar, I'd chose a garden. Not unhesitatingly perhaps, but the decision would be inevitable. Decent wine is easily obtained and clearly and honestly labelled – simply because more of the law-makers' attention is given to alcohol than to food, a great deal of which is neither nutritious nor wholesome.

This fact is fully understood by those companies which manufacture and 'process' our daily diet. Even more fully is it understood by the men who package it so attractively and by those who concoct the advertisements. As a rough rule of thumb it seems that the more artificial colourings, flavourings and preservatives a product contains, the more likely it is to be sold by means of TV adverts showing Arcadian landscapes and bucolic countrymen filmed through those soft-focus lenses.

Worse than the 'convenience foods' for those who have little time or inclination to cook, are the deadly junk foods for people who do not want to cook at all. Last year I watched a dietician display a typical junk food meal. It consisted of (warmed-up) chicken heavily coated in batter, fried potatoes, bread, cheese, ice-cream and coffee. Alongside this meal its ingredients were displayed: a bowl of bleached flour, artificial sweeteners, chemical colourings, various preservatives and a huge slab of animal fat. Anyone about to step into a 'fast food outlet' would do better to look around for a fruit and vegetable shop. That's what I do when I'm in a hurry.

But, as this book demonstrates, the battle is not yet lost. More and more people are discovering that the best way to get good nutritious bread is to make it yourself. Vegetables are not so difficult to grow, even without chemicals, and more and more shops are interested in selling such foods.

Better eating is the result of many very different people pursuing a common interest. Cooks, cranks, dieticians, food writers, gourmets, gourmands, vegans, vegetarians and, most important of all, businessmen looking for a chance to turn a profit, are all playing a part in what will mean better food, better health and better lives for us and our children. Speaking personally, I ask only one thing – don't give me a lecture about the politics of South America every time I buy an avocado.

For Dana, with love and thanks

Editor's Note

I am delighted that Pan Books have taken over from Garnstone Press this publication of *The Health Food Guide*, which now appears in a much enlarged edition. A whole new Part One has arrived; the first chapter is mine, and distinguished contributors have written the others. I hope they offer a coherent view of health food today. Part Two is a revised and expanded version of *The A to Z of Health Food Terms* (1973) by the editors of the previous guide. Part Three is the guide to names, addresses and places; in this part, prices are generally no longer given because inflation would make them less than helpful.

I would like to thank the owners, managers and secretaries of all the shops, restaurants, health farms, organizations and societies mentioned in *The Health Food Guide* for their help in providing the information we needed to make the book as complete and helpful as possible. The marshalling of it all was mainly in the capable hands of Ruby Rae, the assistant editor, and Sara Ellis, who edited the extensive guide to Britain's health food shops; Judy Allen, the joint editor of the last edition, edited the chapter on health farms; Kate Jenney and Anjou Merchant typed the words with great patience and accuracy, and I am grateful to them.

My special thanks go to the contributors of the chapters in Part One: Miriam Polunin on nutrition and equipment; my wife Dana for her delicious recipes; Leslie Kenton on natural beauty preparations; Leah Leneman for her look at health food around the world; Keith Michell for his macrobiotic experience.

Finally I acknowledge with gratitude the help of my wife, who sustained and refined my endeavours in so many ways.

Michael Balfour

Part one

Health Food Today

1 The Last Few Years: Movements, Responses and Necessities

The need for *The Health Food Guide* exists because 75 per cent of all the food we consume annually in the UK has been processed at least once;[1]* because about 45 per cent of total food manufacturing industry expenditure is accounted for by processing and distribution costs;[2] because 10 per cent of what we spend on food in shops is for the packaging; and because about 33 per cent of the grocery trade is controlled by six companies.[3] Whilst the last edition of this book was being prepared, the desire for the information and guidance in it was widely sensed; now the needs are actually known and are being met by efforts which have to bypass the serious effects of the absence of a national and rational food policy.

The standard of living in the UK fell from 1970 to the end of 1977, due to a combination of external and internal economic problems: the national diet was affected as food purchasers were forced to change their habits. The result was that the average household consumption of recommended essential nutrient intakes fell,[4] and this was particularly true of low-income/large family groups. Moreover, during this time, expenditure on food (in real terms) declined, whilst at the same time the price of foods increased faster than all other categories in the retail price index. The result is that food as a subject is becoming ever more politically sensitive. The realization will hopefully spread into the minds of career politicians that the determining factors in eating patterns really are social and economic, and that they are the people responsible for bringing about vital changes in diets. Unfortunately the annual National Food Survey, which is always widely noticed in the media, is a survey of expenditure and not of nutrition, and, as such, fails to show up the decrease in the national expenditure on actual food (as opposed to the packaging, etc.) since the proportion of processed and manufactured foods increases (as do the risks from the chemicals involved, to which I shall come).

The nation's food requirements for the future are at present calculated on the basis solely of either expected or existing *economic* demand, or a

* See references at the end of this chapter.

combination of both: the supply of food is not linked to the factors that affect health by its consumption. It is easy to see that the food *supply* in the UK can be influenced by import controls, and by grants and subsidies to the well-lobbied processing and manufacturing industries; *demand* is affected by the changing levels of disposable income, by the food subsidies which may, and should, influence retail prices, and by the usual market practices, such as television advertising. At food policy-making levels, the health of the UK citizen (baby, school-leaver or pensioner) is, generally speaking, ignored. So true is this that known levels of intake for certain nutrients (iron, for example) have become undesirably low when compared with officially recommended levels.[5]

Short-term food subsidies are frequently used for purely political ends. One of the bargaining counters in the negotiations in 1975 between the TUC and the Labour government was the rate of increase in food prices. Foods representing some 20 per cent of consumer expenditure (bread, butter, natural cheese, flour, milk and tea) received subsidies between 1974 and 1976.[6] When the subsidies were removed, by the end of 1976, the price index for these foods for the year rose by 25 per cent against 16 per cent for all foods. It was solely the expedient and much promoted economic impact upon the lower income group's food budget which guided these political actions; a parliamentary debate of the nutritional aspects of these policies cannot be recalled. The food policy makers knew (and chose to ignore) that awareness of the importance of nutrients in diets is directly associated with income group (the lower, the less). Butter and milk together contribute about 30 per cent of total national dietary fat intake; this percentage is much too high for the national health; and yet until March 1979 imported New Zealand butter received a subsidy of 8½p per pound from the British government (and this benefit continues under an equivalent Common Agricultural Policy import levy reduction); furthermore, the massively subsidized Milk Marketing Board increased its advertising expenditure between the beginning and the end of 1976 by 100 per cent (in the event there was a 2.3 per cent fall in liquid milk consumption in those two years; the board stated that their additional expense 'helped to prevent more severe damage to the market'.[7]) A leaflet urging a greater consumption of fruit and vegetables was planned by the Fresh Fruit and Vegetable Information Bureau; the British Farm Produce Council refused to allow it, on the grounds that it might adversely affect the national consumption of butter.[8] Movements for the reintroduction of school milk can be seen only as an attempt to assist with the disposal of the EEC's 'milk lake'. The Medical Research Council announced not long ago that older people might delay the natural ageing process by lowering their intake of protein; this was just at the time when a special bonus was announced for pensioners who increased their consumption of meat and veal beyond a certain quantity! My attention has been drawn by the Farm and Food Society to two sharply contrasting and yet authoritative statements, which serve to summarize the running farce in which the general public is the bewildered victim:

It is reasonable to accept the strong consensus of opinion on diet in the reports of the eighteen national committees. They are: to reduce the total fat intake of 30 – 35 per cent of the energy [diet], to restrict consumption of saturated fat, cholesterol, sugar and salt, to increase unrefined carbohydrate and polyunsaturated fat [*Dr R. W. D. Turner MD FRCP FRCPE*].[9]

Mr Gundelach [EEC Commissioner for Agriculture] made it clear that lower prices alone were not enough to reduce production and surpluses. In the dairy sector, for example, where stocks currently stand at 400,000 tonnes for butter and 665,000 tonnes for milk, the Commission is planning to continue charging a levy on surplus milk which will be used in a massive campaign to encourage consumers to drink more milk and spread their butter more thickly [*Euroform*].[10]

The second of these statements is purely political; for no data exists with which to calculate accurately the impact of such policies among different income groups, either within the EEC or in the UK. We must recognize only that UK governmental considerations of pay and price restraints consistently ignore the realities of essential nutrient intake in a healthy diet. The fact that the established health food or wholefood production and distribution industries will shortly have unusual challenges to meet should surprise no one; the Farm and Food Society's demand in August 1979 for a minister of nutrition was timely.[11]

The authors of the valuable report *National Food Policy in the UK*, published in November 1979,[12] write as follows:

Significant changes in attitudes both to the procurement of national food supplies and to the nutritional and health status of the population as affected by the consumption of food are emerging in the UK. These changes appear to be based on concern with the inflationary effect on the level of food prices and a growing awareness of nutrition and health among the concerned professions and among consumers. So far, these concerns are not well coordinated or interrelated, and in many cases are neither well articulated nor generally held.

I agree with these observations, but add that I suspect the editors of the report were perhaps not aware of the tremendous growth in the last few years of alternative means of production (by pure and humane methods) and distribution of essential and nutritious foods, on a fair profit or break-even basis, and for the spiritual as well as economic benefit of those involved in the enterprises. The emergence of wholesale and retail cooperatives, collectives and buying groups all over the country is a natural response to a natural need, and it binds into the health food or wholefood business a political ingredient which is necessary, has become urgent, and will become fashionable.

It might be said with some justification that the next edition of this book is in fact going to have to be divided into two separate volumes. A Health Food Guide (I will come to a Wholefood Guide presently) will concentrate upon faddist demands for remedies, prevention and patented cures, with their attendant profit-orientated shops and organizations. The nationwide Holland & Barrett chain of shops (a large proportion of the total listed in the present edition) perhaps exemplifies this prognostication. They are of modern design and layout, with speedy stock turnovers of a wide range of manufactured goods (some of which other companies in the Booker McConnell group produce) and are a separate profit centre of a huge food-orientated financial conglomerate. Though each has less than one-tenth of the number of their stores, Holland & Barrett are being challenged by at least three small chains of stores; Country Life Wholefoods in the South; Hillstart (Happy Nut Houses, etc.) in the Midlands and North; and Goodness Foods in the Northamptonshire area.

Franchising has arrived in the health food trade since the last edition of this guide, in the shape of Nature's Way shops. There are several large wholesale organizations now operating in different parts of the country (Suma, Infinity, etc.), and several of them will supply buying groups which have been informally assembled by members of the public on a cash-and-carry basis if the order amounts to over £40. To provide for people who feel that health foods and remedies make sense but don't know why, there has also been an explosive increase in the number of books published; the two largest publishers of books in this sector are Thorsons and Rodale, whose catalogues are available on request.[13] Since the mid 1970s the press and television have reflected the ever increasing interest in the generalities of health food, and one of the predictable yet pleasing results has been the arrival of a great number of new health food shops run by the newly converted. They tend to have exotic names: Funny Foods (Ossett), Beans 'N' Herbs (Cardiff), Bonkers (Princes Risborough), Healthy, Wealthy and Wise (London W1), Nut 'n' Meg Wholefoods (Manchester), Eggstasy Wholefoods (Spalding), What Comes Naturally (Hull), On the Eighth Day Cooperative (Manchester), Nitty Gritty Grain Store (Leominster), Truefoods (Felixstowe, Goodness Gracious (several shops), Honeysuckle (Oswestry), and many more which are accounted for in Part Three.

The buyer of this book and the occasional customer of a health food shop (as opposed to the committed and constant customer of what I will be terming a wholefood shop later in this chapter) is probably bothered and uneasy about the possible effects upon him/her of several aspects of the food processing industry today. It is instructive as well as alarming to examine briefly a few of the important subjects about which this 'average' customer is probably concerned, because when added together they surely contribute to some sickening statistics, revealing that the quality of life is very often not what it should be. One such statistic is that in 1967 sixteen million tranquillizers were prescribed in Britain; in 1977 the figure reached twenty-five million.[14]

Every doctor is familiar with the fact that changes in daily family routines disturb balances in metabolisms. Reexamining the daily or monthly food eating routines in our own lives soon reveals the shifting of some behaviour patterns: our eating habits are being Americanized – we are now in an era of convenience/junk/fast food with all its frightening implications. The day of the druidburger on sale at Stonehenge cannot be far away.

Economics have everything to do with these shifting patterns. Compared with ten years ago, Britain now has fewer of the vital *Fs*: farmers, farmworkers, fishermen, food companies and food shops. As I have already mentioned, about 33 per cent of our grocery trade is now in the hands of the six largest supermarket companies; as margins on foods in the other shops are remorselessly squeezed by these companies, the smaller shops are forced to look to non-food items for survival. A combination of social and economic circumstances has, at the same time, seen to it that we spend less time preparing meals and less time eating them (and we rarely meet for tea at all!). The cafés have gone; the gleaming counters of the burger houses and take-aways have arrived. Over the past decade it has been made more convenient for us to eat, and probably cheaper – but the counting of the *real* cost has yet to come.

The hapless chicken now features prominently in our diet – and perhaps it is the terrible story of this bird which is creating customers for the health food trade. There was an eightfold increase in the consumption of chickens between 1957 and 1978,[15] and the number of chickens expected to be produced by factory farming methods in the UK in 1979 was 360,000,000.[16] The only daylight the birds ever saw was from the transporters taking them to the slaughterhouses at about seven weeks old. Their brief, fish-meal-fed lives were led on battery farms in cubicles which, according to the 1965 Bramble Committee's government report, should allow for one square foot per bird. Often the stocking density is down to one-half of a square foot per bird, 'to prevent pecking-order-type stresses', according to a television programme on factory farming.[17] In farmer Derek Humphrey's chicken production facilities, artificial light regulators are used for all-year-round egg laying; this way he gets 2,000,000 eggs per week. Unfortunately for him he loses 12,000 chickens a year, due perhaps partly to the fact that, according to the programme, some birds are confined five at once in a cage five square inches in size. We should not be surprised that 4,000 die annually through cannibalism – although of course this, it was said, can be cured by removing part of the chicken's beak. And debeaking would save a little space too, wouldn't it? Miss Brigid Brophy has said, 'Intensive farming is simply a synonym for a concentration camp;[18] Mr Nevile Wallace, director-general of the British Poultry Federation, said in reply that 'members were always ready to consider better methods'.[19]

The ghastly tale of the chicken's journey (perhaps to the Sunday lunch table, where five million households are now accustomed to seeing it regularly)[20] is by no means over. Nearly one-half of all UK frozen chickens receive injections of polyphosphate solution so that they retain their moisture and acquire improved flavour and texture – and the government's Food Standards Committee is worried about it. Whilst we, the customers, await the committee's report on hazards from microbiological aspects of food, we receive cold comfort from buying frozen water at chicken prices. The addresses and details of the Free Range Egg Association and the Vegetarian Society are given in Part Three.

The scandals of the factory farming of chickens, pigs and veal are now out in the open, but it is as well to remember that it is now seventeen years since the publication of Ruth Harrison's famous book *Animal Machines*. I agree with David Williams, deputy chief veterinary officer for the RSPCA, when he says, 'The price of cheap protein is becoming too great for the cruelty suffered by these animals. It is a time for very drastic measures.'[21] The health food faddist is bound to be disturbed when he hears of the 25,000 calves turned into veal on one British farm each year in the UK, after they have spent four months in cubicles only $2\frac{1}{2}$ feet wide (the Bramble Committee said they should be a minimum of $3\frac{1}{2}$ feet wide), being fed on skimmed milk with ingredients added to substitute for sun and grass. And he won't be happy with the fact that over 250,000 sows spend their pregnancy tied down in their pens, without light: this is illegal for dogs, who are no more intelligent. The drastic measures that David Williams, Compassion in World Farming, the National Society against Factory Farming, and every sensitive being are calling for have to fight fierce parliamentary lobbying by an industry that turns over more than £700 million per year. Moreover, the law concerning battery farming is so vague in the UK that no prosecutions have been brought during the last five years;[22] campaigns are in progress here, as well as in Germany, Holland and Switzerland, to phase out the battery systems of farming. Two interesting questions among many for discussion are: do animals have rights? And, are there debasing effects on the operators/farmers of such systems?

The only real beneficiaries of factory farming methods are the investors in profitable convenience/fast/junk food (which I will summarize under the term junk food) enterprises. Junk equipment worth £121 million was sold in 1977; by 1983 the turnover is forecast as £283 million.[23] A nutrition expert has claimed:

Nutritionally, fast foods are not as unsound as a lot of people imagine. A hamburger and chips a couple of times a week isn't going to do you any harm, if you're eating a normal balanced diet the rest of the time. It's how your diet balances out over a week that counts, not what you eat at one meal.[24]

The trouble with this argument lies with the 'normal balanced diet the rest of the time', as against the realities of all those sales in all those supermarkets of all that junk food. Food manufacturers are food modifiers, and they have unusual victims. In 1979, Dan White, a fanatical anti-homosexual living in San Francisco (where a large percentage of the population is gay), murdered Mayor George Moscone and Harvey Milk, two of his City Hall colleagues who were homosexual; his courtroom defence was that the crimes were committed in a moment of diminished mental capacity, which was 'induced largely by excessive consumption of junk food'.[25] Changing eating habits in America eventually arrive here, so non-health-fooders in the UK should note that, at present, America is 1.444 billion pounds overweight[26] – some convenience! Some junk food!

Attached to the horrors of the modifying of food in the UK is the final indignity: the way it is now most often cooked in public eating places. It is 'cooked' in a microwave oven. Microwaves are a form of electromagnetic energy; they have a wavelength between those of visible light and medium-wave radio. These microwaves do not have enough energy to break chemical bonds – unlike ionizing radiation (x-rays, nuclear radiation, etc.). At one time the effects of microwaves were assumed to be limited to heating skin tissues – and so it was that the safe levels for exposure in both America and Britain were set at no less than one thousand times higher than in the Soviet Union, where only 0.01 milliwatts per square centimetre is the limit for occupational exposure. It is the central nervous system that is held to be at risk by the Soviets.

The microwave scare in this country started with the use of early radar; it was found that birds colliding with radar masts were cooked by their beams. We now know more: the patient receiving blood heated by microwaves will die; microwaves can penetrate four inches into human flesh and inflict internal burns which are not sensed by skin nerves; they heat by vibrating molecules and causing friction between them; they can affect the lens of the human eye because it has no blood supply to carry away excess heat, and so the membranes lose their elasticity; they can stop frogs' hearts; they may interfere with scarcely understood electrical properties of the body and the brain (possibly a kind of radio receiver, so some eastern bloc scientists believe), and so affect growth and reproduction systems. To quote Professor Edward Grant of Queen Elizabeth College, London, 'Only a few milliwatts of microwaves were found capable of altering the amount of calcium entering and leaving the brain tissue. Experiments like this show that we ought to be very cautious about the effects of microwaves'.[27] Professor Abraham Lilienfeld of John Hopkins University, Baltimore, knows this; in the 1960s he discovered a link between radar workers and mongolism in their children.

So far, pressure by public health authorities to improve microwave exposure and leakage standards is virtually unknown in Britain. Paul Brodeur is one of America's leading microwave experts and the author of *The Zapping of America*; he says, 'If microwaves were a food additive, a chemical or a drug, they would be banned.' This should be remembered by potential buyers of apparently convenient microwave ovens – and some 20 per cent of UK households will have them within ten years according to Merrychef Ltd,[28] unless pressure groups do their work effectively.

The greatest of all physicians, Hippocrates, used to urge upon the citizens of Athens that it was essential that they should pass large bulky motions after every meal and that to ensure this they had to eat abundantly of wholemeal bread, vegetables and fruit [*Sir Arbuthnot Lane*].[29]

If it is possible to condense 3,000 years of history into a few lines of print, the best way to do so is perhaps to consider the various groups who have pressed for bread to be brown or white [*R. A. McCance and E. M. Widdowson*].[30]

The recent trial and conviction of sliced white bread in Britain is only the most recent event in the long history of 'the staff of life' to cause controversy, but it has at last removed for ever the old notion that white bread is posher. The First World War started the movement. The *Baker and Confectioner* for 14 December 1923:

Now of all the minor evils of the Great War, the loaf of this period may fairly claim front rank. It is more depressing than the plague of darkness and infinitely more dangerous than the air raids. One could hope to escape being hit by German bombs, but who could dodge the bulletproof crust and sour soggy interior of a war loaf?

A Hull journal of the time had this confident headline: 'White bread's best, brown won't digest.' On the other hand, the *Daily Mail* screamed: 'The whiter your bread, the sooner you're dead.' Bread is no longer the single most important element in the national diet and therefore no longer claims the constant attention of politicians; the Luddite, Chartist and Anti-Corn Law movements all originated during times of rising bread prices. Sliced white bread is the victim of common health sense, to the great concern of the two great companies (Rank Hovis McDougall and Associated British Foods) who supply 70 per cent of the country's factory-made bread. Not only is there a great swing towards wholemeal bread and home baking, but in fact the average home consumption of bread is dropping quite fast. In 1966 each of us ate 51.08 ounces per week; in 1978 we ate 32.13 ounces.[31] The Chorleywood baking process is a mechanical/chemical process which makes factory bread quicker and cheaper than we can at home, and 70–75 per cent of us still buy it. What needs to be discovered is whether, as I suspect, the lower income groups are the main purchasers; I discussed the political implications of this question earlier in this chapter. Sliced white has fallen from favour because the fact is becoming ever more widely recognized that wholemeal bread provides extra vitamins, dietary fibre and (in the wheat germ) essential fatty acids. Bran is the fibrous shell surrounding the wheat grain and is now seen as a vital combatant of some of the diseases of civilization; a herd instinct is among us that we need more cereal fibre, and it is fascinating to learn that more than one of the giant supermarket groups are planning to cater for this demand by installing their own branch bakeries. The smell of baking bread would be an enterprising loss leader.

Health food hunters should beware of 'wheatmeal' bread, incidentally, which might contain up to 20 per cent of white flour; in some 'wheatmeal' loaves the brownness of the flour is due to the original flour being refined and whitened, and then dyed brown again with caramel colouring.

'Bread, peace and land' was the cry of the Russian revolutionaries; our own standard-bearer in the matter of better bread is Mr A. Mackie, former director-general of the Health Education Council. If the trial of sliced white bread really started with the bread strike in December 1977, Mr Mackie had warmed to his correct task of public education in a speech prepared for delivery in the autumn of 1979 to a nutrition conference organized by the Flour Advisory Bureau (sponsors? guess who?); in fact it objected to his text; he refused to change it, and the speech was never delivered. Mr Mackie's text attacked the 'just add water and heat' food manufacturers: 'It is they who have created generations of little fatties, whose daily waddle between bed, school and the telly is interspersed only rarely with visits to the dentist for the dentures that will enable them to resume their munching to the greater glory of the gross national product.' Professor Brian Spencer of the Flour Milling and Baking Research Association, representing the plant-baked bread industries' development efforts, has handed out a typical response on behalf of sliced white – its arguments are of course economic: 'Plant-baked bread makes vast savings; it costs the country £80 million a year in subsidies to the EEC to import the hard North American wheats used in the traditional loaf. The Chorleywood bread process uses more soft European wheats, and saves the government millions.'[32] He is speaking on behalf of a lot of dough – the plant-baked bread industry has an annual turnover of more than £500 million. This is probably more than ten times the business of all health food merchandisers. The address of the Flour Advisory Bureau is 21 Arlington Street, London SW1A 1RN; in February 1980 they issued a booklet called *Bread in Britain*. It is fascinating reading.

One section of the public that has long been involved with so many of the subjects I have discussed so far is the vegetarian population. Whether the committed vegetarian is born or converted I don't know (although a survey has shown that a lot more of them are left-handed than the rest of us), but the movement is strong and growing everywhere. A recent poll in America found that there are about 1.2 million strict vegetarians there (and about 6 million more are 'fish and chicken vegetarians').[33] Of the strict ones, 56 per cent became so for health reasons, and 16 per cent for ethical reasons. A vegetarian society in Holland has 30,000 members and thirteen vegetarian restaurants (there were only three in 1974); even in Paris there are some twenty-five such restaurants. The Vegetarian Society in the UK (for details see Part Three) has about 7,500 members, and its calls are becoming more strident – we are eating 'innocent' animals. We are certainly spending about a third of our food budget on meat; in fact, during each of our lives we get through about 10 cattle, 29 pigs, 530 poultry and 38 sheep. The Vegetarian Society has called for a weekly V-day, and this well-organized education and research charity is getting its message widely heard. Gone are the prewar days of its association with Esperanto, naturism and pacificism.

One reason for this is the clear meaning of the word *vegetarian*. A recent report by the Federal Trade Commission in Washington proposes a ban on the expression *health food*, and neither does it care for the words *natural* and *organic*. In response, the US National Consumer's League welcomed the proposals: Most of us equate '*natural* with goodness, and companies will frequently use the word *natural* as an excuse to raise prices, when the food is not markedly different from earlier versions.[34] Britain's Consumers Association's magazine *Which?* has expressed a similar opinion:[35] 'It [the term *health food*] should be legally defined, or, better still, banned . . . You can get a good, healthy diet without ever visiting a "health food shop".' *Which?* pointed out that 'Alfonal, probably the most common soft margarine, has just the same list of ingredients as most other brands of soft margarine – including emulsifiers, flavouring and colour.' Alfonal margarine is produced by an associate company of Holland & Barrett.

In the 1980s the committed health food addict will prefer the expression *wholefood* – it stands for pure and nutritious food, purchased away from the high street at fair prices, and much more besides. Under the heading of *wholefood* a bandwagon is gathering speed fast, and it is easy to understand why.

Wholefooders don't want to know about the six thousand million tins of food used in Britain each year; they are interested only in pure, unrefined food, from which no more is removed from the original source than is necessary to make it edible; food which is unprocessed, and free from any of the two thousand or so approved chemical additives and preservatives – food, in fact, which is in accordance with the definition of health food given at the beginning of this book. The two hundred or so wholefood shops, cafés and groups in this country have a quite different character and appeal: the word *natural* is both the obvious and the correct adjective to describe the goods in stock *and* the people who serve you there. Knowledge and friendliness abound; food labelling is honest as well as helpful; prices are fair, and elastic if you want to buy in bulk (lowering your food bill by up to 30 per cent sometimes); and the choice of the foods is based on food values rather than money values. You will generally find the following in a good wholefood shop: wholewheat bread and flour, grains, pure honey, molasses, pulses (beans and peas), organically grown vegetables, fruits (mainly dried) and yoghurt. You will also probably find information on local and community activities and perhaps modest facilities for holding meetings. The really interesting thing about such a shop is that it is probably a cooperative or collective.

This is a very important development with, I believe, considerable implications for the future of what I have been calling the health food trade, and also for the food processing (modifying), distribution and retailing industries in the UK. It means that politics has entered the food business from the proper end – the marketplace. The cooperative/collective (co/co for short) movement is well established, expanding very fast with its alternative methods, and it is very serious about its intentions. The highly effective Friends of the Earth have spent years acting as the so-called alternative lobbyists, and how right they have been in so many of their campaigns. The co/co movement has taken the problems it sees into its own hands and is solving them very effectively. As I said earlier in this chapter, the food giants must beware, and national food policy makers must investigate, listen and learn from the co/co movement. Already, for instance, there is one wholesale co/co in the North of England (where the movement is particularly strong) which turns over in excess of £1 million a year of wholefoods – and only twelve people are involved.

With the kind permission of the Down to Earth Wholefood Collective, Sheffield (of whom further details can be found in Part Three), I reprint below the leaflet they make available to explain themselves.

What we sell

Down to Earth sells basic foodstuffs which have undergone the minimum of processing, preserving or artificial enrichment, and which retain as much as possible of the nutrient content found in their natural state. We sell, where available, produce grown under organic conditions, free from artificial fertilizers or chemical sprays. We sell only vegetarian foodstuffs, believing that the consumption of animal products can be grossly inefficient use of the Earth's natural resources. We do not sell 'health' foods.

We sell as cheaply as we can, passing on all price reductions to the customer. Packaging is kept to a minimum, and recycled materials are used where possible. We buy from other cooperatives and collectives where we can, avoiding reliance on large distribution chains or capitalistic enterprises.

Our structure

Down to Earth is a registered cooperative, with limited liability, and as such is owned equally by its members. It is non-profit-making, and any surplus remaining after overheads are met is distributed to other collectives, campaigns and local community projects.

We are associated members of the Industrial Common Ownership Movement, and attempt to promote the growth and development of small cooperative enterprises run on collective lines. If you are interested in forming a cooperative we would be willing to give any advice and assistance that we can. We would also like to encourage food buying coops in the area. These are groups of people who buy food together in bulk, not only to benefit from bulk-discounts, but also to enjoy the spirit of cooperative work. We offer a coops discount to these groups, as well as our normal bulk reduction. We would be glad to assist and advise anyone interested in forming such a group. (Note: it is our normal policy not to offer any price reduction on large quantities purchased, because this discriminates against people who can only afford to buy small amounts. However, if you purchase an entire sack or box of something, then a reduction is available.)

Our politics

We attempt not to support oppressive regimes (by buying goods from them) but, while no alternatives exist, we sell goods from countries like Argentina, believing that acting in isolation would achieve little, and having the goods in the shop provides a focus for information on those countries. We boycott totally goods from Chile and South Africa, since there already is a mass boycott of their products called by the people of those countries. We cannot be completely aware of conditions of production in countries where our goods come from, but we attempt to find out as fully as possible.

We aim to give out information to our customers about food issues: for example obesity and commercial exploitation. We see ourselves as providing a focus for local activity; we advertise local events in the shop window, and also run campaigns in the shop about issues on which we feel strongly, for example, vivisection, nuclear power, public service cuts.

The collective

We are members of the Federation of Northern Wholefood Collectives, which is an association of shops like Down to Earth and other collective enterprises and communities involved in the production and distribution of wholefoods in the North of England.

Being a collective means that all of the workers have equal say in decisions, and equal responsibility. We participate in all aspects of running the shop, from serving customers, packing the food in the back room to doing the book-keeping and ordering. (This is one reason why the shop sometimes seems crowded and why queues form; we have no separate office or packing room in which to do the work. If we did not do these jobs during opening hours the shelves would soon be empty.) We have no bosses, and all our knowledge and skills are shared between the workers. We see part of our job as teaching people who are not good at figures, for example, to do the accounts. We are attempting to break down sexual stereotypes, and do not distinguish between 'men's' and 'women's' jobs. Our wages are currently £33.00 per week (after deductions) for a single person, plus whatever rent we pay. We are trying to pay ourselves according to what we need. Members of the collective with children, for example, receive proportionately more money.

Please do not hesitate to ask any of us about other things you would like to know. We would also welcome suggestions and criticisms about the shop.

And there are sure to be goods we sell which you are not familiar with or don't know how to use and cook – we will be pleased to try to help you with anything you want to know.

This trading profile is repeated all over the country, and the important point is that co/co enterprises are not to be equated with hot flushes of enthusiasm for a new technique, fashion or trick. That this is so has been helped, as the Down to Earth leaflet mentions, by the Industrial Common Ownership Movement, which is a promoting and coordinating body. ICOM Ltd publishes its own model rules which enable cheap and easy registration for groups of people wishing to start a cooperative enterprise, and through its financial arm (Industrial Common Ownership Finance Ltd) can provide some seed capital for promising ventures. ICOF has so far received £80,000 from the Department of Industry fund which was established under the Industrial Common Ownership Act of 1976; the Scott Bader Commonwealth Development Fund is also a main contributor of funds. ICOM states its aims as follows:

Common ownership is an ancient principle. We are applying it in new ways to produce what is in effect a new form of ownership. Common ownerships are not just successful enterprises. They must have social as well as economic objectives. Our aim is to achieve the ownership and control of an enterprise by those who work in it. We seek to establish common ownerships in order to improve quality of life at work through transforming working relationships.

A co/co differs from nineteenth-century producer coops in two important respects. First, if it dissolves, the residual assets go, not to members, but to other co/cos or to charity; this ensures that the means of production are, in the words of the late Fritz Schumacher (once a director of an ICOM firm), 'effectually neutralized'. Second, only people employed in the co/co can hold shares in it, and only one share each; this helps ensure that labour employs capital and not the other way around.

The address of the Industrial Common Ownership Movement is Beechwood College, Elmete Lane, Roundhay, Leeds LS8 2LQ (telephone: 0532 651235). ICOM now has well over 200 co/cos registered, and it is worth emphasizing that it is not associated with any political party.

Another group of people who have discovered a need, and gone on to provide for it, are the initiators of the Wholefood School of Nutrition. This is a national charity and an ICOM co/co which was established about five years ago to teach basic wholefood cookery and nutrition around the country. It trains teachers to hold lessons and courses in their own areas, under the aegis of the WSN, and it publishes a bi-monthly magazine, *Wholefood*. The address of the Wholefood School of Nutrition is 54–7 Allison Street, Digbeth, Birmingham 5 (telephone: 021 632 6909). Lessons in wholefood and nutrition can also be enjoyed at the Community Health Foundation (East West Centre); see Part Three for details.

The Henley Centre for Forecasting, which looks ahead on behalf of industry, has recently published its view of the next decade.[36] Robert Tyrell, the chief analyst, foresees that social class differences will remain a feature of the British way of life; at the same time he sees the arrival of 'polyocracies' – groups which defy traditional social and economic classifications and whose memberships find common cause around a single issue. I hope I have explained how this is true for food already. Robert Tyrell also foresees a biomedical revolution which will affect mental and physical welfare, and not always for the best.

Again, I am confident he will be proved right, and there is already in the early 1980s a wide sense of this. At the National Vegetable Research Station, Wellesbourne, Warwickshire, a £300,000 world seed bank has been set up, to save 12,000 varieties of vegetables which will otherwise become extinct. At the same place, vaccinations for vegetables have been developed by government scientists; three-week-old sprouts and cauliflower plants can now be innoculated; the liquid evolved to fight mosaic virus includes abrasive carborundum to penetrate the plant skins.[37] It doesn't sound right. Nor do the experiments being carried out at the Nikolai Vavilov Plant Breeding Institute in Leningrad to show that ultrasound processing can help ensure successful cultivation of cereals in extreme conditions. I prefer wholefood, produced in accordance with the definition at the beginning of *The Health Food Guide*.

Michael Balfour

References

1 *National Food Policy in the UK*, Report 5, Centre for Agricultural Strategy, Reading, 1979
2 Marsh, J. S., *UK Agricultural Policy within the European Community*, Report 1, Centre for Agricultural Strategy, Reading, 1977
3 *Social Trends* 10, HMSO, 1979
4 Ministry of Agriculture, Fisheries and Food HMSO, 1977
5 ibid.
6 ibid.
7 ibid.
8 *Farming for a Future*, Farm and Food Society, 1979
9 'Perspectives in Coronary Prevention' opening address to a symposium, *The Prevention of Coronary Disease Starts in Childhood*, Central Middlesex Hospital, London, June–July 1977; Dr Turner is Senior Research Fellow in Preventative Cardiology at Edinburgh University and Chairman of the recently formed Coronary Prevention Group
10 February 1979
11 Milk Marketing Board, Annual Report, 1977
12 Centre forAgricultural Strategy, Reading
13 Thorsons Publishers Ltd, Denington Estate, Wellingborough, Northamptonshire NN8 2RQ; Rodale Press Inc., Chestnut Close, Potten End, Berkhamsted, Hertfordshire HP4 2QL
14 *Social Trends* 10, HMSO, 1979
15 ibid.
16 Colin Wilson, marketing director, Suffolk Sovereign Roasting Chickens, quoted in *Daily Telegraph*, 4 July 1979
17 'Down on the Factory Farm' BBC *World About Us*, 24 June 1979; written and produced by Tony Edwards
18 News report by Hugh Clayton, *The Times*, 12 December 1979
19 Report by Simon Kinnersley, *Daily Mail*, 26 June 1979
20 Colin Wilson, *Daily Telegraph*, 4 July 1979
21 Simon Kinnersley, *Daily Mail*, 26 June 1979
22 Letter from Peter Roberts, general secretary, Compassion in World Farming, *Country Life*, 18 October 1979
23 *Financial Times*, 19 June 1979
24 Elisabeth Morse, senior scientific officer, British Nutrition Foundation, *Daily Mail*, 10 November 1978
25 Report by William Scobie, *Observer*, 9 December 1979
26 Review of *Getting Into Shape*, *Daily Mail*, 30 January 1980
27 Report by Martin Weitz, *Telegraph Sunday Magazine*
28 Report by Pamela Judge, *Financial Times*, 23 March 1978
29 *The Prevention of Diseases Peculiar To Civilization*, London, 1929

30 *Breads White and Brown: Their Place in Thought and Social History,* London, 1956

31 *Sunday Times Magazine,* 20 January 1980

32 *Sunday Times Magazine,* 27 January 1980

33 Report by Jane M. Friedman, *International Herald Tribune,* 28 August 1978

34 Report by Larry Kramer, *International Herald Tribune,* 1 December 1978

35 June 1978

36 Robert Tyrell, *Planning Consumer Markets,* Henley Centre for Forecasting, February 1980

37 Report by David Brown, *Sunday Telegraph,* 17 December 1978

2 Cooking: Nutrition and Equipment

'But health foods are so *expensive*!' That's one of the commonest cries of those who don't eat them. But is it their reason for not doing so – or is it an excuse?

There are two separate questions to be asked: firstly, expensive compared with what? How extravagant something is depends on how important it is to you. Some people in affluent countries spend remarkably little on food – and look for cheap prices above all. They don't feel a need to bother about the food value of what they eat. But experiences of the last fifty years have shown that having plenty to eat is not the same as being well nourished. In fact, we are seeing increases in several diseases directly related to a poor choice of foods: tooth decay, diverticular disease, gallstones, heart disease, maturity-onset, diabetes, overweight, and others. Food has a direct effect on the whole experience of life – how you feel every day, immediate and the long-term achievements, your health and looks now and in later life. 'Economy above all' might have seemed a reasonable principle once – but now the big effect of food quality has been recognized, so the second question is: does eating healthily cost a lot more than eating unhealthily? What *is* the best possible fuel for human beings?

We know that we need the following food substances: proteins, fats, carbohydrates, fibre (or roughage), minerals and vitamins. The first three, together with alcohol, supply calories. (A calorie is a measurement of the amount of energy released when food is digested.) Fibre is really part of the carbohydrate family, but it does not supply calories, as the body does not break it down. It is essential, however, as it stimulates the intestinal muscles and prevents constipation. Exactly how much is needed of each group varies from person to person. But there are tables of average requirements which show that humans need between 2,000 and 3,500 calories a day, the lower requirement being that of a lightweight woman living a sedentary life, and the higher that of a large-framed man taking heavy physical exercise. Teenagers of both sexes need most calories of all – but again, one person will have a much bigger appetite than another. As people grow older they need fewer calories per day.

Proteins

About ten per cent of our calories need to come from protein; that is, 200 to 350 calories per day. Approximately four calories are produced by every gram of protein, so we need between 50 and 80 grams of protein per day. Teenagers need more than anyone else, even than large adult men. There are not any wholly protein foods. Natural foods are never more than 40 per cent protein – the rest of their weight being other elements such as fat, carbohydrate, fibre or water. The most concentrated protein foods are not really natural foods, but those with the water removed – like brewer's yeast powder or skim milk powder. Both are about 40 per cent protein. After that are foods like peanuts (about 25 per cent protein), wheat germ (about 25 per cent), meat and fish (15–20 per cent) and cheese (the hard variety, about 25 per cent). But these foods are far from being the only good sources of protein, for there are significant amounts in the most basic of our foods: grains (about 10 per cent protein), and all the foods – bread, biscuits, even spaghetti – made from them.

Only recently have grains and other vegetable protein sources like nuts, beans and pulses in general become more widely appreciated for their protein content. Although on their own, they do not supply the whole range of proteins which the body needs, they can be combined to do so and are a useful and normally cheap supplement to the more traditionally recognized high-protein foods from animals.

For every hundred grams of cereal we eat, we get at least ten grams of pure protein – and this is why .

Cereals contribute between a quarter and a third of the average person's daily protein. Too much protein is of no benefit to the body; it is turned into energy calories, and in the absence of energetic activity surplus calories are stored as fat. So a healthy diet should not overplay the proteins – 55 to 65 grams a day is plenty.

Fats

About forty-two per cent of an average person's calories come from fat – an immense proportional increase over the past 200 years. All studies of what a healthy diet should contain agree it is too much. We all need some fat, but probably only about ten per cent of our calorie intake should come from it. This is extremely difficult to achieve, as fat is present in most foods, from milk and nuts to the leanest of meat. Most people could cut a vast amount of fat out of their diet if they kept away from just a few of super-fatty foods such as chocolate, pastry, rich biscuits and cakes, mayonnaise, shop-bought pâtés, cream and sausages. A maximum intake of two to three ounces of fat a day is plenty. The most useful fats are those in oily fish – which have lots of vitamins A and D – those rich in polyunsaturates – which are the runny fats, like safflower, corn, soya and peanut oils – and also various other foods – seeds, avocado pears, wheat germ and grains.

Carbohydrates

A gram of carbohydrate produces just under four calories, and a gram of protein produces four calories. So ounce for ounce, they are the same in energy-producing or, in excess, fattening terms. A lettuce, celery or grapefruit is almost entirely free of carbohydrate. What about bread? A really big slice of bread – 50 grams – has the same amount of calories as there are in a rather meagre 25 grams of Cheddar cheese. So much for bread being fattening!

The process of refining takes out of the original food the elements that food manufacturers do not want – but which are often the ones the body needs most: fibre, vitamins and minerals. And refining also makes food more concentrated in calories, by the removal of water along with the fibre, enabling more calories to fit into less weight. Take sugar as an example. Sugar cane and sugar beet are natural foods – complete with vitamins, minerals and other nutrients. Their high fibre- and water-levels make them very filling, so in practical terms they are not particularly fattening. But sugar refined from them has virtually no water and absolutely no fibre or any other useful nutrient. It has got energy value as it is virtually *pure* carbohydrate. The unrefined carbohydrates – the ones in fruit, vegetables, brown rice, wholemeal bread and other whole grains – are the foods which should supply most of our calories (about two-thirds of them or more) in a well-balanced diet, and contain our main source of fibre, minerals and vitamins. Make all the carbohydrates unrefined ones, because they supply lots of nutrients per calorie, but will not make you fat if eaten in reasonable quantities.

Fibre, minerals, vitamins

In spite of increased knowledge of these essential ingredients in our food, nobody knows what the optimum intake of them is, compared with the minimum needed to avoid getting scurvy, for instance. For the person who eats a healthy balance of food, it's an unimportant, if interesting, question. If you eat the correct balance of the right kinds of protein, fat and carbohydrate, your intake of minerals, fibre and vitamins should be sufficient.

Healthy meal patterns

There is nothing wrong with eating breakfast cereal for supper or lunch, or with skipping meals, or never eating cheese or milk, if you don't really like them. A sample day might look like this (quantities should be according to appetite and waistline):

Breakfast
Fresh fruit with yoghurt – no sugar **or** wholegrain cereals, porridge, wheat germ or similar with milk (or water, fruit juice or yoghurt) and fresh fruit – no sugar. If cereals are uneatable without some sweetening, moisten them with unsweetened fruit juice, or soak some dried fruit in the milk overnight, or eat with sweet fruit such as bananas or sultanas, or a minimum of honey; **or** wholemeal bread with little or no butter or margarine, or with low-fat cheese; with poached, boiled or scrambled eggs. Fresh fruit.

Main meal
The main family or social meal of the day is usually the evening one – just when we don't need a lot of calories, because they are not going to be burnt up in energy. There are two ways in which to cope with this trend. Either try to eat a light lunch and eat dinner before eight; or make the main evening meal important in social rather than calorie terms. From a health point of view, one meal a day should be a predominantly salad one. This is simply because uncooked fruit and vegetables are the best source of vitamins and minerals, and possibly of other unrecognized elements. It may be convenient to make a habit of salad suppers, which don't have to be unsatisfying or boring; they can

be made with dozens of ingredients, in many combinations. Salads can be different every day of the week; based on shredded carrots and root vegetables on one occasion; on beans, another; on apples and celery, another.

Good salads are light-years away from the curling, vinegary offerings which frequently appear in restaurants. Everything in a salad does not have to be raw; brown rice, beans and cooked grains add appeal and goodness. You can make the rest of the meal hot with soup or baked potatoes. And the protein content of the main meal (cheese, nuts, eggs, fish or meat) can either be built into the salad, or served with it. If the main meal is not salad, it should still offer a generous amount of vegetables, which should be fresh and given a minimum of careful cooking. There's nothing wrong with 'meat and two veg' provided the meat (or fish) is not greasy, or coated in refined carbohydrate and fat, in the form of batter, coating or pastry; and always provided the vegetables are not 'ghosts' boiled to death, or tinned, dried or kept hot, when all the goodness evaporates away.

When it comes to desserts, there are many which are as good for you as they taste. Examples are fresh fruit salad, cheesecake made at home without additives and using low-fat cheese, baked apples stuffed with dates, stewed dried apricots puréed with plain yoghurt, fruit fool made with yoghurt rather than cream, cheese and an apple, or a handful of nuts and raisins. To sum up, this main meal should provide a lot of protein, vitamins, minerals and roughage; and it should not contain much fat, processing additives, or sugar.

The next chapter contains a whole range of appetizing recipes from which to choose.

The third meal

Today few of us live sufficiently energetically to use up the calories from a second large meal in a day. Therefore the third meal, whenever we eat it, should be light. It could be another salad (one can't eat too many!) or it could be whatever is convenient given the basic guidelines above. If you eat lunch at work, get into the habit of taking it with you. Thinking ahead in the morning quickly becomes automatic and saves a little aggravation and a lot of money. If you have working lunches, make the most nutritious choice of what is there. In a pub, a salad, brown bread sandwich, baked potato or a packet of nuts and raisins are usually the healthiest options. In a chips-with-everything establishment, avoiding the frying pan is the main aim. As always, what you are out to avoid is fat, sugar and refined stodge.

Snacks

Some people are natural nibblers. Despite the aura of immortality with which our Victorian forebears have surrounded 'eating between meals', all the evidence shows that this is a perfectly healthy thing to do, provided that the total daily intake of food is no higher. Indeed, as studies have shown, snacks are less fattening than eating food in two large sessions, because it is not difficult to burn up small amounts of food as energy but larger amounts are not so easily burnt off and what is left will quickly be stored as fat. And from a physical point of view, it's sensible, since snacks help us refuel our bodies more consistently with the continual use of energy.

There is no doubt that most of us are less choosey about what we eat when we are very hungry. If you know that you won't be able to resist the scent from the fish and chip shop you pass on the way home, or the bakers on the corner, when your light lunch was hours ago, carry emergency rations to divert your attention. This may seem to be carrying nutritional thinking to extremes, but what is the difference between a packet of raisins in the pocket and carrying a packet of cigarettes around? Such emergency rations have the advantage that they keep fresh, and are handy for these moments when hunger coincides with temptation.

What to drink

There are not many healthy drinks. Fruit drinks, squashes, cordials and crushes are mainly sugar, with flavourings and additives – read the labels! Coffee, tea and chocolate all contain stimulants like caffeine and tannin, which not only keep one awake, but also artificially stimulate many body processes. If this sounds depressing, it should not. All these drinks are relatively new, and certainly not necessary to the quality of life.

Instead, you can benefit from drinking fresh, unsweetened fruit juices, herb teas in all their variety and, with different good effects, milk, skim milk and skim buttermilk (provided you are not allergic to milk, or liable to catarrhal symptoms which it exaggerates), and mineral or plain water.

Real value for money

This section on how to eat opened with the question: does eating healthily cost a lot? Well, fresh fruit, wholemeal flour, unprocessed meat and other healthy, natural foods may seem expensive to buy, but there is good food value for every penny. And you are saving money on all these poor-food-value items you do not buy:

soft drinks	cream	sugar
potato crisps	chocolate	tinned fruit
ice cream	pies	tinned vegetables
sweets	convenience	dessert mixes
cakes	meals	sweet biscuits

Equipment

A liquidizer is probably the most useful kitchen machine for the cook who wants to make soups and desserts. It will purée fruit and vegetables in seconds, whereas rubbing them through a sieve can take fifteen minutes or more.

If you want to make your own wholemeal bread, you may also want to grind your own flour from wheat. Health food stores sell small hand-grinders; larger, more efficient hand and electric stone-grinding models are also available. (SAMAP flour mills are made in France and imported to Britain by Springhill Enterprises, 38 Buckingham Street, Aylesbury, Buckinghamshire.) Flour grinders are expensive, but give you the freshest flour, and you can save in the long run by buying wheat cheaply in bulk. The makers of the Kenwood Chef have recently developed a new mill attachment for grinding flour from wheat and other grains. Otherwise, making your own bread requires no special equipment. You don't have to bake it in bread tins – any cake tin or baking sheet will do. You don't need a thermometer. If you measure your recipe liquid out as one-third boiling and two-thirds cold, you will get the right temperature.

Most people possess a grater, which is certainly very helpful for making salads. The best is the four-sided kind which gives a choice of different surfaces, and sits firmly while you grate. Always choose a stainless-steel one. Some people prefer plastic graters which sit like lids over bowls, and all the shreds end up where you want them. Just check that the cutter is sharp. If you like finely grated salads, using the Spong kitchen set is the fastest way of producing an immaculately even-cut bowlful. It has two drums, one for fine shreds, the other for thinly slicing cucumbers, carrots and so on. It is an expensive gadget to buy just for shredding, but it also has a mincer head, which means you can make your own pâtés, nutmeats, mincemeats, etc., and it is good for shredding fruit too.

If you like fruit and vegetable juices, a juice extractor will give you the best ones. Juices are an easy way of obtaining a concentrated amount of vitamins and minerals, but remember that the goodness starts to decrease within half an hour. Having your own juicer also enables you to make many different varieties not available in shops. A continuous juicer, which ejects the squeezed pulp as it goes, is preferable, although it will cost more. Moulinex make the cheapest one, and a juicer attachment is available with the Kenwood Chef. Both models have aluminium interiors, which will react with the acid in fruit, and are therefore not nearly as good as stainless metal. AEG and Nature's Bounty are two models with non-reactive interiors, but they cost more. Juices are a source of vitamin and mineral supplement in food form – a terrific booster when you're not eating properly, convalescent, slimming or run down. But normally you do just as well by eating the whole fruit or vegetable and thereby including some fibre.

The argument about aluminium versus stainless-steel saucepans has been going on for decades, with no conclusive evidence for either side. However, to be safe choose the stainless. You may wince at the price tag, but they will last far longer and age gracefully too. Enamelled pans are fine, provided that the quality is good; those that chip easily allow food to touch the reactive metal underneath. Heatproof glass is perhaps the best of all – if you can find it.

A wide-mouthed glass-lined vacuum flask can be a great help in the nutrition-minded kitchen. It can carry home-made soup for healthy packed lunches and it can be used to simplify cooking foods like beans or barley. Just put them in, top up with boiling water, and leave for several hours. Porridge made from oatmeal is difficult to wash off saucepans, but cooks wonderfully in a vacuum flask overnight for a super-quick, no-pan-washing breakfast. Dried fruit tastes very good after being soaked for a few hours in boiling water (but first simmer off any liquid paraffin that may be on the fruit and throw that liquid away).

Miriam Polunin

3 Health Food Recipes

Soups

Carrot soup

340 g (¾ lb) carrots, scraped and shredded
1 large potato, diced
1 onion, chopped
salt and pepper to taste
25 g (1 oz) butter
550 ml (1 pt) vegetable stock
½ tsp lemon juice
1 tbs fresh parsley or chervil, chopped

Sauté the carrots, onion and potato in butter. Season with salt and pepper. Leave the pan covered over a very low flame for about 15 minutes. Pour over the stock and simmer for another 15 minutes. Purée the mixture, using a hand potato masher to get rougher texture. Correct the seasoning, add lemon juice, sprinkle with herbs. Serves 3.

Iced avocado soup

2 large and 1 small avocados
½ tsp grated onion
425 ml (¾ pt) chicken stock, cold
140 ml. (¼ pt) double cream
280 ml (½ pt) natural yoghurt
140 ml (¼ pt) fresh tomato juice
salt and pepper
Fresh chives, chopped

Peel the large avocados, remove the stones and cut the flesh into small pieces. Put them into a blender with the salt, pepper, onion and half the cream and blend until smooth. Then add the remaining cream, stock, yoghurt and tomato and blend again until mixed. Taste and adjust the seasoning. Chill in the refrigerator for 1 hour. Peel the small avocado, remove the stone and cut it into thin slices. Add these to the soup together with the fresh chives. Serves 4–6.

Cold cucumber soup

1 clove garlic
1 large cucumber, coarsely grated
280 ml (½ pt) natural yoghurt
140 ml (¼ pt) cream
2 tbs chopped fresh dill
2 tbs tarragon vinegar
salt, pepper and pinch of sugar

Rub a bowl with 1 clove of garlic – then mince the garlic into the bowl. Add the cucumber and sprinkle with salt and pepper. Leave in a cool place for 5 minutes. Mix together the yoghurt, cream, vinegar, sugar and dill, then pour the mixture over the cucumber. Stir well and chill for at least 1 hour. Serves 4–6.

Cream of cauliflower soup

1 kg (2 lb) cauliflower, divided into
   small stalks
1 large onion, coarsely chopped
1 potato, diced
1 stalk celery, coarsely chopped
1 tbs butter
2 tbs olive oil

2 sprigs parsley, chopped
1 litre (2 pt) boiling water
140 ml ($\frac{1}{4}$ pt) cream
40 g ($1\frac{1}{2}$ oz) grated cheese
salt and pepper
grated nutmeg

Sauté the cauliflower, onion, potato and celery in the butter and oil. Don't allow them to brown. Add the water, parsley, salt and pepper. Cook till the cauliflower is soft. Rub the soup through a sieve. Reheat the soup, but do not allow it to boil; whisk in the grated cheese and add the cream. Correct the seasoning with salt, pepper and a little grated nutmeg. Serves 4–6.

Tomato and potato soup

1 tbs celeriac, finely chopped
white parts of 2 leeks, diced
340 g ($\frac{3}{4}$ lb) potatoes, diced
225 g ($\frac{1}{2}$ lb) tomatoes, skinless
1 tsp brown sugar

salt
850 ml ($1\frac{1}{2}$ pt) boiling water
50 g (2 oz) butter
3 tbs cream
1 tbs parsley, chopped

Sauté the celeriac, leeks and potatoes in the butter very slowly until medium-cooked. Do not allow them to brown. Add the tomatoes and cook until they start to give out juice. Season with salt and sugar. Add the water. Simmer slowly for about 30 minutes. Purée in the blender, and reheat. Add the cream and sprinkle with parsley. Serves 4.

Potato soup

3 tbs butter
1 tbs olive oil
3 medium potatoes, peeled and diced
2 tsp celery, chopped
2 tsp onion, chopped
1 litre (2 pt) boiling water

1 tsp caraway seeds
1 bay leaf
1–2 tbs flour
salt and pepper to taste
2 tbs parsley, chopped

Heat a third of the butter and the oil together. Add the onions and celery and sauté for about 10 minutes. Add the potatoes and continue cooking for about 2 minutes. Add the boiling water, salt, pepper, caraway seeds and bay leaf. Melt the rest of the butter in another saucepan, add the flour and blend until smooth. Add this to the simmering soup, stirring until smooth. Reduce the heat, cover and simmer for 1 hour. Before serving add the chopped, fresh parsley. Serves 4.

Haricot bean soup

170–200 g (6–7 oz) beans, soaked
   overnight in cold water
3–4 potatoes, diced
1.5 litres (3 pt) water

3–4 tbs sour cream
1 tbs tarragon vinegar
salt to taste

Cook the beans in half of the water until tender. Add the potatoes, the rest of the water and salt and cook until the potatoes are soft. When the soup is ready add the cream and vinegar. Correct the seasoning. Serves 4–6.

Autumn soup

2 medium carrots, chopped
6 stalks celery, chopped
1 medium Spanish onion, chopped
⅓ cup safflower oil
½ cup cabbage, shredded
2 small courgettes, chopped
5 ripe tomatoes, skinned and blended
250 g (9 oz) fresh tomato juice

2 cups water
5 sprigs watercress
1 cup cooked brown rice
3 tbs soya sauce
1 tbs butter
fresh basil, dill or parsley, chopped
6 tbs cottage cheese

Sauté the carrots, celery and onion in the safflower oil until they are medium-cooked (10–15 minutes). Add a mixture of tomatoes, tomato juice and water, blended together. Add the courgettes, cooked rice, cabbage, watercress, soya sauce and herbs. Cook for about 10 minutes. Serve immediately. Add 1 tablespoon of cottage cheese to each bowl of soup. Serves 6.

Gazpacho

1 small clove garlic
6 large ripe tomatoes,
  peeled and quartered
1 small Spanish onion,
  finely chopped
1 large green pepper, seeded
  and finely chopped

½ cucumber, peeled and
  finely chopped
550 ml (1 pt) fresh tomato juice
4 tbs lemon juice
2 tbs parsley, finely chopped
salt and demerara sugar to taste
5 tbs single cream

Blend the tomatoes and garlic in an electric blender at *low* speed stirring occasionally. Strain the mixture into a large, chilled serving bowl, add the tomato juice and the rest of the ingredients. Chill for about ½–1 hour. Gazpacho is nicer served with garlic *croûtons*:

2 tbs butter
2 slices bread, diced
1 clove garlic

Heat the butter with the garlic; toss in the diced bread; fry until crisp and golden and put into small serving bowl. Serves 4.

Summer borsch

1 litre (2 pt) sour milk (to make sour milk: put the milk in shallow glass or
  china dish, add 1 tbs sour cream, cover with some paper or cotton sheet
  and keep for 2–3 days in a warm place till it gains the consistency of light
  jelly; before use, take off any excess fat on top of milk)

140 ml (¼ pt) single cream
1 bunch young, small beetroots,
  peeled and finely chopped
  (together with leaves)
280 ml (½ pt) water
½ fresh cucumber, peeled and
  finely chopped

2 tbs fresh chives, finely chopped
2 tbs fresh dill, finely chopped
1 tbs tarragon vinegar
1½ tsp sugar
salt to taste
4 eggs, hard-boiled

Blend the sour milk and cream together until smooth. Cook the beetroots separately (together with the leaves) in the water with little salt for about 10–15 minutes. When soft, allow the beetroots to cool, then add them, together with the remaining liquid, to the milk and cream mixture. Add the rest of ingredients, except the eggs. Mix well, season and put into the fridge for about 2 hours. Place 1 egg halved into each bowl and pour the soup on top. Serves 4.

Spring borsch

370 g (13 oz) young beetroots, peeled and finely cut in strips (together with leaves)
1.5 litres (3 pt) chicken or vegetable stock
2 onions, chopped
2 carrots, diced
3 sprigs of parsley, chopped
2 bay leaves
5 balls allspices

1 tbs flour blended in ½ cup cold water
salt, demerara sugar to taste
4 tbs sour cream
1–2 tbs tarragon vinegar or lemon juice
2 tbs dill, parsley and chives, chopped and mixed

Cover the onions, carrots, parsley, bay leaves and allspices with the stock and cook till soft. Then add the beetroot. Cook for 15 minutes. Add the flour blended in the water. Boil for 1 minute more. Then add the cream and salt, sugar, vinegar or lemon juice to taste. Sprinkle with chopped herbs. Serves 6.

Pearl barley soup

115 g (4 oz) pearl barley
850 ml (1½ pt) water
3 dried mushrooms, diced and soaked for ½ hour in ½ cup of hot water
2 litres (3 pt) any white stock
1 leek, cut into slices
1 large onion, chopped

2 carrots, cut into slices
2 stalks celery, cut into small pieces
3 sprigs parsley
3–4 medium potatoes, diced
salt to taste
5 balls allspices
2 tbs fresh dill or parsley, chopped

Simmer the barley in the water, salt, allspices and mushrooms for 1½ hours, stirring occasionally. Half an hour before the end of the cooking time add the stock and vegetables. When cooked taste for seasoning and add the dill or parsley. Serves 6.

Quick cold tomato soup ('just in love' soup)

4 large ripe tomatoes, peeled and quartered
1 cup sour cream
1 cup natural yoghurt

salt, pepper, demerara sugar to taste
5 sprigs fresh dill

Purée the tomatoes in a blender, strain the mixture into a chilled serving bowl. Stir in the sour cream and yoghurt (or twice as much of either). Add salt, pepper, and sugar to taste. Keep the soup in the refrigerator for 1–2 hours. Pour it into bowls and serve topped with dill. Serves 4.

Fish

Cod baked with lemon

1–1.5 kg (2–3 lb) whole cod
3 tbs olive oil
2 tbs lemon juice
1 onion, finely chopped

1 tbs fresh parsley, finely chopped
sea salt and freshly ground
 black pepper

Rub the fish with a mixture of salt and pepper, and place it on a large piece of baking foil. Mix together the oil, lemon juice, onion and parsley, and pour over the fish. Wrap up the fish in foil and leave it in a cool place to marinate for 1 hour. Bake in a moderate oven for 45 minutes, basting from time to time. Serve with plain boiled potatoes or French beans or green peas. Serves 4–6.

Grilled trout

4 medium-sized trout, cleaned
2 large garlic cloves, halved
4 sprigs rosemary

salt and pepper to taste
3 tbs olive oil
1 lemon, quartered

Rub the fish with salt and pepper. In the cavity of each fish put one sprig of rosemary and half a garlic clove. Make two cuts on each of fish, lightly coat them with oil and put them on a preheated, lined grill pan. Grill for about 5 minutes then turn the fish over, brush with oil and grill for another 5 minutes. Remove the garlic and rosemary, and serve with lemon, plain boiled potatoes and green salad. Serves 4.

Baked mullet with herbs

1–1.5 kg (2–3 lb) whole mullet,
 gutted but with head left on
2 tomatoes, sliced
1 lemon, sliced

4 sprigs of rosemary
2 tsp powdered cumin
oil, salt and pepper

Put the fish on large piece of foil, and brush the fish with oil. Stuff the rosemary into the cavity. Mix together the salt, pepper and cumin and sprinkle this all over fish. Place the lemon and tomato slices over and around fish, and wrap up completely in the foil. Bake in preheated oven at 200°C, 400°F, or gas No. 6 for 40–50 minutes. Arrange the fish on a serving dish, garnish with the lemons and tomatoes and cover with the cooking juices. Serve with new potatoes boiled in their skins. Serves 4–6.

Gratin de Colin

1 kg (2 lb) fillets of cod,
  skinned (keep skin)
2 carrots, sliced
3 medium onions, quartered
5 stalks fresh parsley
2 bay leaves
5 balls allspice

salt, freshly milled black pepper
1 litre (2 pt) chicken stock
1 tbs flour
1 tbs butter
4 tbs cream
5 tbs cheddar cheese, grated

Put the onions, carrots, bay leaves, parsley, allspices and pepper into the stock, and cook gently adding the skin from fillets. When the vegetables are cooked, remove the fish skins and put in the cod fillets. Cook gently for 15 minutes. Correct the seasoning. Remove the fish from the pot together with the carrots and onions and put these into a warm, flame-proof medium-sized shallow dish.

Make a sauce from the butter, flour, some of cooking stock and cream and cover the fish with it. Sprinkle the top very generously with cheese, and put the dish under the grill till the sauce reaches a pale golden colour. Serve with plain boiled young potatoes and green salad. Serves 4–6.

Cod with dill butter

1–1.5 kg (2–3 lb) fillets of cod
2 carrots, sliced
20 small pickling onions
5 stalks fresh parsley
75 g (3 oz) butter
1 tbs fresh dill, chopped

juice from 1 lemon
sea salt, pepper
1 bay leaf
5 balls allspice
280–550 ml (1–2 pt) hot water

Soften the butter to room temperature, add the dill and a little salt. Mix well then roll the butter into a cylinder. Wrap it in greaseproof paper and chill well for 3 hours. Cut it into 6 slices before serving with the fish.

Cut the fillets into 6 portions, sprinkle with a little salt and half the lemon juice. Leave for ½ hour in a cool place.

Put the carrot, onion, bay leaf, allspice, pepper and salt into a pot. Cover with the hot water and cook till the vegetables are tender. Then add the portions of fish and the rest of the lemon juice and cook over a moderate heat for 20 minutes.

Remove the fish to a serving dish, garnish with the cooked carrots and onions. On top of each portion of fish place one piece of the dill butter. Serve with new potatoes cooked in their skins and green salad with oil and lemon dressing. Serves 6.

Meat

Spicy chicken kebabs
4 chicken breasts skinned, boned and cut into 2.5 cm (1 in) cubes
Marinade

1 cup natural yoghurt
5 cm (2 in) long fresh ginger,
  finely shredded
5 cloves garlic, crushed

1 medium onion, grated
2 tsp chilli powder
1½ tsp ground coriander
salt

Mix together well all the marinade ingredients, add the chicken cubes and keep in fridge overnight. Place the chicken onto skewers and put them on a lined preheated grill rack. Grill for 8–10 minutes basting with the rest of the marinade and turning them occasionally. Serve with a mixed salad of tomatoes, cucumber, onion and French dressing. Serves 4.

You can also serve the kebabs with special spicy yoghurt sauce.

Spicy yoghurt sauce

2 tbs sunflower oil
4 garlic cloves, crushed
1 small onion, chopped
1 tsp turmeric powder

½ tsp chilli powder
2 tsp coriander leaves, finely chopped
1 cup yoghurt
salt to taste

In a saucepan gently cook the garlic, onion, turmeric and chilli in oil for 1 minute. Remove the pan from the heat. Stir the mixture into the yoghurt, add coriander and salt to taste. Mix well in a blender and chill before serving.

Grilled chicken with herbs

1 chicken, skinned and cut into
 4 portions
2 large garlic cloves, crushed
salt, pepper
4 tbs sunflower oil

4 tbs butter, melted
2 tbs lemon juice
1 tsp dried sweet basil
1½ tbs fresh parsley, chopped

Rub the chicken with salt and pepper. Mix the butter, oil, lemon juice and herbs together and brush the mixture all over the chicken. Place the chicken pieces into a bowl. Cover them with the remaining marinade and keep covered in the fridge for 3 hours, turning the chicken pieces occasionally. Then place them on to a lined preheated grill rack and grill for 10 minutes on each side, basting with the marinade mixture. When the chicken is cooked serve with seasonal salad. Serves 4.

Beef fondue

1 kg (2 lb) fillet or rump steak, cut into 2.5 cm (1 in) cubes
550 ml (1 pt) sunflower oil
1 large apple, finely chopped and sprinkled with little tarragon vinegar
½ cucumber, finely chopped (skin left on); sprinkled with tarragon vinegar, mixed with a little brown sugar and salt
2 tbs fresh herbs (chives, basil, parsley), finely chopped

Three side sauces

Tomato sauce
1 tbs butter
5 tomaties, peeled and halved
2 tbs sunflower oil
1 onion, finely shredded
1 clove garlic, crushed

1 bay leaf
1½ tsp dried basil
- salt and pepper
touch of demerara sugar to taste

Melt the butter, add the tomatoes and simmer for 10 minutes, stirring occasionally. Put through a sieve to make a purée. Heat the oil, add the onion and garlic, and simmer gently till soft. Do not allow them to brown. Then add the tomato purée, bay leaf, basil and seasoning. Simmer for 10 minutes. Remove the bay leaf, allow to cool and transfer to a shallow serving dish.

Aioli sauce

8 fat garlic cloves
2 egg yolks
280 ml (½ pt) olive oil

salt and freshly ground black pepper
lemon juice

Crush the garlic with a little salt to a smooth paste in a mortar. Add the egg yolks and mix well. Whisk using a balloon whisk and gradually add the oil, drop by drop, as for mayonnaise. The sauce will eventually reach its proper stiffness and firm consistency. Season with salt, pepper and very little lemon juice, and serve chilled in a bowl.

Strong horseradish sauce

2 tbs horseradish, very finely grated; dowse with boiling water and drain
  immediately to remove excess strength
2 tbs apples, peeled and finely grated   1 tbs lemon juice
3 tbs single cream   salt and sugar to taste

Mix well all the ingredients, serve chilled in the bowl.

Put the onion, apple, cucumber and herbs in small individual serving dishes to accompany the beef.

Arrange the beef on a large serving dish.

Heat the oil until it is very hot. Pour it into a fondue dish and put over lighted spirit burner. The oil is now ready for cooking the meat. Have ready one or more of the accompanying sauces for dipping the cooked meat into. Serves 6.

Grilled fillet steaks with rosemary
4 fillet steaks
salt and freshly ground black pepper
2 tbs olive oil
4 sprigs fresh rosemary

Rub the steaks with salt and pepper. Brush each side with oil. Put the steaks on a lined preheated grill rack. Place a sprig of rosemary under each piece of meat. Grill for few minutes on each side, as required, remembering always to keep the rosemary under the meat. Serve with grilled tomatoes and potatoes, boiled in their skins and tossed with dried mixed herbs and a piece of butter. Serves 4.

Veal in lemon sauce

4 thin slices veal escalopes –
  about 100 g (4 oz) each – cut in half
salt and pepper
4 tbs olive oil
2 tbs butter

juice from 1 lemon
½ tsp soya sauce
touch of sugar
1 tbs fresh parsley, chopped

Sauté the slices of veal in very hot oil for 4 minutes. Remove them from the pan, sprinkle with salt and pepper. Melt the butter, add the parsley and cook for few seconds. Add the lemon juice, soya sauce, sugar and salt to taste. Cook for few seconds more. Put the veal into the sauce and baste it well over the heat for a few seconds. Serve immediately with cooked broccoli, fresh beans or cauliflower. Serves 4.

Minced veal with vegetable sauce

680 g (1½ lb) minced veal
2 tbs breadcrumbs
1 egg
1 medium onion, minced
850 ml (1½ pt) water
2 carrots, sliced
2 stalks celery
2 onions, quartered
2 leeks, sliced in 1 cm (½ in) pieces

2 bay leaves
5 balls allspice
salt and pepper
5 sprigs fresh parsley
1 tsp arrowroot powder,
  dissolved in little water
3 tbs soured cream
2 tbs fresh dill or parsley, chopped

Mix the meat, breadcrumbs, egg, minced onion, salt and pepper very well, kneading for 10 minutes. Form the mixture into little balls (ping-pong ball size). Boil the water in a shallow pot. Add the rest of the ingredients except the cream arrowroot and herbs. When the vegetables are cooked, add the meat balls. Boil moderately till cooked. Strain the stock from the meat and vegetables, keeping them hot. Add the arrowroot mixture to the strained stock and when it has boiled, remove the liquid from the heat and add the cream. Correct the seasoning. Cover the meat balls and vegetables with the sauce and sprinkle with herbs. Serve with plain cooked rice and green salad. Serves 4–6.

Vegetable dishes

Bean, tomato and potato stew

1 kg (2 lb) runner or French beans, broken into two halves
450 g (1 lb) tomatoes, peeled and chopped
450 g (1 lb) very small potatoes, scrubbed and cut in halves
3 big onions, coarsely chopped
4 cloves garlic, crushed
5 tbs sunflower or corn oil

280 ml (½ pt) hot water
1 bay leaf
1 tbs sugar
1 tsp sweet basil, freshly chopped or dry
1 tbs freshly chopped parsley
3 tbs Gruyère cheese, finely grated
salt and a pinch of chilli powder

Heat the oil in a large pan, then put in the onions and sauté them very gently till they are transparent. Add the tomatoes and cook them gently till soft. Then add the beans, potatoes, bay leaf, salt, sugar, chilli powder and water. Simmer until tender; five minutes before finishing cooking add garlic, sweet basil and parsley. Serve warm or cold, sprinkled with cheese. Serves 4.

Hummus

225 g (½ lb) chick peas, soaked over previous night
140 ml (¼ pt) lemon juice

2–3 cloves garlic, crushed
140 ml (¼ pt) tahini paste
sea salt

Garnish
1 tbs olive oil
1 tsp paprika
1 tbs finely chopped parsley

Boil the peas for 1 hour till soft. Drain them, retaining the soaking liquid. Purée the peas in an electric blender adding some of the soaking liquid to thin it. The purée should have consistency of thick cream. Beat crushed garlic into the paste then add the tahini paste and lemon juice. Blend well, correct the seasoning, and pour the purée on to a flat dish. Mix the red paprika with the olive oil and pour this over the hummus. Sprinkle it with parsley. Serve with pitta or Arab bread. Serves 4–6.

Lentil kedgeree

225 g (½ lb) rice, soaked in cold water
  for an hour
225 g (½ lb) lentils, soaked in cold
  water for an hour
5 lbs sunflower oil

1 medium onion, thinly sliced
¼ tsp peppercorns, crushed
1½ tbs fresh ginger, finely chopped
2 bay leaves
salt

Drain the rice and lentils. Heat the oil in a heavy, deep pan. Add the onion and sauté slowly till it is golden. Remove the onion, pressing all excess oil back into the pan. Keep the onion warm. Put the drained lentils and rice into the pan; add the ginger, bay leaves, pepper and salt to taste, and fry, stirring constantly till the oil has been absorbed. Then add enough warm water to cover ingredients by 2 inches. Bring to the boil, then cover and simmer on lowest heat. It should take about 1 hour to cook the lentils and rice, and the water should evaporate. Add the onion and mix well. Serve the kedgeree with green salad and a mild French dressing. Serves 4.

Lentil curry

225 g (½ lb) lentils (English orange
  lentils are not suitable)
2 medium onions, sliced
3 tbs olive oil
1 banana, thinly sliced
1 tsp lemon juice
1 tsp garam masala

½ tsp mustard seed
½ tsp turmeric, ground
½ tsp ginger, ground
¼ tsp chilli powder
½ tsp coriander, ground
50 g (2 oz) seedless raisins
salt

Put the lentils in a saucepan with a scant litre (2 pt) of cold water. Bring to the boil and then simmer until the lentils are soft, by which time the water should almost have evaporated; halfway through the cooking add some salt and the raisins.

Fry the onions in oil in a pan until soft and golden, then add the spices, raisins and banana, and cook for another 5 minutes. When the lentils are ready (their consistency will be thick and 'soupy'), stir in the onions and spices mixture and the lemon juice. Reheat, stirring all the time. Serve with sliced hard-boiled eggs. Serves 4.

Tahini

1–3 cloves garlic
juice of 2–3 lemons
½ tsp cummin, ground

550 ml (1 pt) tahini paste
6 tbs parsley, finely chopped
salt

Crush the garlic in the salt and mix in a little of the lemon juice. Add the tahini paste and mix well. Then add the remaining lemon juice and enough cold water to make a thick smooth cream. Season with the cumin and salt, and garnish with parsley. Serves 2.

Tahini Parsley Sauce

140 ml (¼ pt) tahini paste
juice of 3–4 lemons
140 ml (¼ pt) fresh yoghurt

1 tsp Dijon mustard
4 tbs parsley, chopped

Put the tahini paste, lemon juice, yoghurt and mustard into the electric mixer, and mix well, adding a little water if the mixture thickens too much. Pour into a bowl and spoon in the parsley. Serve with chicken or fish. Serves 4.

Buckwheat (Kasha) stuffed marrow

1 medium marrow
1 teacup – about 125 g (4½ oz) –
  buckwheat
salt
pepper
1 leek, diced
1 onion, sliced
1 carrot, diced

4 tbs olive oil
soy sauce
3 cloves garlic, crushed
2 ripe tomatoes, skinned and chopped
1 tbs dill, chopped
1 tbs parsley, chopped
25 g (1 oz) cheddar cheese, roughly
  grated

Lightly salt two teacupsful of boiling water in a saucepan, add the buckwheat and boil briskly for 1 minute. Then lower the heat as much as possible, cover the pan and leave for 15–20 minutes or until the water is absorbed.

Meanwhile, scrub and wash the marrow, but do not peel it. Cut in half lengthwise and scoop out the seeds and pith. Boil the marrow in another pan for about 5 minutes, then strain off the water.

Stuffing

Cook the onion in the oil until it colours. Parboil the carrot for a few minutes, then drain and add to the onion, together with the diced leek. Cook until the ingredients are soft and golden; then add the buckwheat and mix well. Now add salt, pepper, soy sauce, tomatoes, herbs, and the garlic, and mix well again.

Grease a large fireproof dish and place in it the two halves of the marrow, side by side. Fill them with the stuffing mixture, and top with the grated cheese. Bake in a moderate oven (180°C, 350°F or gas No. 4) for about 30 minutes. Serve with fresh tomato sauce. Serves 4.

Fresh tomato sauce

1 kg (2 lb) tomatoes, skinned and
  chopped
4 tbs olive oil
1 tsp unrefined sugar

2 cloves garlic, crushed
2 tbs mixture of fresh parsley,
  basil and marjoram, finely chopped
salt and pepper to taste

Heat the oil in a heavy pan and stir in all the ingredients. Cook until the tomatoes have dissolved or for 10 minutes, and serve. Serves 4.

Brown rice, Indian style

2 cups brown rice
2 large onions, finely chopped
4 tbs olive oil
4 tsp brown sugar
3 × 5 cm (2in) piece of cinnamon

12 cardamoms, slightly opened
12 cloves
1 tsp salt
6 peppercorns

Soak the rice in water for half an hour, drain and fry lightly. In another pan fry most of the onion in oil until it is brown, and then remove from the heat. Put the sugar in a small pan and heat until it is dark brown. Add a cup of water, cook for a few minutes until the caramelized sugar has melted, and set aside.

Add the drained and fried rice to the fried onions, and fry further, stirring frequently. Add the caramel water and cook for 5 minutes. Then add the cinnamon, cardomoms, cloves, salt, peppercorns and 3½ cups of warm water. Cover, lower the heat, and continue cooking until the water is absorbed and the rice is cooked. Sprinkle the remaining chopped onion, crisply fried, over the rice before serving. Serves 4.

Brown rice with vegetables

170 g (6 oz) brown rice
4 tbs olive oil
1 medium onion, roughly chopped
115 g (4 oz) carrots, sliced
1 leek, white part only, roughly
  chopped

2 stalks celery, chopped
115 g (4 oz) mushrooms, chopped
360 ml (¾ pt) chicken stock
sea salt and black pepper

Heat the oil in a pan, add the onion and cook gently until it starts to colour. Add the carrots, celery, leek, and mushrooms, then stir in the rice. Fry the mixture for 2 minutes, then add the heated chicken stock and seasonings. Bring to the boil, cover and simmer gently for about 45 minutes or until the water has evaporated. Serves 4.

Raw vegetable pâté

2 medium carrots, finely grated
2 tsp spring onion, chopped
2 tbs celery, chopped
2 tbs raw mushrooms, chopped
2 tsp lemon juice

2 raw egg yolks
1 tbs fresh herbs (such as basil,
  garlic and parsley)
4 tbs mixed nuts, grated

Mix all ingredients together, form into a pâté, leave for 15 minutes in a cool place, then serve.

Cauliflower curry

1 cauliflower, broken up into sprigs
5 cloves garlic, crushed
2.5 cm (1 in) piece ginger, ground
1 tsp salt

juice of ½ lemon
2 tbs olive oil
1 tsp marsala

Mix the ginger, garlic, salt and lemon juice. Heat the oil and fry the mixture for a few minutes. Add the cauliflower, cover and cook on a very low heat, stirring occasionally. When the cauliflower is soft sprinkle with the marsala. Serves 2.

Stuffed marrow

1 medium marrow
3 tbs olive oil
¼ cup ground maize
¼ cup soya beans, cooked
½ tsp fresh or dried tarragon

8 black olives, with pips removed
1 small green pepper, chopped
  and dried
25 g (1 oz) butter
salt

Wash and scrub the marrow but do not peel it. Cut it lengthwise in half and scoop out the seeds and pith. For the stuffing, mix together the rest of ingredients, add enough water to make a thick dough and cook for 5 minutes.

Boil the marrow separately for about 5 minutes and strain it.

Grease a fireproof dish and put in the marrow halves side by side, fill them with the stuffing mixture and top with small pieces of butter. Boil in a moderate oven (180°C, 350°F, or gas No. 4) for about 30 minutes. Serves 4.

Baked potatoes with avocado topping

4 baked potatoes
1 large avocado, halved, stoned
  and peeled
115 g (4 oz) cream cheese
1 tbs chives, finely chopped
salt and pepper

Mash the avocado with the cream cheese. Season with salt and pepper. Make a cut across the tops of the baked potatoes, squeeze them open and put a generous spoonful of the mixture into each and sprinkle with chives. Serves 4.

Ratatouille

7 tbs sunflower oil
3 courgettes, sliced
2 aubergines, diced
2 onions, sliced
1 big green pepper, diced
1 small red pepper, diced
6 ripe tomatoes, peeled and chopped
2 cloves garlic, crushed
salt
pepper, freshly ground
½ tsp demerara sugar
½ bay leaf
1 sprig thyme
½ tsp sweet basil, freshly chopped
  or dry
1 tbs freshly chopped parsley

Heat the oil in a frying pan, add the onion and sauté it gently till it is transparent. Add the garlic and green and red peppers. Sauté for a further 5 minutes. Add the aubergines and courgettes. Sauté again for 5 minutes. Add the tomatoes and season with the salt, pepper and sugar.

Put the whole mixture into a baking dish. Add the bay leaf, thyme, basil and parsley, then cover with foil and bake for about ½ hour at 200°C, 400°F, or gas No. 6. Serve hot or cold. Serves 4.

Cheese and egg dishes

Courgettes à la Milanese

8 small courgettes, thinly sliced	sea salt
2 tomatoes, skinless, roughly chopped	touch of demerara sugar
3 cloves of garlic, crushed	3 tbs olive oil
1 tbs fresh parsley, chopped	4 tbs Gruyère cheese, finely grated

Put the oil and crushed garlic in a frying pan and heat. When the oil is hot add the courgettes, and cook them gently for 5 minutes, turning them often. Then add the tomatoes, parsley, salt and sugar. Cook gently for 2–3 minutes. Serve sprinkled generously with cheese. Serves 4.

Wholewheat pancakes with mushroom filling

115 g (4 oz) wholewheat flour	2 eggs
115 g (4 oz) white flour	280 ml (½ pt) milk
½ tsp salt	280 ml (½ pt) water

Mix all the ingredients together in a blender. Stand the mixture in a cool place for about 30 minutes. Make large, thin pancakes.

Mushroom filling

340 g (12 oz) flat mushrooms	salt
65 g (2½ oz) butter	pepper
280 ml (½ pt) sour cream	

Sauté the sliced mushrooms in butter till soft. Stir in the cream and season well.

Aubergines in yoghurt dressing

3 large aubergines, sliced	salt
2 tbs olive oil	5 small leaves of fresh mint,
1¼ large carton natural yoghurt	finely chopped
3 cloves garlic, crushed	½ tsp paprika

Fry the aubergines in oil till they are just soft. Put them to cool on a shallow plate. Smoothly blend the yoghurt with the garlic and salt to taste. Pour this over the aubergines. Sprinkle with mint and paprika. Serves 4.

Mixed herb omelette

The individual 2–3 eggs omelette is always the tenderest. So it is better to serve separately made omelettes for each person.

3 eggs
½ tbs butter
1½ tbs fresh mixed herbs (e.g. tarragon, chervil and parsley), chopped
salt and pepper

Beat the eggs, with salt and pepper and two-thirds of the fresh herbs, beating just enough to blend the whites and yolks thoroughly. Place the butter in a frying pan over high heat, and when the butter is very hot pour in the eggs. When the omelette is ready sprinkle it with the rest of fresh herbs. Serves 1–2.

Cheese soufflé

3 tbs butter
3 tbs flour
1 cup hot milk
1 cup strong cheddar cheese, grated

4 egg yolks
½ tsp salt, touch of pepper
6 egg whites
knob of butter to grease soufflé dish

Melt the butter, add the flour and stir well for 1 minute. Add the hot milk and cook gently until smooth and thick. Remove from the heat and add the cheese. Return to the heat and cook gently till the cheese is melted. Remove from the heat again and add the egg yolks one by one, beating vigorously. Add salt and pepper, and allow to cool. In a separate bowl beat the egg whites till stiff and then fold them in to the warm cheese mixture with wooden spatula. Pour the mixture into soufflé dish and bake in a preheated oven at 180°C, 350°F or gas No. 4 for 35–40 minutes. Serve immediately. Serves 4.

Stuffed eggs

4 hard-boiled eggs
3 tbs mayonnaise
few drops lemon juice

2 tsp dried chervil
salt and pepper
4 leaves of green lettuce

Mayonnaise
2 egg yolks
280 ml (½ pt) olive oil
2 tbs lemon juice
salt

Keep all the ingredients at room temperature. Beat the egg yolks, add half the oil drop by drop, whisking continually. When it gets really thick add the lemon juice. Keep on whisking, adding the rest of the oil, again drop by drop, till thick once more. Add some salt to taste.

Cut the eggs in half. Remove the yolks, chop these finely and add them to mayonnaise. Add the herbs, lemon juice, salt and pepper to taste, and beat together well. Spoon the mixture into the cavities in the egg whites. Serve 2 halves per person on a lettuce leaf. Serves 4.

Eggs Florentine

1 kg (2 lb) young spinach	salt, black pepper and nutmeg
2 tbs butter	6 tbs double warmed cream
6 eggs	6 tbs grated Gruyère cheese

Cook the spinach carefully for about 5 minutes, only using the water which is clinging to the spinach after washing. Strain it well and toss in melted butter. Add the salt and nutmeg. Spread the spinach over the bottom and sides of 6 cocotte dishes, then put on top of each portion a lightly poached egg sprinkled with salt and pepper. Pour a tablespoon of warmed cream over each egg, sprinkle each with a tablespoon of grated cheese and grill until the top is well browned. Serves 6.

Eggs stuffed with caviar

4 eggs, hard-boiled, skinned, halved lengthwise	2 large tomatoes, thickly sliced
115 g (4 oz) caviar	2 tbs freshly chopped chives
	2 tbs home-made mayonnaise

Remove the yolks from the egg halves and stuff the whites with the caviar. Place the slices of tomatoes on a serving plate. Pipe a mayonnaise ring on to each tomato slice leaving the edges of tomatoes visible. Inside each mayonnaise ring put half an egg sprinkled with chives and grated egg yolk. Serves 4.

Fromage blanc

90 g (3½ oz) low-fat cottage cheese
130 g (4½ oz) yoghurt
3 tbs lemon juice

Mix all the ingredients well in electric blender till the result is as smooth and thick as lightly whipped cream.

Cream cheese

Fill a muslin bag with yoghurt and hang it from the kitchen taps overnight, so it drips into the sink. The cheese is ready the next morning. You can vary its taste by adding:

1 Sliced fresh banana and honey.
2 Chopped chives and sea salt.
3 Sliced radishes, chopped chives, sea salt and little cream.
4 Fresh strawberries or raspberries, or any other soft fruit purée, sweetened with a little honey.

Stuffed tomatoes with cottage cheese and herbs

6 large tomatoes	1 tbs fresh chervil, chopped
8 tbs cottage cheese	1 tbs fresh chives, chopped
3 shallots, finely chopped	6 fresh lettuce leaves
2 cloves garlic, crushed	salt and pepper to taste
1 tbs fresh tarragon, chopped	

Cut the stalk ends off the tomatoes, then cut slices from the other ends to use as lids. Scoop out the seeds and cores, sprinkle the insides with a little salt, then leave the tomatoes upside down to drain out the moisture. Mix the cottage cheese with the herbs, shallots, garlic and salt and pepper. Fill the tomatoes with this mixture, cover each tomato with its lid and chill well for $\frac{1}{2}$–1 hour. Serve each portion on lettuce leaf.

Curd cheese

550 ml (1 pt) milk
1 tbs sour cream, or yoghurt or 2 tbs sour milk

Pour the milk into a shallow glass or china dish, add the cream, yoghurt, or sour milk. Cover the dish with a piece of paper or cotton sheet and keep for 2–3 days in warm place till it becomes the consistency of light jelly. Before use of this curd, remove any excess fat from the top. To make cheese put the soured milk into a muslin bag and hang from kitchen taps for 12–20 hours. Then sieve or beat the curd to get a cheese of smooth consistency.

Or alternatively: put the dish with the soured milk into a shallow saucepan with hot water, over very gentle flame. Be careful not to warm it too much, avoiding a solid consistency. Then put it into a muslin bag, and hang it up to drain off excess liquid. The liquid obtained from the cheese is rich in vitamin B, especially vitamin B_2, minerals such as calcium, and phosphorus. Use the liquid mixed with your freshly made juices, especially carrot, and with the addition of a little honey.

Curd cheese with black radish, tomato and onion

1 large black radish, peeled and coarsely grated	1 tbs onion, finely chopped
2 tomatoes, cut in thin quarters	5 tbs curd cheese
	sea salt

Dressing

6 tbs yoghurt	touch of demerara sugar
2 tbs cream	few drops of lemon juice
1 tbs sunflower oil	

Put the grated radish on a plate, sprinkle it with a little salt and cover it over with another plate. Leave for $\frac{1}{2}$ hour and then drain any liquid from it. Mix the radish with the onion and dressing, put it into bowl, sprinkle with cheese and tomatoes. Serve as a spread on rye or any brown bread, and as canapés.

Cream cheese squares

4 square wafers	2 eggs, lightly beaten
225 g (8 oz) cream cheese	rind from 1 lemon
25 g (1 oz) sugar	2 tbs mixed, toasted nuts
3 tbs honey, warmed	little butter

Beat the cream cheese, sugar and honey together. Add the eggs and beat for a few more minutes. Fold in the lemon rind and nuts.

Place 2 pieces of wafers on the bottom of a square baking tin, brush them with a little butter and pour on the cream mixture. Bake in a moderate oven 190°C, 375°F or gas No. 5 for about 30 minutes, then cover each with another wafer and bake for 5 more minutes. When cool cut the wafers into squares.

Cauliflower cheese

1 large cauliflower	13 g ($\frac{1}{2}$ oz) wholemeal breadcrumbs
25 g (1 oz) cheddar cheese, grated	salt
25 g (1 oz)butter	touch of demerara sugar

Cook the cauliflower whole in a minimum of water with a little salt and sugar, covering the pan with a lid (taking care not to overcook). Drain the cauliflower and place it on a warmed plate. Mix the cheese and breadcrumbs and sprinkle over the top. Finally heat the butter and pour it over the top. Serves 4.

Salads

Tahini salad

1 cucumber, diced
1 lb tomatoes, skinned and diced
3 tbs tahini paste
1 clove garlic, crushed

juice from 1 lemon
3 tbs cold water
sea salt

Mix the tahini paste with the water, lemon juice, garlic and salt and pour this dressing over the cucumber and tomato mixture. Serves 2–3.

Dandelion salad

Dandelion leaves should be gathered before plants start to flower, from ground which has not been chemically treated. Also, avoid places near motor traffic.

2 handfuls young dandelion leaves
1 small clove garlic, crushed
2 tbs fresh chervil, chopped
2 tbs sunflower oil

1 tbs lemon juice
salt
touch of demerara sugar

Place the dandelion leaves in a salad bowl. Mix the oil, lemon juice, salt, sugar and garlic; pour this dressing over the salad, and sprinkle with chervil.

Or use *yoghurt dressing*
4 tbs natural yoghurt
a few drops of lemon
touch of demerara sugar
sea salt

Home-made sauerkraut

Sauerkraut, as a rich source of vitamin C, is a valuable raw vegetable, particularly in winter. You can use it in salads, and its juice is a healthy drink.

small wooden barrel or wide-necked earthenware jar
small wooden or china plate
few whole cabbage leaves to put on the bottom of the jar

3.5 kg (8 lb) white cabbage, finely shredded	7 g ($\frac{1}{4}$ oz) dill seeds
	2 bay leaves
115 g (4 oz) carrots, roughly shredded	50 g (2 oz) sea salt
	few juniper berries
15 g ($\frac{1}{2}$ oz) caraway seeds	

Put the cabbage into the barrel or jar in layers, each layer sprinkled with a mixture of the spices. Press each layer down with a wooden masher until risen liquid covers the cabbage; continue until barrel or jar is full. Cover the top of the cabbage with a clean cloth, and then the wooden or china plate (smaller than the barrel or jar neck); press this down with a clean heavy stone. Keep in warm place (about 18°C, 65°F) for 1–2 days. Then move to a storage in a cold and dark place (about 4–6°C, 39–46°F).

Cloth, lid and stone must be washed weekly. Make sure that the cabbage is always covered with juice. This will keep the sauerkraut for up to five months.

Riccota cheese salad
2 cups low-fat riccota cheese
2 tbs natural yoghurt
1 ring fresh pineapple, finely chopped
2 tbs mustard and cress sprouts

Mix together all the ingredients, avoiding too smooth a consistency. Serve with wholegrain biscuits.

Spinach salad

3 cups spinach, roughly chopped	4 tbs sunflower or corn oil
1 small bunch radishes, diced	2 tbs lemon juice
1 tbs onion, finely chopped	demerara sugar
1 clove garlic, crushed with a little salt	salt

Mix all the ingredients together. Stand in a cool place for 10–15 minutes.

Pumpkin, apple and horseradish salad with mayonnaise

170 g (6 oz) pumpkin, roughly shredded	1 tbs freshly chopped parsley
	4 tbs mayonnaise
170 g (6 oz) skinless tomatoes, roughly chopped	salt, demerara sugar
	a few drops of fresh lemon juice
1 apple, roughly chopped	

Mayonnaise
1 egg yolk
280 g (10 oz) sunflower or corn oil
1 tsp lemon juice
salt

All ingredients for making mayonnaise should be at the same temperature. Mix the yolk with a little salt. Whisk the yolk adding the oil drop by drop. Stir vigorously and flexibly all the time. As the sauce thickens, gradually stir in the lemon juice. This makes sufficient for about 6 portions. Mix the salad with the mayonnaise. Season with salt, sugar and lemon juice.

Carrot, apple and horseradish salad
1 big apple, roughly grated
1 carrot, roughly grated
1 tsp horseradish, finely grated
1 tsp parsley, freshly chopped

2 tbs natural yoghurt
a few drops lemon juice
salt
demerara sugar

Mix all the ingredients together. Serve immediately.

Sweet and sour salad
3 tbs white cabbage, finely chopped
2 tbs green pepper, finely chopped
3 tbs celery, finely chopped
3 tbs carrots, finely shredded
1 cup sweet apple, finely chopped
  and sprinkled with lemon juice to
  prevent discolouration

1 tbs seedless raisins
2 tbs peanuts
1 tbs fresh parsley, chopped
2 tbs lemon juice

Mix all ingredients together. Allow the flavours to blend in cool place for half an hour. The salad can be served with a yoghurt dressing (first beat the yoghurt with a few drops of lemon juice until it is a smooth, creamy consistency).

Chicory and fruit salad
3 heads chicory, sliced
1 orange, peeled and cliced
1 sweet apple, diced

115 g (4 oz) fresh green walnuts,
  chopped
white part of one small leek,
  thinly sliced

Dressing

2 tbs sunflower oil	touch of mustard
1 tbs lemon juice	demerara sugar
salt	

Mix all the ingredients, pour the dressing over them and allow the salad to stand for five minutes before serving.

Red pepper salad

1 large red pepper (seeds removed)	2 tbs lemon juice
2 pears, peeled	salt
2 tomatoes	demerara sugar
3 tbs sunflower or corn oil	

Cut the red pepper into thin slices. Roughly chop the tomatoes and pears. Make a dressing from the oil, lemon juice, salt and sugar. Mix it in with the salad.

Avocado salad

2 large avocados, peeled	½ celery stalk, finely chopped
1 tsp lemon juice	4 lettuce leaves
2 medium tomatoes, peeled and finely chopped	2 tbs sunflower or corn oil
1 small onion, finely chopped	1 tbs tarragon vinegar
1 green pepper, finely chopped	salt and freshly ground black pepper
	½ tsp dry or fresh basil

Cut the avocados into thin slices and sprinkle with lemon juice. Add the tomatoes, green pepper, onion and celery. Mix together the oil, vinegar, black pepper and basil. Pour the dressing over the salad, mix well and let it stand for 5 minutes. Serve each portion on a lettuce leaf. Serves 4.

Celery salad

2 cups celery, finely chopped	4 tbs natural yoghurt
1 cup apple, finely chopped	1 tbs single cream
1 tbs fresh parsley, chopped	salt, pepper, demarara sugar
1 tbs white of 1 leek, diced	

Mix the celery, apple, parsley and leek together in a bowl. Blend together the yoghurt, cream, salt, pepper and sugar. Stir this into the salad. Serve after 10–15 minutes.

Spring salad

2 young green salads (small)
small bunch radishes
½ small cucumber
2 tbs fresh chives, chopped

2 tbs sunflower or corn oil
1 tbs lemon juice
salt, pepper, demerara sugar

Separate the salad leaves, slice the cucumber and radishes. Make a dressing from the oil, lemon juice, salt, pepper and sugar. Mix this with the rest of the ingredients and sprinkle with the chives.

Winter salad

2 cups savoy cabbage, thinly shredded
⅓ cucumber, thinly sliced
2 tomatoes, thinly sliced
white part of 1 leek, thinly sliced, or
  1 tbs onion, finely chopped

salt, pepper, demerara sugar
1½ tbs tarragon vinegar
3 tbs sunflower oil
2 tbs fresh parsley, finely chopped

Put the cabbage and cucumber into a bowl, sprinkle with a little salt, mix and squeeze a little by hand. Leave for 20 minutes. Add the tomatoes and leek or onion. Mix together the vinegar, oil, salt and sugar, pour this on the salad, add some pepper and mix well. Leave to stand for 10–15 minutes.

Sweets

Strawberry muesli

4 tbs oat flakes, soaked overnight
2 large apples, roughly grated
1 cup fresh strawberries,
  roughly chopped

4 tbs yoghurt, mixed with
  4 tbs honey
4 tbs grated mixed nuts

Pour yoghurt mixture over the oats. Add the apples and strawberries mixed. Spread the nuts on the top and serve at once. Serves 4.

Yoghurt surprise

2 cups natural yoghurt
2 tbs chopped mixed nuts
  (lightly browned in the oven)
4 tbs raisins (washed, then dried
  on a paper towel)

2 tbs soft brown sugar
1 tbs rosewater
½ tsp vanilla essence

Combine all the ingredients in one bowl. Chill for ½–1 hour, then serve in individual portions. Serves 4.

Date dessert
170 g (6 oz) dates, chopped
280 g (10 oz) rolled oats
115 g (4 oz) butter
2–4 tbs honey (vary quantity to taste)

Soak the dates in warm water for 3 hours, to soften. Wash the oats under cold running water and drain them. Mix the softened butter with the honey. Drain the dates and mix all the ingredients together. Serves 4.

Sweet avocado
2 large avocados, halved, stoned and peeled
2 tbs lemon juice
4 tbs fine brown sugar
2 tbs chopped mixed nuts and almonds (lightly toasted in the oven)

Put the avocados in the blender. Stir in the lemon juice and gradually add the sugar. Beat until the mixture reaches the consistency of lightly whipped cream. Stir in the nut mixture and spoon the sweet into small cold glasses. Chill for 5 minutes. Serves 4.

Dried apricot fool
225 g (8 oz) dried apricots, soaked overnight in cold water
1 carton natural yoghurt
2 tbs almond flakes (lightly toasted)

Purée the apricots in an electric blender, adding yoghurt to produce a creamy consistency. Chill well and pour into small glasses. Spread some almond flakes over each portion. Serves 4.

Wheat germ ice cream
2 whole eggs
2 egg yolks
65 g (2½ oz) vanilla sugar
280 ml (½ pt) milk

280 ml (½ pt) double cream
50 g (2 oz) Kretschmer's sweetened
  wheat germ

Beat the eggs and the egg yolks together in a bowl, gradually adding the vanilla sugar. Heat the milk just to boiling point; pour it on to the egg mixture, and continue to beat until well mixed. Place the bowl over a pan containing boiling water and stir the mixture until it thickens to the consistency of egg custard. Then pour it through a strainer into a clean bowl placed in a dish of cold water. Stir occasionally to prevent skin forming.

Whip the cream, fold it into the cooled mixture, and then pour the mixture into a sorbetière, or put the mixture into a plastic container and freeze for two hours (mashing the mixture with a fork occasionally). Then allow it to freeze solid for one further hour, stir in the wheat germ and continue to freeze until required. Serves 4.

Wholewheat bread

First a word about bread: since it is a food product whose formula is standardized by regulations, it is unnecessary for manufacturers to list ingredients on packaging – but it does enable them, in general, to produce a very poor-quality bread.

For example, low-grade white flour with the addition of caramel colouring, plus added bran, chalk, synthetic vitamins and minerals may be sold as 'brown' and 'wheatmeal'. That is why the only way to be confident about the nutritious quality of your bread is to bake it yourself.

Although the developed nations' diets are richer than ever, and one can get all the carbohydrates, proteins, mineral salts and vitamins from many sources, bread is still the most convenient portable form of food. People should never cut out bread (an important source of carbohydrates) from their diet for fear of growing fat otherwise some of the valuable proteins they eat (meat, fish, eggs, cheese, nuts, etc.) will be converted by the body into carbohydrates instead.

To get the fullest nutritional value out of baking your own bread it is better to use wholewheat flour rather than white flour. The reason for this is simple: the extraction rate of white flour is 70 to 75 per cent (see Part Three, *Flour*), which means that all traces of bran and germ are automatically sifted out from the ground grain. Also most of the commercially sold white flour is chemically bleached by using chlorine dioxide and benzoyl peroxide, and although these two chemicals are believed harmless, they do destroy the vitamin E in the flour. It is sensible to argue that not whiteness, but nutritional content, should be most important.

Wholemeal flour, milled from wholewheat grain, contains not only starchy white endosperm found in the white flour, but also the germ and the husk. The germ gives the flour more protein, fat, fatty acid, iron, thiamine, riboflavin, phosphorus, calcium, vitamin E, and the B vitamins. The husk is a valuable source of roughage, which is a very important part of the diet.

In many countries, particularly in the East, bread-making is a central part of domestic activity. Some people even consider it a spiritual exercise. It will take a little time to succeed in making 'real loaves', as bread baking is a very individual matter. Even if you are following the recipes exactly you still have to work out your own ideal timing, kneading techniques, amount of water, etc. Once you have succeeded, your bread-making becomes a simple and quick routine.

Recipe

680 g (1½ lb) stoneground
  wholewheat flour
1 tbs brown sugar
1 tbs sunflower oil

1 tbs sea salt
25 g (1 oz) fresh yeast *or*
13g (½ oz) dried yeast
3 cups warm water

Pour the water into a bowl. Add the yeast and sugar, stir a little but do not leave to stand. Start adding the flour, stirring continuously till a thick creamy consistency is obtained. At this stage you should have used about half the total amount of flour. Beat for another couple of minutes to incorporate as much air as possible. Cover the bowl with a dampened cloth and let it stand in warm place for 30 minutes. The dough should double in size.

Now sprinkle salt over the dough, add the oil and fold in the rest of the flour gradually. Take the dough out of the bowl, place it on the floured table and knead until it no longer sticks to your hands or the table yet remains moist. Too much kneading will damage the gluten (the gluey nitrogen part of the flour). Return the dough to the bowl, cover with the damp cloth and leave it in a warm place for a further 30 minutes. Then put the dough in a well-oiled baking tin and bake in a freshly lit oven at 200°C, 400°F or gas No. 6, for about 1 hour. The bread is baked when it is brown and crisp, and sounds hollow when tapped underneath. Cool the bread on the top of the oven rack after taking it out of the tin. Do not cut it till it is cool.

Dana Balfour

4 Natural Beauty Preparations

Some people don't think often or much of their skin. They give it up to the modern devils of unnatural ingredients. Toxic substances, heavy perfumes and harsh additives crush it into submission. But don't be taken in by your skin's meakness; that can soon turn to acne, allergies and early ageing.

This is why so many women now turn to natural skin-care products. For they contain pure, simple ingredients – perhaps a base cream, plus plant extracts and a preservative which itself is natural – all of which respect the delicate balance of your skin. Synthetics arrived on the scene only in the last half-century or so; plants have been appreciated since Culpeper's day and so far no natural cosmetic has ever needed to be 'withdrawn' from the market.

But a natural cosmetic is not a pseudo-natural cosmetic. You can tell the difference by reading the back of the packet. A good one will clearly list its ingredients and simply state what each one is for; for example, orange juice is antiseptic, so orange flower water will help acne. It is an honest approach. You know what you are getting and you are not encouraged to have false expectations – quite different from the cream with aspirations to being a natural cosmetic; that will keep a modest silence over the ingredients but will claim miraculous results. It likes to emphasize one 'star' component of the cream – usually a slightly exotic herb or fruit. So beware: a so-called strawberry hand-lotion may not even contain a single real strawberry.

This is a pity, because strawberries have the same pH value as our skin, and pH is important. Skin is covered with an emulsion, formed by secretions from the sebaceous glands and called the hypolipidic film. It should be slightly acid; it will then repel attacks from microorganisms and keep the skin moist and pliable. So what then is this mysterious and all-important pH value? pH means the potential of hydrogen. The pH scale runs from 0 to 14. pH0 to pH7 is acid and pH7 to pH14 is alkaline. Our skin is 4.5 to 5. A strawberry is 5 while most cosmetics and soaps go up to 9 or 10. Harsh chemicals used in cosmetics can destroy the skin's natural resistance to harmful bacteria and can also cause dryness, flakiness and irritation. One such chemical is glycerine.

Glycerine is a humectant used as a base cream in most synthetic cosmetics, it draws moisture from the atmosphere – but also from the skin, and you can't tell that this is happening. As long as you keep applying it your skin looks soft and elastic, but once you stop using it the skin can turn hopelessly dry and raspy. And sadly *any* alkaline substance used on the skin, although enriched with oils and proteins, will only damage this vital acid barrier.

Most natural cosmetics, however, aim to bolster up this barrier. Two companies whose products place particular emphasis on this are Yin Yang, and Faith Products. Faith Products label everything with its pH value. There is nothing over 5.5. Yin Yang use soya flour as the basis of their cosmetics. Soya flour is not only rich in protein but alkaline. So plant enzymes are introduced. They have a natural chemical effect which turns the alkaline to acid. The cream can then be used on the skin to readjust it to its proper acidic level and strengthen the acidic mantle. These plant enzymes have another important function. They break down the molecular structure of the proteins in the cream. Only then can the skin absorb them. Ordinary proteins have too large a molecule structure. They will feel lovely to put on but all their goodness will sit on the skin's surface – nothing gets absorbed. The things that will penetrate your skin and really do it good are extracts from plants, plant hormones and creams with an oil base. So using natural products made with plants will give your skin the opportunity of a feast.

The same principles work for your hair. Plant extracts have a better chance of being absorbed by the hair shaft than chemical conditioners and shampoos. Your hair, like your skin, also has a pH value of 5 to 5.5. Most shampoos are nothing more than fierce detergents of pH 11 or 12. Conditioning creams are there to do a cover-up job. They are simply an attempt to hide from you that the shampoo has washed the acidic barrier and natural oils down the plug-hole! It is best to be sceptical of these repair jobs, especially henna conditioners. If it is in liquid form then it's not pure henna. There is no such thing as liquid henna. To make it so, a mineral or chemical has been added and that is going to damage the acidic barrier of your hair. Nor is a balsam rinse the complete answer. In December 1974 the Federal Trade Commission in America challenged the advertising claims of a company making such a rinse. They admitted that although the advertisement said the conditioner could be used as much as desired, it would be harmful to do so. The hair would gradually become limp and over-soft. It is far better to treat your hair to a natural, slightly acidic shampoo, such as organic seaweed shampoo made by Faith Products, and two tablespoons of apple cider vinegar in the final rinse.

Natural cosmetics have another great advantage. The philosophy for making them, whether it's you doing the making or a company, is if it is edible it can be a cosmetic. This is very sensible, because the skin and hair demand the same proteins, vitamins and minerals as the body. If the body can get them from good natural food so can the hair and skin. The only difference is that these proteins, vitamins and minerals must be of a molecular structure small enough for the skin to absorb.

You are what you eat, and a good diet will make for a lovely complexion. But if you can put on your face what you eat the result can be very spectacular. Honey for instance is full of essential nutrients and minerals. Dab some on to your skin to increase the circulation without any drying effects, or try a honey and cucumber clay mask supplied by mail order from the Body Shop, 1 Crane Street, Chichester, West Sussex PO19 1LH. It might be your answer if the stickiness of raw honey puts you off.

You can't apply the principle of edibility to synthetic cosmetics. Your unhappy choice might be kaolin. It is the base for many make-up products because it is so cheap. But what you brush on your face happens also to be the drying agent in plaster of Paris. And if you eat it, what will clog up your pores will certainly turn into cement in your stomach!

It is much better to stick to a natural product like the New Era tissue salts. They are the essential minerals in each of our cells which combine with organic matter to create new tissue. From the tip of the tongue in pill form they take thirteen seconds to be absorbed into the blood. Once absorbed they stimulate cell growth and can be useful in treating many ailments. Combined in a make-up – the New Era's Hymosa range – they help stimulate the production of new cells. This is vital, when you realize that the day that cell production grows sluggish is the day ageing begins.

But when is a natural cosmetic not a natural cosmetic? Often when it is packed with preservatives rather than nutrients. Often the way the problem of preservation is dealt with will tell you how good the product is. Faith Products for instance use natural preservatives. The Body Shop frankly warns you its products may go off after two or three months. The cosmetics of the future will be deep-frozen the moment they are made. Already some companies are experimenting with this idea. The cosmetics will keep indefinitely in the freezer, and for about thirty to forty days in an ordinary refrigerator. Once opened the jars will have to be used within twenty-four hours if kept at room temperature. But the makers hope this trouble will pay off. For the make-up will contain no preservatives, antioxidants or stabilizers, which are often the cause of allergies. There should then be little decay of the active components in the make-up. Most exciting is the possible use of new ingredients which cannot be used now as they deteriorate so quickly; amino acids, vitamins B and C and enzymes will remain active in deep-frozen products. This will be an effective way then of keeping elsuvie goodness alive in cosmetics.

Inquiries as to where the cosmeticians find their supplies will also show how genuine the product is. Some grow their own; Weleda and Faith Products both have their own gardens in which all the ingredients are grown. Boy Scouts collect seaweed from the nearby beach for Faith Products. The Body Shop (address on page 68) has two suppliers who grow everything except the more exotic oils. They invent a recipe, test it for three months and then send it to the shop. They can afford to experiment. There are no big production lines to be rescheduled, just an initial outlay to make a few gallons for the customer to try. Often it is indeed a case of 'small is beautiful' when it comes to imaginative and good products at low prices.

One exception to this is the Boots Original Formula range made on a big commercial scale. The recipes are based on those found in books from as far back as the sixteenth century. But, where the old recipes each centred on one herb, Original Formula combine two or three with complementary properties for a concentrated effect. The range tries to use natural preservatives like aromatic vinegar.

To make quite sure of the naturalness of your cosmetic of course you can always make your own. The formula for a basic cream is easy – rather like preparing a hollandaise sauce.

Basic Cream

40 g (1½ oz) solidifier and/or solid fat, such as:

25 g (1 oz) lanolin and 14 g (½ oz) beeswax

75–115 g (3–4 oz) skin oil, e.g.

50 g (2 oz) sweet almond oil and 50 g (2 oz) wheat germ oil or hazelnut oil

25 g (1 oz) her water e.g. rosewater

5 drops essential oil; e.g. oil of geranium

Melt the solidifier and/or fat together in the top of an enamel double boiler or in a small enamel pot. Add the oils slowly, stirring all the time. Switch off the heat. Then mix in bit by bit the herb water. Stir all the time until the cream has cooled. Add the essential oil and stir it in completely. Ladle into 170 g (6 oz) jars. The glass should be opaque or you may use clear glass but only if you cover the jar with some kind of coloured paper. Different ingredients can then be added depending on whether you want a night cream, face pack, shampoo or whatever.

Why not decide to have 'natural make-up day'? Start by making your teeth really white with three tablespoons of bicarbonate of soda mixed with two tablespoons of salt. Keep it in a pot and use it as you would any commercial tooth powder. Or simply rub a strawberry over your teeth; it will polish them beautifully.

Be converted to the soap and water principle by making your own gentle soap.

Herbal Soaps of Lettuce, Cucumber, Strawberries or Carrots

Cut 115 g (4 oz) of fruit or vegetable into pieces and crush it in 1 cup of water then simmer it in the top of a double boiler for 10 minutes. Infuse until cool, then remove any particles by straining through a cheese cloth. Pour this liquid into 115–450 g (4–16 oz) of shaved castile soap (depending on how solid a piece of soap you want). Heat until the soap has melted and the juices absorbed. Let it cool and cut into bars.

If your skin is oily use your own astringent, which is just as effective as a commercial one, but not nearly so damaging. Simply wipe your face with a slice of pear; it has a disinfecting action especially good for those with acne. Or simmer a handful of sage leaves in a cup of water. Remove from the heat and let it steep until it reaches room temperature.

Those with dry skin might prefer to use a cleanser instead of soap. One of the best is milk. Warm a couple of tablespoons with a few drops of vegetable oil and shake them hard so they mix well. Use cottonwool (not paper tissues) to wipe the skin clean. Finally apply a thin film of oil to seal in the moisture. Then you are ready for the day.

To refresh you after a day's work before an evening out you can make a mask from two tablespoons of cucumber and mint mixture with four drops of peppermint extract and a pinch of alum powder if your skin is shining. It will make you feel fresh and alive again. To do the same for your eyes relax with a couple of tea bags over them. If your hair bores you, change the colour with a natural dye. Make sure your hair is free of chemical colour by washing it and rinsing it well. Check that your face and body will be protected from any drips from the dye. Pour and re-pour a strained infusion through your hair, rubbing it into the roots. Rinse lightly but completely and towel dry.

There are a great variety of herbs to turn your hair different colours. They also condition the hair as well. Henna mixed with cloves gives the hair a gleam and turns it a dark red brown, with indigo the hair turns a blackish brown, and with camomile it will go a lovely golden red. Camomile brightens 'mousey' hair. Black or blue malva will add a blue tint to white hair. Marigold, lemon peel and camomile will condition and brighten brown and blonde hair. All these dyes are subtle and the best for you are the ones closest to your real hair colour. To condition the hair use anything that is full of protein, like mashed avocados or egg yolks. If your problem is a last-minute invitation and dirty hair, simply brush through a few drops of lavender water or cologne. It will take away grease and grime and, just as important, leave your hair smelling lovely.

Last thing at night indulge in a really delicious face mask that will enrich your skin. Mash half a banana with two tablespoons of cream and a spoonful of honey. Put this on your face and prepare a herbal bath to relax in. They can be easily made. Bring to the boil half a cupful of herbs barely covered with water in a non-metallic pot. Leave to simmer for 10 to 20 minutes, then strain through a piece of material. Use the herbs in a cloth bag for a scrub and put the liquid in the bath. Stay in the bath for at least 20 minutes to get all the goodness from the herbs. Instead of a herbal bath you could have used bath oils, or rub oils into the body after the bath. There is a tradition in southern India that massaging essential oils into the body builds resistance to disease and helps cure many problems from acne to frigidity. You won't find the tradition in England but you can buy less complex versions of oils made by New Age Creations. Their carrier oils, in which the essences are mixed, have been cold-pressed to preserve their essential goodness. They do not have preservatives, and they are so pure they could almost be used for a salad dressing. Essential oils are added to these to give each a certain characteristic. A rub with the New Age rosemary massage oil will improve circulation while stimulating the mind and relaxing muscles. Gomorrah massage oil contains Moroccan jasmin, honeysuckle bark and other floral essences, and is supposed to have aphrodisiacal qualities. Or you can visit the Body Shop, who will supply you with a basic bath oil into which you can add your own essences, also available from them. Remember that each essence will do something different for you. It's up to you at the end of your natural make-up day whether you chose the Gomorrah massage oil or an essence such as bergamot, which has sedative qualities. Finally, to sleep well – the best of all beauty treatments – fill your pillow with sweet-smelling herbs. Take 25 grams each of the following: cinnamon, cloves, coriander seeds, lavender flowers, rose leaves and orris root; pulverize all the ingredients and put into small bags; tie them securely and slip as many as you want into your pillow. They will be much nicer than a sleeping tablet, and like all natural products will do a great deal of good *without* doing a great deal of harm somewhere else inside you.

Recommended Stockists of Natural Beauty Preparations

Aromatic Oil Co 12 Littlegate Street, Oxford (0865 42144)
This company stock over forty-five essential oils, pure vegetable oils, and their own aromatherapy products such as preblended oils and creams. They can also supply every book currently available on the subject of aromatherapy. Although they sell to some health food shops, everything can be ordered by mail directly from the company, who will be pleased to send you their catalogue and price list.

L'Artisan Parfumeur 194–6 Walton Street, London SW3 (01 584 9632)
An elegant shop, with rows and rows of the beautiful eaux-de-toilette displayed on glass shelves in open dark-wood cases, and the air filled with perfume. There are fragrances and potpourris of flowers, herbs,

leaves, fruits, vegetables – gardenia, iris, hyacinth, orange, grapefruit, celery, pepper, cloves, vanilla. There are *parfums d'environnement* too, sprays with different fragrances – a summer garden after the rain, an autumn garden, a bouquet of roses, a forest glade. L'Artisan have shops worldwide. Callers only, Monday to Saturday 10.00–18.00.

G. Baldwin & Co 173 Walworth Road, London SE17 (01 703 5550)
Baldwin's, herbalists since 1844, look like dingy, old-fashioned English chemists. They sell sarsaparilla from plastic buckets on the counter, have tall jars of lozenges on the shelves, and faded posters proclaiming that 'Balsam of lungwort is the great winter remedy.' They have drawer upon drawer of herbs, from celery seeds for rheumatism to mistletoe (a sedative), and a large stock of fluid extracts of herbs and perfumed oils. As well as their original herbal medicines, they will make up prescriptions, and though not strictly consultants they are extremely knowledgeable. Callers only, Tuesday to Saturday 9.00–17.30.

The Body Shop 1 Crane Street, Chichester, West Sussex (0243 785810)
The Body Shop are a convenient mail order company, since their range is so extensive. They do over fifteen treatments for the hair, from nettle shampoo to a hot oil hair treatment, and have herbal henna in four colours plus clear. Among the cleansers available are a honey and oatmeal scrub mask, thyme cleansing cream, and jars of beauty grains; there are also the four basic flower waters as toners: rosewater, orange flower, elderflower and honey. There is a selection of creams, oils, lotions and soaps, too, and a long list of essential perfume oils and essences, although only a few herbs – camomile and marigold flowers, rosemary, lavender and orris root. Open 9.30 to 17.30, Monday to Saturday. There are Body Shops to visit in various parts of the country, addresses listed on the company's form for ordering by post.

Crabtree and Evelyn 34 Savile Row, London W1X 1AG (01 734 1513/4)
Crabtree and Evelyn combine exceptional presentation with pure and natural ingredients in all of their soaps and toiletries. The Swiss-made soaps are triple milled, beautifully shaded, and contain a natural base with extracts and essential oils of herbs, fruits and flowers. There are almond, maize and oatmeal soaps, Scotch heather, English lavender and Devon violets, the buttermilk and goat milk, and 99.9 per cent pure glycerine soaps with rosewater, or extract of wood strawberry and

mountain raspberry, for example; also toilet waters like honey and elderflower, shampoos, and bath gels.

Creighton Laboratories Ltd Water Lane, Storrington, Sussex (09066 3452)
Creighton's have a pretty booklet to illustrate and explain the many herbs and flowers used in their preparations – the rosemary in the toner, the linden flower in the cleansing lotion, the fruit in the oils and creams of their apricot range, and the various plants in their herbal baths and shampoos. There are also cactus oil or honey and vinegar moisturizers, seaweed or protein shampoos, and a sunscreen lotion based on horse chestnut, a natural screen. Some health food shops and department stores stock the range, but you can write for the booklet and price list, and order by mail.

Culpeper 21 Bruton Street, London W1 (01 629 4559)
Named after the seventeenth-century herbalist Nicholas Culpeper, this is another shop where the herbs are prettily presented. You can buy them from tall cartons crowded underneath a table loaded with gift packs of soap and tea, or find them in linen bags or simply wrapped in cellophane packets. Culpeper's also keep most oils, herbal shampoos, scented cushions and potpourris. They have their own range of herbal cosmetic; these include a lovely milk of lilies cleansing milk, an elderflower toner, orange cream for normal skins, red elm for dry, and Velvet cream, made of cold-pressed almonds, for very dry skin. There are spices such as nutmeg, and a few packets of more unusual plants such as dried marigold flowers and juniper berries. Lastly, you can buy many books about herbs there, amongst them *Culpeper's Herbal*, first published in 1653. Callers only, Monday to Friday 9.30–17.30, Saturday 10.00–17.00.

Daniel Galvin Colour Salon 59 George Street, London W1 (01 486 8601/2)
An excellent London *salon* who specialize in colouring and treating hair with their own natural products. They use henna, saffron, rhubarb roots, marigold, orange and bilberry, as well as other vegetable dyes, for colouring hair; cactus oil, avocado, mint, and organic protein in their conditioning treatments. Conditioners may be bought for home use. They also have a witch-hazel antistatic lotion for use before blow-drying. Open daily 9.00–17.00.

L'Herbier de Provence 351 Fulham Road, London SW10 (01 352 0012)
A delightful shop, the floor covered by blocks of sacks containing over 200 varieties of aromatic, culinary and medicinal herbs. The shelves are stacked with bars and cubes of soap – honey, olive or palm oil, olive, and lavender for dry skin; oil of bitter almonds for normal to oily skins. They also stock pure olive oil, spices, syrups, honey and vinegar, and, for presents, cotton lavender bags, nougat, and *calissous*, a delicious sweet made of orange blossom water, vanilla, egg whites and almond paste. Callers only.

Martha Hill 39 Marylebone High Street, London W1 (01 935 4050)
After successfully using natural ingredients on her skin, Martha Hill now produces cosmetics from herbal extracts, vegetable oils, and vitamins. They are professionally formulated in laboratories in Switzerland. The skin care products consists of a cleanser, gel, and day and night creams, designed to work together. Martha constantly improves her products; for example, by next year there will be a new range incorporating a sunscreen, and she has recently brought out a seaweed shampoo, scalp tonic and conditioner. These contain coconut oil to stop dryness and camomile for the shine, and they are scented with lavender and rose. There are nearly forty different preparations for make-up, skin care and hair available at Martha Hill's shop or by mail order. Open Monday to Saturday 9.45–18.30.

Liberty & Co. Ltd Regent Street, London W1 (01 734 1234)
The beauty department at Liberty have quite a selection of high-standard natural products: the Crabtree & Evelyn and Truc of Switzerland ranges, the New Age bath and body oils, including their Egyptian Massage oil, jasmine and rosemary, and their boxes of Old English lavender flowers. Liberty are also the only shop keeping the small Molton Brown range, containing camomile and rosemary shampoo, seaweed setting lotion, aromatic hair oil, and elm and lemon hand cream. On the other side of the store you will find the herb shop. Here you can buy a selection of the more unusual herbs under the name of L'Herbier de Provence, sold from sacks and packaged for you; these include cornflower, flower tea, and orange blossom. Liberty also stock spices, honey, teas, sea salt, their own packets of herbs, and a chart listing various uses for herbs. Open Monday to Wednesday and Friday 9.00–17.30, Thursday 9.00–1900, Saturday 9.30–17.30.

Leslie Kenton

5 Health Food around the World

For the adventurous traveller, one of the most enjoyable aspects of visiting other countries is the opportunity to sample local cuisine. But what about health or wholefoods – is it possible to eat healthily abroad while at the same time sampling regional specialities? The answer to that question is a resounding yes! Health food stores and restaurants are a common feature nowadays in many countries; and while conforming to wholefood standards, the products to be found in these establishments possess the distinctive flavours of the country.

To begin at the nearest point to Great Britain, health foods have long been a fact in FRANCE. This may come as a surprise to some, considering the French reverence for *haute cuisine*. For many years only a minority insisted on the benefits of wholefoods (but a minority large enough to sustain health food stores in all parts of the country); recently this concern has become far more widespread. Whereas at one time it was only the well-known long *baguettes* ('French bread') that could be found at local bakeries, now *pain complet* (wholemeal bread) is quite readily obtainable and is very popular. Health food shops stock a large variety of goods: all the basics like wholemeal flour, grains, nuts and dried fruit, and also specifically French specialities made with unrefined ingredients like *crème de marrons* (sweetened chestnut purée), containing only raw sugar. It is a pity that more of these high-quality health products are not imported into Britain but their retail prices would be prohibitively high.

Health food restaurants have never flourished in France to the extent they have in some other countries, though there are well-established exceptions, not only in Paris but in Lyon and Nice. The big craze in Paris some years ago was for macrobiotic food. Many of the restaurants established at that time are still running, and the wholefood-minded visitor to Paris will always find satisfactory meals at any of them.

There are a number of *pensions* in France catering for those who prefer wholefood, and also a few health hydros. The emphasis tends to be on the hearty, keep-fit, spartan approach that one associates more with the Germans than with the supposedly sybaritic French, but then perhaps it is a form of reaction against the latter tendency. Up-to-date information on restaurants, *pensions* and hydros can be obtained from Mrs L. B. Doro, General Secretary, Association Végétarienne de France 8 bis rue Campagne-Première, 75014 Paris (telephone 633 43 25).

BELGIUM and HOLLAND also have plenty of health food shops, and Holland in particular has a large number of natural food restaurants. In proportion to its population, Amsterdam must have more of them than any other city in the world. When the macrobiotic craze swept Europe, the liberated young Dutch took to it in a big way. There are in fact two distinct strains in the Dutch health food movement: the older one falls in the hale, hearty, clean-living tradition, while the newer one is associated in people's minds with marijuana and pop music (not necessarily justifiably). In any case, with the many wholefood eating places there, the visitor can take his pick. An up-to-date list can be obtained from Mrs Eikeboom-Brokeman, Secretary, De Nederlandse Vegetariersbond, President Kennedylaan 146/11, Amsterdam (telephone 020 446132). One of the world's leading names in health food is Lima of Belgium; their fine products are available in most health food stores.

The spartan strain is very much in evidence in SCANDINAVIA. There is a thriving manufacturing industry of health food products, and many health food shops and restaurants; but it is the health hydros that are really significant in Scandinavia, many of them doubling as guest houses. Sweden alone has well over a dozen, which, in a country with its population, is amazing. A full list is available from Svenska Vegetariska Foreningen, Radmansgatan 88, 113 29 Stockholm (telephone 08/32 49 29).

The most health-food-minded nation of them all on the Continent is undeniably GERMANY. Even the smallest town is likely to have two or more health food shops, known as *Reformhausen*, and most chemists also have a health food section. Germany is unique in having an organization which vets all new health food products before they appear on the market, giving its seal of approval. This is then incorporated in the packaging designs of those which conform to its high standards of what constitutes a genuine health food. The seal of approval satisfies customers that they are getting real value for their money and is therefore much coveted, particularly as these products are exported to other countries, notably Austria and Switzerland. (Attempts have been made from time to time to introduce such a system in Britain, none of them so far successful.) Wholefood restaurants can be found in most German cities, and the food is usually very tasty (in the traditional rather than the macrobiotic style). German eating habits on the whole leave much to be desired health-wise – saturated animal fats are heavily relied on, and refined cakes and pastries with mounds of whipped cream regularly indulged in – but at least the Germans have never given up eating genuine black, sour-dough rye bread.

In Germany, apart from the German Vegetarian Union, there is a health food organization, the Bund für Lebenserneuerung, eV, Herr Rudolf Meyer, D3000 Hanover, Munzeler Strasse 18 B. There is also a separate health food organization for young people, the Deutsche Reform-Jugend, Secretary, Professor Dr Klaus Detering, D23 Altenholz-Stift, Posener Str. 26.

Germany is another country with a great number of health hydros, some of them also guest houses. Many of them publish glossy promotional leaflets and are very professional. A similar situation exists in AUSTRIA, with many such establishments situated in very beautiful, peaceful mountain areas.

The Austrian health food organization is the Österreichische Vegetarier Union, Secretary, Dipl-Ing. J. Fleischanderl, A8010 Graz, Lechgasse 2.

SWITZERLAND is in many ways another Mecca for the health food enthusiast. It was here that the great dietician Ralph Bircher-Benner formulated the recipe for the well-known Bircher muesli, the combination of oats, nuts and fruit which has now gained such popularity in Europe and America. Many ordinary restaurants in the German part of Switzerland have Bircher muesli on their menus – not just for breakfast but also for lunch, tea or as a dessert. Apart from the renowned Bircher-Benner Clinic in Zürich, there are not a large number of hydros in Switzerland; but there are plenty of health food shops. Zürich, for example, has three vegetarian restaurants, each offering an enormous and varied menu, catering for everyone from the raw-food fan to the devotee of Indian food. There are also some vegetarian *pensions* in Switzerland, and they can be found in all three parts of the country, German, French and Italian. The Swiss contact address is Schweiz, Gesselschaft für Vegetarismus, CH-5614 Sarmenstorf, Postfach.

Turning to ITALY, one cannot really speak of a health food movement as such. There is a very active vegetarian society, and they publish a list of restaurants willing to cater for vegetarians, but the emphasis is on the ethical rather than the wholefood angle. There is one macrobiotic restaurant in Rome, but in general the whole idea of a shop or restaurant devoted to wholefoods is quite foreign to the Italian character. On the other hand, one cannot claim that all the recent discoveries about the importance of fibre in the diet have simply passed them by, for there has been a slow but steady growth in the demand for wholemeal spaghetti and macaroni.

There is nothing really comparable to our own health food movement in SPAIN or PORTUGAL either – although, rather surprisingly, local vegetarian societies exist in most Spanish cities. Barcelona has a long-established vegetarian restaurant, and there is a health hydro in Valencia. Lisbon has no less than four vegetarian restaurants. One reason why there is not more talk about health foods in Spain (and this would certainly hold true of a country like Greece as well) is that in the Mediterranean climate fresh salads and fruit are a mainstay of the diet anyway, and much locally grown produce is free from chemicals. It is only when an affluent society strays too far from the ideal diet that it is necessary to form organizations to promote a simpler style of eating.

The country which perhaps best combines natural conditions and enlightened knowledge is ISRAEL. A superabundance of natural produce is readily available all the year round, and the people take full advantage of what they have. Visitors are astonished to find that Israelis breakfast on fresh salads and cheese, and that freshly made fruit juices are sold on street corners. Given this wealth of natural health foods it might seem unnecessary to have special health food shops, but Israel is very much a western country with some bad western habits – like eating refined foods. Fortunately there has always been an enlightened minority aware of the hazards of this way of eating, and there have always been shops and restaurants catering for them. In recent years there has been an increase in diet awareness, so that health food stores are more numerous and better stocked than ever before, and so-called 'dairy restaurants' proliferate. There are also three fine natural health centres in the country. Further information from Israel Vegetarian Movement, David Nahum, Secretary, PO Box 40026, Tel Aviv.

It comes as rather a surprise to find that there is a small health home in EGYPT; on the whole, Arab countries of the Middle East are in no way geared to ideas of natural health. In the case of the majority of the population the diet is basic and on the whole healthy enough, with *légumes* and olive oil as staples.

IRAN is a similar case, though in the cities it would be very difficult to find a meal that was without meat; meat is regarded as a symbol of affluence. However, fresh salads are readily available; fruit juices are made with a juicer on street corners, and pistachios and other nuts are easy to find. One curiosity in Teheran which might interest wholefood-minded visitors (if it still exists) is a Raw Vegetarian Society and Restaurant: 2 Kamkar Avenue, Sanai Place, Teheran (tel. 828878). It is the brainchild of an Iranian, Arshavir Ter Hovannessian, who is perhaps one of the world's most forceful exponents of a diet consisting entirely of raw foods.

Raw foods are all too seldom eaten by those living on the Indian subcontinent. Of course, there are tropical fruits available, and these are cheap enough to be eaten by virtually anyone, but salads are a comparative rarity in INDIA. Part of the reason for this is that so many vegetables grown there are very tough and require much cooking to make them palatable; this is also one reason why everything is curried. Curry is not just one flavour. A typical Indian vegetarian meal would have three different vegetables, each of them flavoured with a distinctive combination of spices chosen to suit the particular vegetable. A constant diet of highly spiced fried foods served with white rice could hardly be considered ideal; on the other hand, saturated animal fats are not often eaten (clarified butter, known as *ghee*, always used to be the cooking medium, but this has now largely been replaced by vegetable *ghee*). *Chapaties, puris* and *nan*, in fact all Indian forms of bread, are made with wheatmeal flour, so that at least the germ if not the bran is eaten by most people. The biggest problem there, of course, is that the poor do not have enough to eat, while the rich are liable to overeat; under these circumstances it will be a long time before western ideas of health foods are relevant to the East.

In the FAR EAST, too, patterns of eating are such that one could hardly expect to find health foods on the western model. However, even in a prosperous country like Japan, traditional eating habits are to a large extent adhered to. And while the use of refined white rice throughout the East is deplorable, the diet on the whole is much closer to a health or wholefooder's ideal than anything found in the West. Vegetables, when cooked, are usually only lightly stir-fried; various nutritious seaweeds are a regular part of any eastern meal; and *tofu* (soya cheese) is a much more common form of protein than meat. It is well worth trying to break past the language barrier in the Far East in order to eat the same foods as the inhabitants of the country.

In THAILAND, BURMA and other tropical countries, fresh fruits like mango, papaya, pineapple and watermelon are sold on street corners, chopped into convenient sizes ready for eating. For a popular drink, the top of a green coconut is hacked off and a straw thrust in to drink the milk out of the naturally cooled interior. For another thirst-quenching drink, sugar cane is pulped and the juice served with ice; it is not over-sweet, as it is close to its natural state.

One would have to move still further east – to AUSTRALIA – before finding western-style affluence and, as a corollary, western-style health foods. Aside from local vegetarian societies there is a Natural Health Society of Australia: 131 York Street, Sydney 2000 (tel. 29 8656). Health food stores and restaurants can be found in all the major cities. One of the biggest manufacturers of health foods in Australia has connections with both the USA and the UK; called Sanitarium Health Foods in Australia, Loma Linda in America, and Granose in Britain, they are all Seventh Day Adventist organizations. This religious sect advocates a vegetarian diet, and in order to make this easier for their members they developed, earlier this century, a wide range of tinned substitutes for meat. (Tests carried out on large numbers of Seventh Day Adventists have proved that they are in general healthier than the population as a whole.) Some products are manufactured in the USA and then exported to Australia and Britain – soya milk powder being the most notable example – but in most cases the products are made locally, tailored to fit the different national requirements. Browsing round an Australian health food shop will therefore reveal many unusual items to sample.

So will browsing in a health food store in SOUTH AFRICA. Considering the relative cheapness there of exotic fruits and nuts, it is logical for these to be incorporated into many health food products. Johannesburg and Cape Town each have a vegetarian society and restaurant, and there is also a health hydro in South Africa. The address of the South African Vegetarian Union is PO Box 23601, Joubert Park 2044, Transvaal (tel. 46 1078). Most of the African continent eats in traditional ways; it was the absence of cancer of the colon in Africans which led researchers to discover the connection between this disease and a lack of dietary fibre. Maize is the staple in most of Africa, and it is still ground at home by the people themselves.

There could hardly be a greater contrast in diets than between Africa and the UNITED STATES; it is doubtless because the latter has departed so completely from a natural diet that the swing back towards health and wholefoods has been so widespread. It is still true that the majority of Americans live mainly on junk foods – convenience meals, tinned, frozen or dehydrated, artificially flavoured and coloured, pumped full of preservatives. However, the possibilities which exist in a nation with such resources have been avidly seized by an ever-increasing proportion of the population. Nowadays even supermarkets – those arch purveyors of junk foods – generally stock a good selection of health foods as well as wholemeal bread. There is an incredible variety of fresh produce available all the year round at reasonable prices – and the growing demand for fruit and vegetables grown organically is being met by a host of growers supplying health food stores across the nation; virtually every health food store in the USA has a large section for fresh fruit and vegetables grown without chemicals. With the selection of vegetables available, health food restaurants are able to come up with incomparable salads.

Another advantage which health food restaurants have in the USA is the diversified ethnic background which has long accustomed Americans to eat the foods of many different nationalities. On the menu of a health food restaurant one might find Chinese chop suey, Mexican chilli con carne, Italian pasta, Indian *idli-dosas*, etc. The enormous number of such restaurants all over the continent (and including Canada) means that they range from small take-away bars to gourmet restaurants. The latter serve true *haute cuisine*, and a few use only organically grown and unrefined ingredients. The fact that they are able to charge prices comparable to any other highly-quality restaurant is indication enough that they are not patronized by young faddists. For information on these, contact Vegetarian World, PO Box 46187, Los Angeles, California 90046.

American health food stores – some of them as large as British supermarkets – are also a revelation. In recent years the shift has been away from tinned imitation-meat products towards whole cereals and grains. Some of these have even been used for convenience meals; for instance rice, millet or cracked wheat combined with dehydrated vegetables and herbs and spices – easy to prepare, but with nothing refined and no artificial flavouring or chemical additives. Modern American research and technology is here put to exceptionally good use. Similarly, there is generally a frozen food section full of wholefoods – pizzas made with wholemeal flour, macrobiotic-type burgers made from brown rice all ready for frying, Mexican *tamales*. Even those preferring raw food are catered for; there might well be frozen savouries and sweets made entirely from uncooked ingredients. It is difficult to deny that America is way ahead of most other nations when it comes to the quality and availability of health foods.

Leah Leneman

6 A Health Food Experience

Macrobiotics . . . the word has an ominous ring. It might be the name of a new anti-drug or a whiter-washing detergent. The word in fact comes from the Greek and means great (macro) life (biotic) – the *great life*!

When people ask me about my eating habits – which they frequently do – I find talking about it rather difficult. I am not an authority on the subject and have learnt whatever I know mostly by reading books about it. If the question is an idly curious one or a plain conversational gambit, I can't help feeling that the person asking me is disconcerted when I suddenly let fly, however gently, with words like Zen Buddhism, *yin* and *yang* and *unique principle*. He or she must think they've struck a religious fanatic – and an Oriental one at that! Macrobiotics is so simple that it is almost impossible not to make it sound complicated, partly because it is from the East but mainly because we live in a complicated, sophisticated society in which simplicity is often interpreted as stupidity. When explanations are demanded I usually say that some foods are acid, some are alkaline, that it's a matter of balancing the foods we eat just as we try to balance everything else in life, and I leave it at that.

Four or five thousand years ago the Zen Buddhist monks had learnt which foods were acid and which were alkaline. They also ate food grown in their own climate and vicinity, and naturally had no chemicals, preservatives, insecticides, fertilizers, dyes, bleaches, protein supplements, or vitamin additives to cope with, either. We can still learn from their discoveries today. A great deal of our sophisticated western solid and liquid diet is very acid, and most of us are on a perpetual acid trip which, over the years, can become wearying and disillusioning, if not unbearable.

Georges Ohsawa at the beginning of this century was lucky enough (he considered) to be condemned by doctors to tuberculosis. He was saved by the diet of a Japanese healer named Ishizuka and went on to devote his life to promoting, lecturing and publishing books on the philosophy of macrobiotics all over the world. According to Ohsawa:

Macrobiotics is not scientific, religious, mysterious or superstitious. It is the art of living . . . It is not a mystical Oriental cuisine. Some people think it's brown rice and giving up pleasure at meals . . . it is an understanding of the orderliness of nature ... it enables us to prepare delicious meals and achieve a free life. The most important principle of macrobiotics is gratitude ... without it there can be no freedom, no happiness [*Macrobiotics: An Invitation to Health and Happiness*].

Ohsawa passionately believed in his cause. His books are written in English, not his native language; consequently he uses some rather quaint expressions which make macrobiotics sound more folksy than it should. But it is the simplicity of the great concept of *yin* and *yang* that I like, the balance and interdependence of opposites: female and male, cold and hot; dark and light, evil and good – acid and alkaline. Each needs the other; neither is better or worse than the other.

While I was in New York a few years ago, a friend, to whom I shall always be grateful, gave me a book as a joke. It had the funny title *You Are All San Paku*, and was written by William Dufty, who had suffered, it would seem not only from pyorrhoea, migraines, constipation and piles but had an intriguing marriage break-up which resulted in his being sent to a civil jail, where he fasted. He was honest enough to relate in full these troubles, his adventure into macrobiotics which cured him of all the physical ailments, and his later meeting with Georges Ohsawa himself. *San paku* is Japanese for having three whites to the eye instead of two, the third being below the iris and above lower lid. It is a bad way to be. I was, I saw in the mirror, *san paku*.

When I was in my late teens, I first became aware that what I ate affected my health and general well-being when I suffered – and suffered is the word – an attack of quinsy. My throat and tonsils were ulcerated and swollen. I went to hospital and was given the appropriate drug cure of the time and told I'd better have my tonsils out. However, when the onslaught subsided a wise friend of mine sent me instead to a naturopath. He told me that our tonsils are what he called a danger signal indicating when too much 'poison' was being taken into the body for it to be rid of naturally. Recooked fatty foods, dripping and rich sauces, gravies and sweets were, he said, especially dangerous. My landlady, bless her (still alive, a widow in her eighties), was offended, but I stopped eating such foods. I have had no more ulceration of throat or tonsils since. The fact that I came to London quite soon after this and lived sparsely on £2 a week, forced to eat simply because of postwar rationing, probably helped my health and shape a great deal. I was also exercising daily at the Old Vic School – exercise is important too. But once I started working in the theatre, had some success, and made enough money to eat the way I felt I was expected to, I wined and dined my way through life and slowly grew plump on restaurant food, perpetually, consuming supposedly slimming steak and salad, claret and green figs, or gâteau if I felt thin enough. I made the occasional vague exploration into vegetarianism, which of course meant considerable intakes of *yin* or acid fruits and vegetables. In New York this consisted mainly of drinking large glasses of sweet and syrupy prune or orange juice from large plastic cartons, or eating baked potatoes and tomato salads at the health food bars, none of which did my feeling of ennui much good! I sometimes felt that I had a perpetual hangover.

It took me three or four months before I finally got around to reading *You Are All San Paku* from cover to cover. When I did, it all made such startling, simple, thundering sense that I knew I wanted to try this ancient, and rediscovered way of eating. I was by now back in England on a pre-London tour and had taken the book with me to read. The idea of violent death and assassination associated with *san paku* was brought home by the fact that several of my friends had recently had car smashes, and J. F. Kennedy had been assassinated not long before. Ohsawa had apparently predicted this.

When I got back to London and the play I was appearing in settled down to a run, the decks in the kitchen were cleared for action. I didn't think about what would be involved, or even much about the consequences – I seldom do. I'm inconsiderate that way – but I do cook a good deal of my own food. Everyone felt sorry for my wife but, after all, Ohsawa says that love is an essential ingredient of cooking, and she tolerantly brought her own particular expertise and sensitivity to the preparation of a new type of food. In fact, cooking with grains can be extremely satisfying.

Rice and vegetables are really good with *gomesio* (sesame salt) and *tamari* (soya sauce) and no trouble really to prepare. I went through no revulsion period, but then I wasn't strictly on 'regimen seven', which is rice only and as little liquid as possible. I had a few braised vegetables and sometimes prawns with the rice, and within three or four days I was on a high that was really beautiful. I was aware, on stage and off, of a sharp clarity that was new to me. I also noticed that I seemed to lose that awful nagging terror I used to have in front of an audience and which made me sweat. In six or seven days my body, which was being stripped off for a nude scene in *Abelard and Heloise* each night, started to lose all those fatty bulges. In ten days I'd lost twenty pounds, much to everyone's concern, because I really didn't need to lose twenty pounds – ten, possibly. Diana Rigg, who was in the scene with me, said that the audience wouldn't be able to see me at all if I went on like that! I felt so much better in myself, though, that I did go on. I got a really terrible cold which a sympathetic friend generously called a 'recovery crisis'. It was the last really bad cold I've suffered.

Six months later we took the production to California, and I had to admit that, though the body definition was interesting, I was undoubtedly getting too thin for an actor. I was sleeping fewer hours and better – without the fearful waking up at three or four o'clock in the morning. I called in one day to the Macrobiotic Foundation there and spoke to a woman who was sensible and practical. What she said amounted to this: Ohsawa was Japanese; his way of eating was based on Japanese foods – rice, seaweeds, fish, etc.; he made no sacrifice but simply adapted his discovery of Zen principles, which were also Japanese. But it is the Zen principle of *yin* and *yang*, acid and alkaline, that one needs to remember. Eat anything your body craves – steak, roast beef, whatever – but avoid extremes and always balance it. By now I found the flesh flavour of meat unpleasant and wasn't missing it; sugar burnt my mouth; but I was missing other foods – and, when you think of it, most peasant dishes are basically macrobiotic. Japanese *tempura* is delicious, but paella from Spain, for example, balances fish and prawns with rice and vegetables; Arabic couscous (which need not be made with meat), Russian buckwheat dishes, Welsh and Irish lavabread recipes, Italian pastas (watch the sauces, though!); Greek dolmades, Scottish oatmeal foods – they are all balanced. Peasants have always eaten simply and sparingly. We mostly eat too much and too quickly. Macrobiotics fundamentally has come to mean to me eating a frugal but balanced diet of grains and wholefoods. And if anyone suggests that this is to eat dully or monotonously they don't know the right cooks!

We are not Zen Buddhist monks, and few of us are suffering terminal tuberculosis as Ohsawa was. We live in a hectic society and all have hurdles to jump, needing, like athletes, to be able to work up a good healthy controllable aggression (and to enjoy it) every now and then. But we also need to be able to relax of our own accord without drugs or tranquillizers, and that is when the cereal diet helps.

I was told that macrobiotics has been frowned on by the medical profession. Ohsawa said that modern symptomatic medicine treats the effect and not the cause. In America a few young people have, I believe, suffered from dehydration as a result of the diet. Indeed it is very easy to be fanatical at first and go to extremes too early. The whole concept of Zen is, after all, balance. It is dangerous to go suddenly from the over-indulgences of a western diet, with animal fats, white sugar, caffeine, and alcohol, to the basic necessities of frugal 'regimen seven' eating rice only. However, when people are really sick from overindulgence the specialists they consult invariably put them on to exactly such a macro-type diet. Usually it is too late and too extreme to be of use, and the diet gets the blame. However, 'regimen seven' is excellent for a few days if you aren't well. I am always being asked why on earth I want to live to be a hundred. It is not the length of time I live that concerns me, but the quality of life I am living, and this, in my case, is undoubtedly affected in an extraordinary way which I shall try to describe, though there are those who say it is all in the mind!

The chief advantage of macrobiotics, in my opinion, is a change in attitude to life and its problems when you balance your nerve-making acid intake. Let's face it, life is a pretty incredible and exciting experience. It should never be allowed to become frightening, boring or infuriating, as it had tended to become for me. But to keep taking drugs, tranquil-lizers or large quantities of alcohol to cope with it is stupid. Of course your problems don't vanish overnight, but they are easier to manage and even quite enjoyable in perspective. There is far less to fear than you thought. I sometimes think that because animals are killed in fear we must suffer distress in some way when we eat them. Probably only through the study of medicine would I understand exactly why my change of food was so effective.

I love food and I always have. But those dreadful compulsive hunger urges have gone away; also the stupid conviction that, even while one is actually eating, one will never eat again, which makes one scoff the food from the plate without chewing each mouthful enough times to make it comfortably digestible. There are difficulties to overcome as well: eating on tour is not easy; restaurants often cannot cope with a macrobiotic's needs and hostesses at dinner parties can become sullen and offended after they have prepared something 'special' for you which you just can't eat. But it is all worth while because you'll feel much healthier and happier for it.

Just after I became interested in this way of eating, it seemed that all the local health food shops suddenly started stocking the necessaries. Strange-sounding condiments such as *gomasio* (sesame salt), *tahini* (sesame paste), *tamari* (soya sauce), with other ingredients like *umboshi* plums, seaweed, *bancha* and *mu* teas – a wealth of new flavours and food textures were becoming popular enough for shops to stock them. I helped one young American actor I worked with to cure his stomach ulcer; he took to macrobiotics wholeheartedly. Many younger people were becoming interested in the subject, and books on macrobiotics were often sold out. My older friends were eating 'plastic' sausages and chips, and steak and eggs, and suffering from indigestion, migraine, constipation, piles and ulcers, though not necessarily all of them, or all at once!

There is always the problem (which this book is tackling) of finding health food, wholefood or organic food that has not been sprayed with chemicals or subjected to additives, preservatives, dyes or bleaches. Ironically, the food that has had nothing done to it costs more today than the food that has suffered. Why 'white' and refined food has become fashionable over the years I can't imagine. The difference in texture between wholegrain bread and fluffy bleached white bread, between brown rice and white is remarkable – not only the taste but the food content. The removal of minerals and vitamins from our food sources should not be tolerated. There is even brown bread on the market which is dyed! I did not realize until recently that most commercial teas are dyed to make them the right colour, and that they also contain caffeine and tannic acid. *Bancha* tea is a delicious clear tea; it doesn't need to be constantly 'made fresh' because it doesn't go sour, and it has had the caffeine burnt out by long exposure to sun for three years or so. A pot on the stove can produce good tea for several days. There's also *mu* tea, which one journalist thought was the Michell family name for tea with milk, and dandelion coffee.

I haven't gone into detail about *yin* and *yang* or given recipes, but if you are interested any good book on the subject will tell you, and explain which foods are acid and which are alkaline. As I say, macrobiotics is all very simple – which the commercial food produced by competitive manufacturers is not; so much of our food is interfered with and used for by-product sales before we get it. We are constantly being conditioned by some pretty lethal publicity campaigns too. According to psychiatrists in an American article I read recently, food faddists are neurotic, and 'normal' people will eat anything! They may be right.

Keith Michell

Part two

A Dictionary of Health Food Terms

Acids

An acid is defined as a substance which, when combined with a base or a metal, will form a salt. Thus, hydrochloric acid will combine with caustic soda to form sodium chloride, or common salt.

There are two types of acid; strong acids such as hydrochloric and sulphuric, and weak acids such as acetic, citric, tartaric, etc.

All acids have a sour taste, the strong ones reacting much more vigorously on the tongue than weak acids. The weak acids are commonly found in fruit; for instance lemons contain a high proportion of citric acid. Tartaric acid is found in wine.

Most unripe fruits contain malic acid and the amount present becomes less during the ripening process as this and other organic acids are converted into sugars, thus the taste becomes less sour. The body can use the weak fruit acids as a source of energy, these being oxidized to carbon dioxide and water, giving no residue to be retained in the body.

As well as fruits, most vegetables also contain weak acids, but these are usually in the form of salts and therefore vegetables do not have a sour flavour.

Because the weak acids in fruit and vegetables are completely oxidized in the body, any bases with which the acids were combined will be available to neutralize excess acid in the body and in this way maintain the acid base balance. It is for this reason that excess acidity in the body can be alleviated by taking a diet of fruit, which at first sight is acid in nature. The base so released will neutralize purin acids, such as uric acid, and and also phosphoric and sulphuric acid which are formed by the metabolism of protein foodstuffs.

In solution, acids dissociate to liberate hydrogen ions. Weak acids dissociate far less than strong acids. The degree of dissociation of acids is measured in terms of hydrogen ion concentration which is usually written as pH. The higher the pH number, the weaker the acid. Thus a solution of hydrochloric acid may have a pH of 1.0 and a solution of citric acid a pH of 4.0. The bases sodium and potassium present in the body will neutralize these acids and raise the pH to the region of 7.2 to 7.4, which is the normal figure for blood and body humours. pH 7.2 is known as the neutral point, where all acids are completely neutralized by bases (see also *Alkalis*). R.F.M.*

*See List of contributors, page 422

Aduki beans

Known as the 'prince of beans' in the Orient, the small red bean has always been held in high esteem as a valuable food. It owes some of its popularity to the sweet flavour produced when cooked to juicy softness. It is the sweetest of beans and is, in fact, often used as the basis for sweet pies and desserts.

As a healing food, aduki beans have long been recognized as tonic food for kidneys and a regular amount in the diet can have noticeable effect if these organs are ailing. An aduki tea can also be taken for kidneys and is made by boiling a teaspoonful of beans in two cups of water (preferably spring water) until it has boiled down to one cupful.

They should be cooked as most other beans. Soak first for best results. Cook one part beans to three parts water, bring to the boil and simmer in a covered pan for one hour. When the bean is soft, add sea salt to flavour and cook for an additional ten minutes. They can also be pressure-cooked or baked. The cooked aduki beans can be incorporated into breads, soups, and savoury or sweet cakes (see also *Macrobiotics*). G.S.

Alkalis

The bases sodium and potassium are also known as alkalis. Other bases which have an alkaline nature are calcium and magnesium – also common components of biological systems – and ammonia, which substance is formed from the metabolism of protein foodstuffs. Sodium and potassium are strong bases whereas ammonium is a relatively weak base.

In concentrated form sodium and potassium as the hydroxides are very caustic and will burn the skin in much the same way as do strong acids. In foodstuffs and in body fluids sodium and potassium are never found in the free state, but always combined with acids.

The bases sodium and potassium when combined with very weak acids have an alkaline reaction in solution and such solutions would have a pH greater than 7, possibly as much as 8 or 8.5.

An example of the alkaline nature of a combination of a weak acid and a strong base is soap, which is the sodium salt of a fatty acid derived from the hydrolosis of fat. The alkaline nature of soap allows it to be an excellent cleansing agent in dissolving grease and as a surfactant so as to allow the removal of dirt from clothes, the body, etc. Sodium combined with taurochloric acid is one of the salts present in bile, which is like a soap. This substance helps to disperse fat globules to assist in digestion by stomach enzymes.

In the bloodstream the alkali sodium is partly combined with one molecule of carbonic acid to form sodium bicarbonate. As more carbon dioxide is brought into the bloodstream from the tissues as a result of the metabolism of sugars and fats, so the bicarbonate becomes changed to carbonate which contains two molecules of carbonic acid. On transporting the sodium carbonate to the lungs, the second molecule of carbonic acid is dissociated and exhaled as carbon dioxide and water, leaving in the bloodstream the original molecule of sodium bicarbonate, which recirculates to again do the job of transporting more carbon dioxide to the lungs.

Although sodium is present in all fruits and vegetables, it is in concentration much less than that of potassium which is the predominating alkali. The potassium salts are strong diuretics and in consequence a high intake of fruit and vegetables invariably leads to increased loss of fluid from the body via the urine. This is one of the reasons why a diet of fruit and vegetables is so useful for reducing body weight (see also *Acids*). R.F.M.

Aluminium

Although aluminium represents about 8 per cent of the earth's crust, it is almost completely absent from plant and animal tissues, rarely exceeding 10 parts per million. There are certain exceptions to this, e.g. mosses such as club moss and sphagnum moss. The latter may contain up to 30 per cent of its dry weight of aluminium. Although soil contains so much aluminium, it is in a particularly insoluble form and certain acid soils become infertile due to the solubilization of minute quantities of aluminium. Human and animal tissues contain even less aluminium than plants, and the level in blood is less than 1 part per million. There is no evidence to date that aluminium is a positively essential element in human nutrition (see also *Trace elements*). R.F.M.

Amino acids

In the same way that sugars are the building bricks of starches, so amino acids are the components of proteins.

All the tissues in the body other than bones and fatty regions are composed of protein which substance has been formed as the result of the polymerization of amino acids, either brought in in the form of protein foods or synthesized in the body.

Chemically an amino acid is a derivative of a weak fatty acid and ammonia. Thus acetic acid when combined in a certain way with ammonia will become the amino acid glycine.

There are about forty naturally occurring amino acids and these have very considerable variation in their structure. They are mainly composed of carbon, hydrogen and oxygen, as well as the nitrogen derived from the amino group. A few have sulphur in addition, and some have iodine.

In the formation of proteins, the various amino acids combine together via the peptide linkage, that is to say the acid group of one amino acid will combine with the amino group of another. Such a compound will have the power to combine with two more amino acids and in this way a whole complex can be built up to make a protein. Protein molecules can contain thousands of amino acid molecules joined together in this fashion.

It is obvious that because there are at least thirty naturally occurring amino acids, the permutations and combinations which can be achieved by building these acids up into structures many thousand times the original size explain why there is such a wide variety of naturally occurring proteins.

In the same way that proteins are built up from amino acids, the reverse process can occur. Thus when protein foodstuffs are cooked, they become partially broken down to polypeptides in which form they are more readily digested. In the alimentary canal the digestive juices contain enzymes which will break down the proteins and polypeptides to the individual amino acids, and when digestion is complete, these simple amino acids can be absorbed across the wall of the gut and so taken into the bloodstream for utilization as food or for the building of new cells in the body.

When amino acids are metabolized in the tissues, the end products are carbon dioxide and water, the same as in the case of the organic acids in fruit, but in addition the amino group becomes converted to urea in which form it is excreted via the urine.

To a large extent the amino acids found in plant life differ from those which are found in the animal kingdom. In particular, animal foods contain the amino acids tyrosine, tryptophane and phenylalanine, which are not found to any extent in fruits and vegetables. These amino acids are synthesized in the alimentary canal of cattle and other herbivores and turned into tissue protein. The flesh of herbivores becomes the source of such amino acids in the diet of carnivores which feed on the herbivores. One such amino acid, tyrosine, is essential for building up the protein in the hormone substance secreted by the thyroid gland (thyroprotein).

The amino acids so mentioned are sometimes called 'essential amino acids' in that they are not capable of being synthesized in the human body, but must be taken in via protein foodstuffs. The substance adrenalin is also based on the amino acid tyrosine. Diminished intake of these essential acids can result in diminished thyroid function and diminished adrenal tone, resulting in lowered blood pressure (see also *Proteins*). R.F.M.

Apple cider vinegar

Apple cider vinegar contains calcium, chlorine, fluorine, iron, magnesium, phosphorus, potassium, silicon, sodium, sulphur and many trace minerals, and is a valuable food for rebuilding the body, especially during middle age and later. It will render ineffective disease bacteria in the alimentary tract, it helps control the intestinal flora and it is invaluable in improving intestinal disorders. Failure to obtain the acids found in this food, with their accompanying minerals, often results in illness.

Physiology textbooks tell us that the body produces a new bloodstream every twenty-eight days. The blood is slightly alkaline, but if its alkalinity is raised above ordinary levels calcium will be deposited in the tissues. By taking a simple acid like apple cider vinegar, calcium deposits in the blood vessel walls can be prevented.

It is obvious, then, that the composition of the blood is not the result of chance but is determined by the food and drink one assimilates, which forms the bone marrow where red blood cells are prepared.

Apple cider vinegar normalizes and improves metabolism; that is to say, it invigorates the kidneys and helps them to drive out poisons and infections. It is instrumental in the proper oxidation of the blood, prevents loss of blood from the body and tenderizes the tissues. It improves digestion and bowel action and its regular use will benefit or completely cure muscle cramps.

Its influence on the metabolic rate of the body makes it one of the best and safest correctives for obesity because it helps in the natural assimilation of carbohydrates so that they will give energy where otherwise they would simply produce fat. Other conditions which benefit from the use of apple cider vinegar are bleeding haemorrhoids, chronic headaches, dizziness, high blood pressure, nose bleeding, deposits of tartar on the teeth, sore throat and pyelitis. The inclination to bleed is lessened or prevented when an acid like apple cider vinegar is taken, because it changes the alkaline urine to one with the normal acid reaction.

It can be used instead of wine vinegar or lemon juice in salad dressings, mayonnaise and so on and, with the addition of spices and oils, as a marinade. One or more teaspoonsful in a glass of water first thing in the morning, last thing at night, and even at meal times in place of wine is recommended in cases of obesity, gout, arthritis, rheumatism and intestinal disorders. It is particularly refreshing in hot weather and replaces potassiums and other salts lost through excessive perspiration. It can be applied neat to nettle and poison ivy stings, burns, shingles and impetigo. One teaspoonful, neat or with honey, will ease a cough, and in cases of cuts, bruises, night cramps, varicose veins and pimply skins it can be applied both internally and externally. Also, one cup of cider vinegar boiled in one quart of water will remove the calcium deposit from a kettle. M.M.

Ascorbic acid
(See *Vitamin C*.)

Barley
Barley, though seldom eaten whole today, is consumed daily by the majority of the British population in the form of beer. There are, however, many more worthwhile uses for barley and it is for this reason that many forms of it are found in the health food shop today. It is such a whole and natural food that it figured as the basic cereal of the Welsh people for centuries. In Ireland it is the basic ingredient of Guinness and in Scotland barley is the sole ingredient of the finest whiskies. 'Scotch' or pot barley is also the variety found on sale in the health food shops.

After the inedible outer husk of barley is removed, it is usually polished until even the valuable inner skin and germ is removed. The barley is then small and round and is called 'pearl barley'. In the health food trade only the variety known as pot barley is sold, as this is the whole food with only the outer husk removed. By retaining the inner skin we preserve the high phosphorus and sulphur content of the grain;

it remains a wholefood – a food containing exactly the right balance of vitamins, minerals, starches and proteins, as nature provides. Pot barley forms an ideal base for soups and stews, lending a smooth creaminess to the soup. It can also be cooked in combination with brown rice (using the same cooking method as for rice). A soup with barley should be cooked for over an hour. Cooked with rice it takes only forty-five minutes to one hour, and should be simmered in a covered saucepan with two parts water.

Barley flour is a remarkable flour for all baking purposes. As it contains gluten (the nitrogenous gluey substance found in flour), it can be used with equal measure of wholewheat flour to form a light, well-risen loaf. It adds a touch of sweetness and a barley-rich aroma to bread, cakes and biscuits. It handles very much like an 85 per cent wheat flour and can be used almost anywhere in place of, or in combination with, wholewheat flour (see also *Macrobiotics*). G.S.

Beans

All beans are rich in proteins, although the complete human protein requirement is only achieved when beans are combined or served with grains, flour, nuts, seeds, or dairy foods. Most beans are also rich in various vitamins and minerals. They form the staple food for many millions of people in all parts of the world (see also *Aduki beans, Chick peas, Mung beans, Red kidney beans, Soya beans*). R.R.

Biochemistry

This has been defined as the chemistry of life and embraces the scientific study of all the chemical processes which go on in all living forms and indeed of the degradation processes which subsequently take place after the organism has died.

Life processes are almost exclusively dependent upon the chemical reactions which take place in individual cells. The understanding of living things is the object of the science biology, and the study of the activity of living things is the science physiology. The application of the physiological processes which takes place is ultimately dependent upon chemical reactions, the study of which may also be called chemical physiology which is synonymous with biochemistry.

Since the early beginnings of biochemistry, a science which did not really get under way until the beginning of the century, the subject has expanded vastly so that nowadays it is subdivided into very many divisions, and biochemists have become extremely specialized.

The fundamental difference between the biochemical processes which take place in the cell and the normal chemical reactions which can take place in a chemical factory are concerned with the scale of which the reactions occur, and the speed. Thus in each living cell there may be hundreds of chemical reactions occurring at the same time, all of them in a rather leisurely manner.

Oxidation of sugars in the muscles to produce energy is fundamentally the same reaction as the burning which takes place when a spoonful of sugar is thrown on a fire, the products being carbon dioxide and water, and energy in the form of heat. In the cell, however, the oxidation is brought about by a most intricate enzymic cycle in which more than a dozen different processes are involved in a sequential manner. In the fire the reaction is rapid and the heat is intense. In the muscle the reaction is slow and the heat regulated. In this way the cell is assured of survival because it does not become overheated or destroyed by the products of its own metabolism.

Biochemistry, consequently, is concerned with the study of micro quantities and involves the use of micro analysis for this purpose. The controlled chemical reactions which take place in the cell are activated by enzymes, themselves self-proliferating entities, which assist in the reaction and remain virtually unchanged at the end so that they can be used over and over again.

The study of biochemistry has allowed us to begin to understand how life tends to reproduce itself through the mechanism of mytosis or splitting of a mother cell to produce two identical daughter cells, in particular how the nucleic acids present in the chromosomes of the cell are arranged in the form of the helical spiral which allows a form of template in which new material will automatically fit so as to reproduce a new helix, and thus new chromosomes.

Biochemistry also allows investigation into the processes which go on in bacteria, yeasts and so on. Bacteria of the soil will produce nutrients for the growing plant, yeasts initiate fermentation, oxygen is carried round the body attached to haemoglobin, carbon dioxide in the plant leaf is converted into starches; the chemistry of digestion and of nutrition. All these subjects and many more are material for study by the biochemist. R.F.M.

Biodynamic farming

The bio-dynamic approach to farming and gardening is based on the teaching of Rudolf Steiner (1861–1925) whose pioneering work in education, medicine and the sciences is being ever more widely recognized. The basis of biodynamic farming rests on the concept of the farm as a living organism in which crop rotation, including fertility-building grass leys (land temporarily under grass), help to make it both self-sufficient and productive. A harmonious balance has to be found for each particular environment between the various factors such as stock numbers, pasture, arable, ponds, hedges, orchard and woodland. The part played by extraterrestrial influences on the development of plants and animals is recognized and operations are planned to fit in with lunar and planetary rhythms.

Particular biodynamic measures – also applicable to gardening – include the use of special plant preparations for compost and manure heaps which aid the formation of a stable humus, and also field sprays adapted to the various stages of plant growth. The consistent use of these measures is found to build up health and disease-resistance in both plants and animals. The aim is always to remain in the biological sphere rather than take recourse to fertilisers and sprays of chemical origin.

The quality of biodynamic produce has been demonstrated in practice and confirmed by scientific tests. At the present time there is only a limited amount of such produce available in this country, but the number of producers is steadily increasing. Literature and further information can be obtained from the Honorary Secretary, Biodynamic Agricultural Association, Woodman Lane, Clent, Stourbridge, West Midlands DY9 9PX; please enclose a large stamped addressed envelope.
J.R.P.S.

Bread

Bread 'is the most important staple food' of the community, says the Foods Standards Committee, so its nutritional quality is a matter of great importance to the nation. In biblical times, bread was called 'the staff of life' but most of the bread of today hardly qualifies for that name any longer.

The Bread and Flour Regulations list five types of bread: (1) wholemeal bread, which must be made from the whole product derived from milling cleaned wheat. No compulsory additives need be included; (2) wheat-germ bread, which must contain not less than 10 per cent added processed wheat germ, calculated on the dry weight of the bread; (3) brown bread, or wheatmeal bread, which is not the same as wholemeal bread. It must contain not less than 0.6 per cent fibre by weight calculated on the dry weight of the bread; (4) white bread, which must contain added powdered chalk at the rate of not less than 235 mg per 100 gm of flour, and not more than 390 mg per 100 gm of flour; (5) soda bread.

All bread flours, other than 100 per cent wholemeal flour which contains the following nutrients naturally, must contain the following, added if necessary: not less than 1.65 mg of iron per 100 gm of flour; not less than 0.24 mg of vitamin B1 per 100 gm of flour; and not less than 1.60 mg of nicotinic acid per 100 gm of flour.

In bread-making, various traditional ingredients can be used, such as sugar, milk, fats, salt, yeast and, in fancy breads, eggs, fruit, malt and nuts. Apart from these traditional and nutritious ingredients there are various 'permitted additives' of no nutritional value, such as bleaching agents, extenders, improvers, and certain other specified substances. These are added for commercial convenience, or for alleged 'customer appeal'.

One disturbing thing about the Bread and Flour Regulations as they stand at present is that they do not say that brown or wheatmeal breads must contain any of the valuable wheat germ. This may be present, but it need not be, which can be misleading for the public who may think that all brown breads contain it.

It has long been known that bran, the fibre content of the wheat, is an important aid to proper elimination, and that its removal in the course of white bread-making causes many people to suffer from constipation. Recent work has shown that this constipation may eventually lead to serious disorders of the digestive tract, such as diverticulitis, and cancer of the colon. Bran is currently used as a good laxative, and many of the other nutrients taken out of flour before bread is made are now used to treat various forms of ill health, including some forms of mental ill health. Certainly many digestive and intestinal disorders might be avoided if wholemeal bread was widely eaten (see also *Carbohydrates. Flour, Wheat germ*). M.Y.B.

Buckwheat

Buckwheat is a sturdy grain used as a principal food in Russia and many of the Balkan countries, where it is known as kasha. It has recently been introduced on the health food market because of its outstanding nutritional qualities.

Buckwheat contains all the essential amino acids, with a particularly high rating in riboflavin, which is essential for maintaining proper circulation. It also controls the blood pressure. It is rich in potassium, iron and thiamine. Buckwheat is one of the main ingredients of macrobiotic eating, and it is considered to be a very *yang* grain. Buckwheat groats are usually roasted before use to bring out the unique flavour. It can be used on its own or in combination with other grains and vegetables. It should be cooked with two parts of water to one part of grain for about twenty minutes (see also *Macrobiotics*). J.L.

Butter

Butter is made entirely from the fat portion – or cream – of milk. Normally cow's milk is used, but in certain areas it is possible to find local supplies of goat's milk available for butter-making.

Butter is formed by churning or beating cream that has previously been heat treated to 72°C. This breaks down the fat globules which on further churning coalesce to form butter grains and buttermilk. The buttermilk is drained off and the grains washed with water. The grains are then worked or moulded together and the butter is formed. At this stage salt may be added, the amount varying according to local tastes and requirements. The salt acts as a preservative and gives the butter its flavour.

The colour of the butter will vary according to the breed of cow from which it is produced. For instance, Channel Island cattle tend to give a rich yellow butter. The time of year will also affect the colour and stability of the butter; spring grass will give a better colour but a softer fat, whereas winter feeding of hay will give a paler but firmer butter. The addition of a vegetable dye as a colouring agent is permitted in the UK, but its inclusion must be stated on the packet.

The composition of butter is strictly controlled in the UK – it must not contain more than 16 per cent moisture, it is illegal to use preservatives other than salt – and imported butter has to comply with British legal requirements and must always carry the words 'imported butter'.

Butter usually consists of at least 80 per cent fat. It contains varying amounts of vitamins A and D and very small amounts of protein, milk sugar and lactose. The calorific value is about 790 per 100 gm.

It is also possible to obtain ripened cream butter. This is butter made from cream that has had a culture of flavour-producing bacteria added, producing a butter of controlled flavour and free from taints, but with a shorter shelf-life than sweet cream butter.

Butter should be kept cool, and away from foods with strong flavours or smells. It is affected by light and should be stored in a dark place (see also *Buttermilk, Cream, Milk*). O.A.H.

Buttermilk

Buttermilk is the liquid obtained during the manufacture of butter. It will be sweet or sour depending on whether ripened or unripened cream is used for the butter-making. It is only available in small quantities in very localized areas in this country although it is popular abroad. Buttermilk has a composition of 90.6 per cent water, 0.3 per cent fat, 4.9 per cent sugar, 3.4 per cent protein and 0.8 per cent ash.

It is considered valuable in certain diets because of its protein content which is in a finely divided and therefore highly nutritive state. It also has a low fat content. It has a calorific value of 35 calories per 100 gm.

Buttermilk has a very limited keeping quality and should be kept under refrigeration (see also *Butter, Cream, Milk*). O.A.H.

Calcium

Calcium is one of the major elements of the body. It is a major constituent of all bony structure and is also present in all body tissues but to a much smaller extent. It plays an important part in the clotting of blood and is possibly concerned with resistance against infection. Lack of calcium in the diet is met by calling upon the calcium reserves which are present in the bones and teeth, and calcium deficiency produces such conditions as caries and osteoporosis.

Calcium metabolism in the body is interconnected with that of vitamin D, and the lack of one or both of these substances can result in rickets.

Calcium may be taken in large quantities in the diet without causing any harm and excess is mainly excreted via the faeces.

The food which is most rich in calcium is milk, a pint of which contains 0.75 gm. Dairy products made from milk, such as cheese, and yoghurt, have correspondingly high calcium contents. In other foods, unless they contain fine bones, the calcium content is negligible in comparison with that in dairy products (see also *Cheese, Chick peas, Cream, Milk, Sesame seeds*). R.F.M.

Carbohydrates

Carbohydrates are a group of naturally occuring chemical substances which are basically polymers of sugar. Simple sugars or units or building bricks of carbohydrates are called monosaccharides. Carbohydrates may also be termed polysaccharides. There are a number of monosaccharides, the simplest being glucose, mannose and fructose.

Glucose is found in free state in honey and fructose in many fruits, the disaccharides are formed when two simple monosaccharides combine, and many different types are found in nature. The most common of these are maltose, which is obtained by the breakdown of starch; sucrose, present in cane sugar; and lactose, which is in the milk of all mammals. Lactose will break down in the body to a mixture of glucose and galactose, and sucrose (or cane sugar) will break down into two molecules of glucose.

Carbohydrates are the main food reserves of plants and are present in all cereals and seeds in the form of starches and as specific carbohydrates. They are also found in other parts of the plant, e.g. tapioca starch in the roots of the cassava plant, starch in the potato.

In the case of the carbohydrates present in cereals, the natural purpose is to act as energy requirements for the growing seed until it has sufficient root system to obtain its nutrients from the soil.

In human nutrition the carbohydrates are present to a large extent as energy foods. In the same way that they are built up in the plant tissue from simple monosaccharides, so in the human digestive tract they are broken down to this form.

The plant leaf under the action of sunlight and chlorophyl converts carbon dioxide and water into simple sugars which then build up into carbohydrates. These processes are brought about by enzyme activity.

In the human digestive tract the same carbohydrates are broken down again to the simple monosaccharides by similar enzyme systems. In the form of simple sugars they may pass through the gut wall and be carried round in the bloodstream to the muscles when the simple sugars are burnt up to give carbon dioxide and water in the form in which they were originally presented to the plant.

Excess of simple sugars absorbed into the human body are reconverted back to a carbohydrate known as glycogen and this material may be stored in the liver and other parts of the body until required. When a large excess of carbohydrates is eaten, then a proportion of it is converted into fat and stored in the adipose tissue of the body. This fat when required, can be reutilized for energy purposes, but in this case it, is not converted into sugar before it is utilized in the tissues.

Refined carbohydrate in the diet is said by many to be harmful in that it supplies the energy needs without at the same time the necessary vitamins and mineral salts important for correct nutrition. Certainly vitamins of the B group are concerned in the enzyme systems which are necessary for the oxidation of the sugars in the cells, and these vitamins are mainly found in that portion of cereals which is thrown away when they are turned into refined flour (see also *Bread, Flour, Rice, Starch, Sugar*). R.F.M.

Cheese

Cheese is a natural live food that is slowly ripening and, like all natural food, is best eaten when at its ripest and nicest. Soft cheeses and curds escape this description and are best eaten when fresh. They are simply skim milk or milk with added cream drained from the sour junket stage and made with a lactic culture and little or no rennet (in the case of Fordhall cheese, a vegetarian rennet is used). The best of these should be free from any other aids or additives if they are to give the recipient the full benefit to health he is entitled to expect. They have only a short life and are best kept in a refrigerator. Their benefits to the maintenance of good health are tremendous.

A large group of soft cheeses are processed and contain many additives that destroy their ability to ripen or change. These cheeses are very difficult to digest, they make no contribution towards better health or vitality and are what might be called dead cheese.

Cheese, as a protein food, is particularly important in the diets of those who do not eat meat. It is rich in calcium and phosphorus and contains vitamins A and D (see also *Cream, Milk*). A.H.

Chick peas

Chick peas, also known as garbanzo beans, could well be called the soya bean of the Middle East. They are a widely popular bean and are served whole (in sauce or in combination with rice), as a rich cream/pâté called hummus tahini (see recipe in Chapter 3), or fried in rissoles. As a source of protein they rival beefsteak, containing a very desirable balance of the vital amino acids. Chick peas are one of the richest beans in calcium and, in fact, contain almost a quarter more calcium than cow's milk.

Though they cook relatively easily, it is advisable to soak overnight unless using a pressure cooker. Drain the soaking water and bring to the boil in a pot with three times water to chick peas. Cover and simmer from 1 to 1½ hours and then add sea salt to taste. At this point they are delicious thrown in with cooked rice or other cereals. They can also be used as the basis of a bean and vegetable casserole (see also *Macrobiotics, Tahini*). G.S.

Cobalamine
(See *Vitamine B12*.)

Cobalt
Cobalt is an essential element in that it is part of the molecule which makes up vitamin B12. This vitamin is essential in the production of haemoglobin. Deficiencies in cobalt intake will result in marked anaemia. Good sources of cobalt are fruits, vegetables, cereals and légumes. Liver and kidney are the richest sources in common foodstuffs (see also *Trace elements, Vitamin B12*). R.F.M.

Coffee
Most coffee drinkers know that coffee contains caffeine, which is a stimulant.

Students who wish to burn the midnight oil drink gallons of coffee, and others who need that midmorning break drink coffee to keep them going till noon.

Coffee drinking can be classified with other habit-forming stimulants such as opiates, barbiturates, nicotine (smoking) and alcohol. The amount of caffeine in three cups of coffee is equal to a therapeutic dose given by doctors as a heart stimulant. According to James McLester MD in his book *Nutrition and Diet in Health and Disease*, coffee raises the blood pressure slightly, slows and strengthens the heart, stimulates renal activity and raises the blood sugar level, thus temporarily relieving fatigue and depression, and when the effects of this stimulation wear off the blood sugar level plunges far below what it should be. At this stage the victim reaches for another cup of coffee which is then the start of a vicious cycle. Other substances which produce this condition are white sugar, starchy foods and tobacco.

The blood sugar level gets much too low for such fundamental requirements as mental alertness, energy or even general health and well being. This condition is known as hypoglycaemia, and besides being caused by excessive intake of carbohydrates, can also be caused by excessive coffee drinking.

In this condition instead of manufacturing too little insulin as in diabetes, the pancreas manufactures too much. The insulin keeps not only the surplus but the required sugar out of the blood and keeps carrying it to the cells of the body. The result is that the brain, which uses sugar (glucose) as its chief food, sends out panic signals that it is not getting enough. There is a desperate craving for more coffee or sweets or biscuits and so on, consequently too much insulin is manufactured, and the sugar is driven right into the tissue cells and stored as fat, while after a short period of relief the brain is once again getting too little glucose. It is a tragic reality that the body is simultaneously starving and getting fatter.

Constant coffee drinking causes excessive production of stomach acids, and in the case of nervous people is conductive to ulcer conditions.

One of the most serious indictments made against coffee is that it has a possible link with cancer. *Chemical Abstracts* (1944, Vol. 38) says that tar extracted from coffee beans has strong carcinogenic properties.

Coffee, therefore, should be avoided by those who have heart disease, angina pectoris, high blood pressure, skin afflictions, arthritis, gout and allied conditions and a faulty liver. Some years ago research revealed that coffee prevents iron being utilized in the body, and that it also creates a deficiency in the B complex group of vitamins. Another trouble which is strongly suspected is that it prevents the utilization of calcium in the body.

For those who wish to continue the habit of coffee drinking because the flavour is pleasant, there are the decaffeinated coffees on the market (Hag).

Whether the extraction of the caffeine eliminates all harmful elements is not scientifically proven. Alternatives are roasted dandelion root coffee, Kneip coffee, which is a malt coffee invented by Dr Kneip, a natural healer, and cereal and fruit coffees as made by Dr Vogel – who has health centres in Europe, South Africa and Mexico. M.M.

Comfrey

The herbalists' comfrey is *Symphytum officinale*, a perennial member of the order *Boraginacea* (forget-me-not family). The cultivated or Russian comfrey is *Symphytum peregrinum*, a hybrid between the first species and the blue-flowered prickly comfrey, *Symphytum asperrimum*, which rarely sets seed so is grown from divisions and sold as plants. There are a number of varieties of which Bocking Number 14 is the most popular because it resists the rust that attacks *Symphytum officinale*. This variety is highest in alantoin, the medicinal principle, with 0.44 per cent in the dried leaf.

This substance is a diureide of glyoxylic acid, a cell proliferant which increases the speed of internal and external healing, a property which made Dioscorides, the Greek MO to a Roman Legion in the second century AD, name it from *Symphuo* meaning 'I unite'. It is present in both the herbal and cultivated comfreys, in stems, leaves and roots. Tea is made from dried leaves of either herbal or cultivated comfrey. Tablets are made from ground leaves and ground dried root and ointments from extracts of both with various bases, mainly coconut oil and lanolin, and a nut oil for the entirely vegetable 'cream' types.

The tea is drunk mainly for arthritis and as recommended by herbalists and nature cure practitioners, the dried replacing the fresh leaf tea when the plants are dormant from March until the end of October. Leaves are used for compresses and poultices against a wide range of skin complaints, but the ointments are more convenient for most people. These have helped a number of obstinate skin problems, including psoriasis, varicose ulcers, athlete's foot and several fungus infections, corns (applied after cutting), bruises, burns and scalds, removing pain and irritation quickly. The vegetarian cream is especially useful for the face, and has been used against hayfever by insertion inside the nose, with the advantage of avoiding the side effects of the antihistamines. (If the symptoms are due to a vitamin deficiency, however, no ointment will produce a lasting effect without a reform of the diet.)

Cultivated comfrey is the fastest builder of vegetable protein, especially where the day length is equal near the Equator, and in Kenya it has yielded 124 tons an acre in twelve monthly cuts. It is also the only land plant so far known to extract vitamin B12 from the soil. Comfrey tablets and comfrey tea are taken by vegans to supply this vitamin, and a number of recipes are available for using a flour made from ground comfrey.

Comfrey – An Ancient Medicinal Remedy by Dr Charles J. MacAlister MD, FRCP (Henry Doubleday Research Association, Bocking, Braintree, Essex includes a chapter on cultivation, analysis and vitamins. The association also issues reports on the horticultural, agricultural, nutritional and medical uses of comfrey, based on continual research on this crop. L.D.H.

Cooking oils

The best type of cooking oil is that with the highest proportion of unsaturated fatty acids. This means that the fat can be assimilated by the body without any harmful effects and is the main reason for the superiority of cooking oil over animal fats.

The oils which are most readily available, all of which produce better cooking results than animal fats, are olive oil, corn oil and sunflower seed oil. Olive oil has been used in Mediterranean countries since classical times, has a distinctive flavour and is particularly good as a salad dressing. The first cold pressing is the most valuable in terms of nutritional content and contains quite a lot of vitamin A. Corn oil comes from pressed grains of maize and is tasteless. Sunflower seed oil comes from the pressed kernels of the sunflower seeds. It is a very light and also tasteless oil and can be used as a salad dressing though it is improved by the addition of a few herbs. A great many oils are available in health food shops, under various brand names, but these three are the standbys (see also *Sesame seeds*). J.B.L.

Copper

Copper is an essential element in that it is involved in numerous of the enzyme systems in the body, particularly those which are concerned with cell respiration. It is also important in the formation of haemoglobin since without the concomitant amount of copper, iron will not enter into the haemoglobin molecule. Copper is essential in the production of adrenalin, the hormone from the adrenal gland, which regulates muscle tone and blood pressure.

Animal and fish liver, crustaceans, shellfish and yeast contain particularly high concentrations of copper. Quite appreciable amounts are found in tea, coffee, cocoa, chocolate, nuts and currants. Dairy products, refined sugars, fresh fruits and vegetables are quite low in copper, but appreciable quantities are found in whole meal cereals, légumes, dried fruit, unrefined sugar, both cane and beet (see also *Trace elements*). R.F.M.

Cream

Cream is a concentration of the butterfat in milk and is obtained either by allowing the fat to rise to the top of the milk over a period of twenty-four hours and skimming the cream off, or, as is much more usual, by the mechanical separation of the cream from the milk by a centrifugal separator.

Cream is a very good energy food, is easily digested and is therefore useful in certain invalid diets.

Cream standards are strictly controlled by law in the UK, and are now classified as follows: clotted cream should contain not less than 55 per cent butterfat; double cream should contain not less than 48 per cent butterfat; whipping (and whipped) cream should contain not less than 35 per cent butterfat, sterilized cream should contain not less than 23 per cent butterfat; single cream should contain not less than 18 per cent butterfat; half cream (and sterilized half cream) should contain not less than 12 per cent butterfat.

Clotted cream is formed by a process of heating, cooling and skimming the cream off pans of milk. This forms crusts of thick cream. It is permissible to add a preservative called nicin to clotted cream.

Whipped and whipping cream may have certain stabilizers added to them to preserve their whipping properties. It is also permissible to add up to 13 per cent sugar. Whipped cream is available in aerosol sprays.

There are also a variety of long-keeping creams on the market, all of which have been subjected to high temperature treatment. It is permissible to add certain mineral salts to replace those lost during the heating process.

Cream contains not only fat but also water, some protein, milk sugar, minerals and vitamins; for double cream the nutrients are present in the following percentages: protein 1.5 per cent; fat 48.2 per cent; sugar 2.1 per cent; minerals 0.2 per cent; water 48.0 per cent; with a calorific value of 131.

Fresh dairy cream needs careful refrigeration during storage. Sterilized and long-keeping creams will keep for long periods without refrigeration but as soon as they are opened should be treated as normal cream. There are various imitation non-dairy creams on the market, but legislation ensures that these are clearly labelled.

Untreated or unpasteurized cream is also available. This should be clearly marked on the container and, being a completely fresh dairy product, should be kept under refrigeration at all times (see also *Butter, Buttermilk, Cheese, Milk*). O.A.H.

Diet

The word is derived from the Greek *diaita* and means 'a way of life', also what one habitually eats and drinks, the usual manner of feeding. It also means a course of food specially selected or prescribed (according to the *University Dictionary of the English Language*).

'The usual manner of feeding' in Britain and all other 'developed' countries has undergone more far reaching changes through the introduction of food technology during this century than at any other previous time in history. This has been made necessary through rapidly increasing urbanization which removes the fast growing populations from the sources of food. Processed and preserved foods are forming an ever larger part of the staple diet of the average town and country dweller.

'Dietetics is the science and art of feeding human beings' (*Encyclopaedia Britannica*). The emphasis today is on food science, and since 1960 degree courses in food science have been established at four British universities and an additional degree course in food technology at the National College of Food Technology.

This has put the emphasis in dietetics on science. The basic food groups are: (1) cereals and cereal products, (2) meat, poultry and fish, (3) eggs, milk and milk products, (4) vegetables and pulses, and (5) fruits and nuts. The essential nutrients in the diet are supplied in the form of proteins, carbohydrates, fats, vitamins and trace elements. Strictly speaking, both the oxygen from the air we breathe and water are to be regarded as foods, for the body's cell functions depend on their adequate supply.

They are an essential part in the process of nutrition on which all life depends for maintenance, development, output of energy and reproduction. Our daily diet and with it our nutrition are adversely affected by imperfect oxygenation of the blood and tissues, resulting from lack of fresh air (air pollution), faulty breathing, bad ventilation, lack of exercise, insufficient rest and want of sleep, overwork and fatigue, worry and stress (negative emotions); lack of sunshine; insufficient or excessive calories and many other factors. But 'the greatest single factor in the acquisition and maintenance of good health is perfectly constituted food', as stated by Sir Robert McCarrison.

Diets for various diseases are prescribed by the medical profession. The bewildering number of contradictory slimming diets prescribed by all and sundry are confusing. Refined carbohydrates such as white and most brown sugars and white flour and all other refined starches and their by-products (sweets, white bread, cakes, biscuits, etc.) are high caloric and easily absorbed foods which must be left out of all slimming diets. This omission also improves general health (see also *Diet reform*). B.L.

Diet reform

The increasing trend towards the refining and processing of all natural foods has alarmed many medical doctors, dentists and naturopaths in Britain, Europe, America and other parts of the world for more than a century. Through careful observations on their patients, through experimental feeding trials with small animals and more recently through population studies they have been able to confirm and verify the detrimental effects of the highly refined and unbalanced diets of western man. They were and are pioneers who stood out against the prevailing teaching of medical orthodoxy. It is thanks to their far-seeing, untiring efforts that the climate of opinion in food fashions is changing.

To the recent concern about external pollution has been added concern about internal pollution of the body through denatured, devitalized foods with their adverse effect on health. Basic diet reform teaches that our food should be whole, fresh and grown on healthy soil – as stipulated by Sir Robert McCarrison, who was Professor of Nutrition at Oxford. To this must be added that all chemical food additives are suspect and should be declared on labelling.

Diet reform aims at leaving all natural foods as natural as possible. Cereals to be eaten unrefined, i.e. 100 per cent wholemeal bread, brown rice, sugar in the form of fresh and dried fruit and honey; fruit and nuts predominantly raw; vegetables eaten raw or conservatively cooked. Natural proteins are contained in a great variety of vegetable and cereal foods and combine well with the proteins in all dairy products; natural fats are found in the same foods. Diet reform advocates avoidance of factory farmed meats and meat products; herb teas and unsweetened fruit juices instead of strong tea, coffee or spirits; fresh and dried herbs instead of too many strong condiments (see also *Diet*). B.L.

Eggs

Eighty-six per cent of eggs produced in Britain come from battery-houses, where birds are packed into cages, often in darkness and fed on a high-energy diet with additives to 'force' egg-laying. This is no guarantee of health. Birds often live over their own excrement which provides optimum conditions for breeding flies and rodents. Disease rate is high and battery eggs have been found to be low in Vitamin B12 content, low in essential fatty acids and high in insecticide residues. Deep-litter birds (12 per cent) are usually also kept in windowless houses, on built-up litter often infested with insects. The remaining 4 per cent of eggs come from free-range hens with outside runs. These eggs are likely to be of higher quality and the birds themselves far healthier than indoor birds. Battery and deep-litter eggs may be sold as 'fresh farm', 'new laid', 'country farm', etc. They should not be sold as 'free range', although they often are. Sometimes faint parallel lines about an inch apart appear on eggs; these indicate a battery of which thin shells, watery whites and brittle yolks are also characteristic. Yolk colour is controlled by feed additives. Customers, even in health food stores, should insist on a free-range guarantee with the address of the farm. Vendors of eggs with a false description (either verbal or written) are liable to prosecution under the Trade Descriptions Act and should be reported to local Weights and Measures authorties. Battery eggs are used almost exclusively in processed foods (mayonnaise, pastas, cakes, etc.) (see also *Factory farming*). J.P.

Enzymes

Enzymes are the biological catalysts which are present in cells and allow the very many complex chemical reactions on which life is based to take place at body temperature.

All enzymes function as complex living systems and the outstanding property of each is the specificity of its function. Thus an enzyme is capable of bringing about a chemical reaction in a very complex medium, although there may be very many similar substances in that medium to those which are intimately concerned with the reaction in question.

It is considered that in an enzyme reaction the chemical substances in the cell which are about to react initially combine with the enzyme and then split off when the reaction is complete, leaving the enzyme free to start the same reaction anew.

The selectivity of enzymes is necessary because of the complexity of an interdependency of the various slow-rate reactions which go on in cellular tissue. Thus glucose oxidase is an enzyme concerned with the burning of sugars, as is also phosphatase and many others. Various types of enzymic reaction go on simultaneously in all cells. For instance, dehydrogenases which have a reducing effect, catalases which are concerned with oxidation, reductases, bringing about another form of reduction and many others are reacting in the cell at the same time.

Enzyme systems can be poisoned by substances called inhibitors. Thus it is found that minute quantities of fluoride ion can have a profound effect upon a number of enzyme activities and prevent them from taking place.

The process of digestion in the human body is brought about by enzymes. Saliva contains diastases which are capable of bringing about partial breakdown of carbohydrates. This is why it is necessary that saliva becomes intimately mixed with foodstuff by the process of chewing.

In the stomach, together with hydrochloric acid, the enzyme pepsin is secreted, which acts upon protein foodstuffs to break them down to simpler form. In the duodenum this process is further continued in an alkaline medium by the enzymic trypsin, which can reduce protein material down to simple amino acid units in which form they can be absorbed into the body.

Also in the small intestine the enzyme lipase is excreted. This permits the breakdown of fats into fatty acids and glycerol, in which form both substances are capable of being absorbed into the bloodstream.

Numerous other enzyme actions take place in the digestive tract to ensure the breakdown of food into the form in which it is acceptable to the organism. After death the process of decay is also brought about by the enzyme systems excreted by the saprophytic bacteria which live on dead material. Thus the whole life cycle from synthesis to analysis is completely dependent upon the multitudinous enzymes which have become adapted to specific natural circumstances. R.F.M.

Factory farming

The term is applied to methods used by a minority of livestock producers operating in some cases on a massive scale. Capital-intensive, labour-saving systems aimed at a maximum turnover of capital in the shortest possible time and to the production of standardized food for the mass market.

Effect on the animals: most of the animals man rears for food are young, playful, intelligent and with a marked sense of curiosity, yet in the more extreme systems they are made to eke out the major part of their existence confined in buildings, often on slatted floors, often in dim light or darkness, and sometimes so closely that they cannot turn round in their pens. A psychiatrist has described these extreme systems as 'characterized by extreme restriction of freedom, enforced uniformity of experience, the submission of life processes to automatic controlling devices and inflexible time-scheduling . . . and running through all this the rigid and violent suppression of the natural'.

Effect on the consumer: widespread use of antibiotics has caused resistent bacterial strains which can be passed from animals to man. Thousands of incidents of food poisoning each year have been traced back to these practices. The use of hormones as growth promoters, banned in EEC countries, is still used in Britain, also arsenicals. Many other drugs are used both prophylactically and therapeutically, also insecticides. The combined effects of these chemicals is still unexplored. There is evidence of loss of nutritional quality in the meats and a change in the composition of fats resulting in a higher proportion of (hard) saturated fat. This latter factor could be a contradictory cause of the increase in coronary diseases. Many people feel that in degrading our food animals we degrade ourselves.

Effect on the countryside: ugly, factory-style buildings, nuisance from flies, smell, noise, dust, rats and mice, enormous, and as yet unsolved, problem of effluent disposal. contamination of waterways to the point of putrefaction. Withdrawal of animals from the land has led to encouragement of monoculture, loss of fertility and loss of humus from the soil, also to a less varied and attractive countryside.

Contribution to the problem of feeding the hungry: none. These foods are produced by affluent people for affluent people whose main problem is overweight. Where malnutrition exists amongst these people it is due to eating the wrong foods. Where hunger exists in developing countries it can best be met by more efficient and economic production of vegetable protein.

Ecological crisis: these systems are biologically unsound. Since the land we have available on our globe is of finite extent, it can only support a finite population. Adjustment must be made by a population policy and a return to agricultural practices which allow for constant regeneration, for there are no other continuing sources of food for *Homo sapiens* (see also *Eggs, Meat*). R.H.

Fats

The importance of fats as foodstuffs cannot be overemphasized since the calorific value of fat is more than double that of the corresponding protein or carbohydrate. In the body, fat is deposited in adipose and other tissues and acts as a storehouse during periods of starvation. Excess protein and carbohydrate foods are converted in the body to fat and laid down in the tissue.

Chemically, fat is a mixture of glycerol esters of fatty acids. The saturated fats are glycerol esters of saturated fatty acids such as stearic acid. These are mainly in animal tissue and the unsaturated fats based on unsaturated fatty acids such as linoleic and linolenic acid are found more in vegetable oils. Generally speaking, saturated fats have a higher melting point than unsaturated fats, so that they are harder.

In the condition atheroma there is excess fat laid down in the arteries, and this condition can lead to ischaemic heart disease if the coronary arteries are involved.

Together in the body with fats are found cholesterol and phospholipids, and an excess of the former is normally found in ischaemic heart disease. It is suggested that the aim of a good diet is to keep the total fat intake low. Preferably this should include a substantial proportion of polyunsaturated fats. All meats and fish contain fats, although the quantity varies considerably with the species. Even so-called lean meat still has appreciable quantities of fat in the tissue. The fish oils contain considerable quantities of cholesterol as well as the unsaturated fats. Most of the white fish have relatively low oil contents, but herrings, salmon and certain other fish have high oil contents. In vegetables the oil is invariably found in the seed. Seed germ oils, which also contain vitamin E, are indicated in the treatment of heart diseases. It has been suggested that provided there is an intake of polyunsaturated fat in larger quantity than the saturated type found in most meats, then the cholesterol level in the blood can be maintained at a satisfactorily low level. Further experiments have shown that the diet high in polyunsaturated fats tends to reduce the level of blood cholesterol.

Of the other vegetable substances which should be mentioned in connection with fats are soya beans which contain a high proportion of oil, coconut and avocado pears, although strictly speaking these substances should be classed as seeds. In periods of starvation, when the body's reserves of fats are called upon to provide calories, there is a tendency for a condition known as ketosis to result. This is often found in diabetics where there is inability to metabolize the sugars and in consequence the fats become the source of energy. R.F.M.

Flour

The flour for bread generally comes from wheat, and a wheat grain has three main parts: (1) the bran and other covering coats, usually about 28 per cent of the grain, (2) the embryo or wheat germ, usually about 2 per cent of the grain; (3) the endosperm, the white centre, usually about 70 per cent of the grain.

Until about a hundred years ago, wheat was stoneground between heavy millstones with an action rather like a grater, and the whole of each grain went into the flour. Some of the coarser particles of the bran could be sifted out, but the millers of those days could not make our very fine, very white flour. About a hundred years ago roller milling was invented, and the wheat passed between many pairs of grooved steel rollers, and many screens, and it became possible to separate the grains into their various parts.

When white flour is made, the bran and other covering coats, and the wheat germ, in other words between 28 per cent and 30 per cent of the grains, is removed, and only the very white central endosperm is used for human consumption. The rest of the wheat, called 'offals' or 'weatings', is bagged up and most of it sold for farm animals and racehorses.

The 'extraction rate' of a flour is the percentage of the grain used in making it, so that wholewheat flour is 100 per cent extraction, and white flour is 70 per cent or 72 per cent extraction. When choosing whole-wheat flour it is wisest to ask for a 'stoneground' flour, because there are available brown flours of various extraction rates that have been 're-constituted'; that is, they are made from white flour to which varying amounts of the outer coats and bran, and possibly the wheat germ, have been returned, but this is less satisfactory than a wholemeal flour that has been milled all in one process. There is also available a stone-ground flour of 85 per cent extraction which has had most of the bran sifted out but which still contains most of the other covering coats and most of the wheat germ. This is a good flour, and many people find it more suitable for cakes than 100 per cent wholewheat flour, which is used for bread.

To give an idea of what is lost from white bread, the wheat germ, bran and outer coats which are removed contain the following: some very good protein; most of the vitamin B complex, including B1, nicotinic acid, riboflavin, pyridoxine, pantothenic acid, folic acid and biotin; nearly all the valuable vitamin E and essential unsaturated fatty acids present in the wheat germ; much of the salts of calcium, potassium, phosphorus, iron, copper, magnesium and other trace elements; all the bran, so essential as roughage to help proper elimination.

The removal of these valuable nutrients from bread has created new factors in modern nutrition. About five million tons of wheat are milled each year to make about 3.6 million tons of white flour, so that about 1.4 million tons of the most nutritious parts of the wheat are eaten by animals instead of by people.

In order to try to repair some of the nutritional damage done when wheat is made into white flour, the Bread and Flour Regulations stipulate that: (1) all flours, except genuine 100 per cent wholewheat flour, must have powdered chalk (*creta praeparata*) added at the rate of not less than 235 mg per 100 gm of flour and not more than 390 mg per 100 gm of flour; (2) all flours must contain, or have added to them, not less than 1.65 mg of iron per 100 gm of flour; not less than 0.24 mg of vitamin B1 per 100 gm of flour and not less than 1.50 mg of nicotinic acid per 100 gm of flour.

The quantities stipulated restore these nutrients to the level present in flour of 80 per cent extraction, not 100 per cent, and only four nutrients are added although a much larger number are removed, so that the natural balance of the wheat grain is upset.

Nearly forty years ago, Sir Robert McCarrison told us, 'If their [the people's] health and physical fitness are not to suffer, they must spend more money on supplementary articles of diet in order to make good the deficiencies of white flour than if they had begun to build on the surer foundation of wholewheat flour.'

Not everyone can afford to do this, nor does everyone realize the importance of doing it even if they can afford it. The loss of the bran is now recognized as very serious indeed. Like the wheat germ, some of the bran is packaged and sold for human consumption. It is a valuable laxative as well as a nutrient.

Apart from serious losses, one further point about white flours and breads is that they may have various additives, such as bleaching agents, extenders, improvers and so on. The chlorine dioxide, which replaced the damaging agene used for a time after the First World War as a bleaching agent, destroys any of the vitamin E still left in the flour. These additives have no nutritional value, but are included for commercial convenience or for alleged 'customer appeal' (see also *Carbohydrates, Bread, Wheat germ*). M.Y.S.

Fluorine

Very considerable controversy surrounds the role of this element in in human nutrition. Whereas it is alleged that an intake of 1 part per million in water is regarded as beneficial, certainly 7 ppm will cause adverse reactions and in particular mottling of the teeth. Fluorine tends to accumulate in the bones where it is invariably found in all human skeletal material examined. The content of fluorine in human bones varies from 0.05 to 0.21 per cent, higher limits being found with individuals living in districts where there are larger concerntrations of fluorine in the natural environment.

Evidence as to the manner in which fluorine acts as an essential trace element is sparse. It is known to be a particularly powerful enzyme inhibitor, preventing the action of certain lipases, dehydrases, esterases and phosphatases. It has been suggested that the possible beneficial action in bone and teeth formation is because of the inhibition of phosphatase activity.

Fluorine is found in almost all foodstuffs, but in quantities rarely greater than 2 ppm. Spinach is said to be a rich source of fluorine, and sea fish may contain relatively large amounts of the order of 5 to 10 ppm. Tea has a high and variable fluorine content of up to 700 ppm, about $\frac{2}{3}$ of which is leached out into the infusion (see also *Trace elements*). R.F.M.

Fruit

The term includes such exotics as the avocado pear but excludes the tomato (a fruit usually considered as a vegetable). Rhubarb, a true vegetable, is also excluded. Fruits contain important quantities of vitamins and are the principal dietetic suppliers of essential minerals required for the good health and general well being of the body; and they aid digestion. They also, apart from the avocado pear and olive, contain negligible quantities of vegetable fats. Citrus fruits are noted for their high vitamin C content, as are such tropicals as guavas and papayas. Among our own garden fruits, the strawberry and the blackcurrant are vitamin C rich. Some dieticians support the popular belief that a large proportion of the vitamins in fruits is just beneath the skin. Nowadays, because of dangerous-to-health chemical residues from the poisonous sprays used in the commercial cultivation of fruit, the skin of such fruits as apples and pears should always be removed before the fruit is eaten.

The peeling may well lead to a loss of vitamins because it removes a proportion of the pulp.

Fresh, raw fruit is a vital element in a balanced diet, but it is not always possible to know what one is absorbing along with the vitamins and minerals. An inspection at British ports leads to the detection of arsenical residues on some fruit consignments from abroad, but port inspection cannot detect residues of several other dangerous pesticides. Commercially grown British fruit is not subject to routine inspection for dangerous chemical residues on its arrival at wholesale markets. Certain chemical pesticides are systematic. They enter the plant system and remain in the sap. These poisons are not recommended for use on food crops. If and when they are used in the cultivation of fruit, only complicated laboratory tests could detect their presence.

People have been conditioned to select fruit on eye appeal. They probably do not know, for example, that a Beauty of Bath apple picked from the tree lacks the flavour of a windfall. To most people a Bramley is a Bramley – they cannot know how the tree may have been chemicalized to produce the apples. 'Morning-gathered' is a popular tag, particularly on shop strawberries. The date of the morning is never specified . . . and strawberries deteriorate rapidly and lose their health-giving vitamins.

The commercial grower is unlikely to take advantage of modern research which proves that the vitamin C content of fruit is higher when fruits are harvested late on a sunny day and from branches exposed to full sunlight. Garden owners can take advantage of this and of course may grow their fruit organically and harvest wholefood crops unadulterated by the products of big business out to make profit and with no regard for the health of the planet's soil or of the health of the consumer.

Pseudo-scientists, who are in fact simply workers in chemical laboratories, have (with a lot of financial backing) misled most farmers and many back-garden fruit growers into believing that the products of the chemical industry, however poisonous, are most necessary to fruit production. The Ministry of Agriculture has aided and abetted them by publishing information suggesting that the regular spraying of fruit plants, bushes and trees is most necessary.

In cases where books are not readily available to guide the home fruit grower intent on producing unchemicalized crops, the Soil Association

(Walnut Tree Manor, Haughley, Stowmarket, Suffolk) will be pleased to send a list of helpful publications. Please enclose a stamped, addressed envelope with your letter. s.f.

Garlic

Garlic (*Allium sativum*) is a bulbous plant of the onion family, found mainly in Central Asia and Southern Europe. The leaves are long and grass-like and the bulb, from 3 to 8 centimetres in diameter, contains compound bulbels which vary in quantity from 8 to 15 per bulb.

The bruised and crushed bulbels have an intensely pungent odour, similar to onion, and the volatile oil contains oil (*Ol. allii*), mucilage, albumen, starch and water. The flowers, if any, are pink and whitish.

Garlic is valued in cooking, salads and so on for its aromatic as much as its health-giving properties. Medicinally, the oil may be used neat, as a syrup or in tincture form. The oil is best for local and external application, the tincture in adult internal use (although the oil is often taken in capsules of gelatin which dissolve in the stomach and reduce the offensive odour from the breath) and a syrupus preparation for child treatment.

Medicinal properties are antiseptic, stimulant, expectorant, diaphoretic, hypotensive, anthelmintic and rubefacient. In brief, this means that garlic will ease bronchitis, pulmonary congestion, coughs, colds and stubborn phlegm; will lower temperatures and reduce fevers by encouraging sweating (diaphoresis) and it has been found invaluable in the treatment of children's catarrh and recurrent colds.

In small doses (5 to 10 drops) it will reduce indigestion and clear gastric and colonic gas, thus easing flatulence, eructation and the abdominal discomfort thus caused. High blood pressure also responds gently to the oil, and its anthelmintic properties have been used for hundreds of years in the treatment and irradication of intestinal worms and convulsions in children.

External applications are extremely valuable. Abscesses are cleared by alium poultice, which is also antiseptic and gently drawing. Atonic deafness and earache have responded following the use of a few drops of the oil in the affected ear, and the stimulant, warming and tonic actions of the plant can be utilized in embrocations for rheumatic afflictions, sprained and bruised musculature and sluggish circulatory conditions (see also *Herbs* (*medicinal*)).j.d.h.

Glucose

The cheapest sugar, glucose was termed 'the champion adulterant' by Dr Harvey Wiley, the first head of the Federal Food and Drug Administration in America.

It is made by the action of sulphuric acid on corn starch, a natural product containing vitamins and minerals. The sulphuric acid, however, destroys all its nutritional values – phytates of calcium and magnesium, phospholipids, and fractions of the vitamin B and E complexes.

At the University of Pennsylvania in 1912, Drs Luken and Dohan showed that glucose was the only sugar to cause diabetes in test animals. Other experiments demonstrated that glucose had destructive effects on the pancreatic tissues.

The serious danger to the human digestive system is its 'quick-energy' factor, all right, perhaps, for the hospital patient who cannot take food in the normal way, but all wrong for the requirements of natural digestion. This 'quick-energy' factor is bad enough in beet or cane sugar, but it is far worse in glucose. Natural sugars, as in fruit, take four hours to enter the bloodstream whereas glucose takes only about fifteen minutes. This puts a very great strain on the pancreas, which has the work of keeping the sugar in the blood at a safe and constant level.

Glucose, or cornstarch as it has been misleadingly misnamed, is used in the processing of many foods today, such as jams, jellies, soft drinks, sweets, tinned vegetables, ice cream, sweetened condensed milk, canned fruit, fruit juices, syrups, and bakery products. Even dried fruits can be saturated with glucose to increase their weight.

As the presence of glucose in foods need not be declared, the unsuspecting consumer may eat quantities of it. The answer is to be as independent of factory-produced foods as possible (see also *Sugar*). D.G.

Herbs (culinary)

Culinary and medicinal herbs were in great use hundreds of years ago, from the ancient Egyptians who passed their knowledge and experience on to the Greeks and Romans. The herbs were brought to this country by the Roman army who knew that they could live without doctors if they had herbs, but not without doctors and herbs. They brought all the culinary as well as most of the medicinal herbs with them, and we owe most of our present herb plants to their occupation, but they were lost in the Dark Ages and only picked up again by the monasteries who planted them for their hostels and their hospitals. In this country there were a number of people who had knowledge on the use of herbs and became very famous, such as Gerard and later Culpeper, who left important books behind.

It is interesting to note that in 1550 the Chinese scientist Li Shi Zhen devoted thirty years to the study of herbs and published fifty-two volumes on the subject, and even today in China there are two recognised schools of medicine, the old-fashioned traditional herbal treatment and the western orthodox medicine.

Many people prefer natural remedies, partly based on medicinal herbs. Unfortunately, the interest in herbs went out of fashion with the development of modern medicine and the many life-saving drugs, but people realize now that some of these drugs have unexpected and disturbing side effects. The interest in herbs has revived during the last ten years to an unforeseen extent in England, influenced largely by the tourists travelling to France, Italy and Germany where herbal tisanes or teas have been in use without a break for a long period.

The main quality of herbs is that they are non-toxic. Herbal remedies are only an extension of a general natural diet, and in a natural way they strengthen our bodies' resistance to disease, promote growth and aid the functioning of individual organs.

Most English people only know thyme, sage, mint and parsley, which of course are very important herbs, but there are many other culinary herbs which improve dishes very much. Fresh or dried herbs for use in the kitchen stimulate the appetite, the gastric juices, and digestion and assimilation generally. They are also beneficial to the nerves, and thus are in themselves remedial in character. The following are a few which, on their own or finely chopped and mixed, may be added to vegetables or salads with good effect: basil, summer savory, borage, dill, tarragon, chervil, marjoram, chives and hyssop.

Dill is used on the Continent regularly for all fish dishes, and it is not possible to imagine a French kitchen without tarragon. Basil, the king of herbs, is used in Italy for all tomato dishes and is altogether a tempting herb. Chervil is used in France as much as parsley is here, although the taste is more subtle. Though chervil has its problems when commercially produced, it can always be grown and dried at home if one can retain the green colour, and chervil soup is an excellent starter to any meal. Green knotted annual marjoram helps many dishes. There are three onion-flavoured herbs. For raw dishes, chives are the best, and for cooking, the Welsh onion-green and the green of the so-called 'tree' onion. Welsh onions have the advantage that you can divide them at any time of the year and they increase enormously, giving a supply of onion-green for a large part of the year.

It is not easy to dry herbs satisfactorily at home. Chiltern Herb Farms Limited, Tring, Hertfordshire, has evolved a new way of green-drying which retains the aroma, colour and vitamins in an acknowledged way. (Elizabeth David in the *Sunday Times* called them 'the best English dried herbs on the market'.)

Further details can be found in *Herbs for Health and Cookery* by Claire Loewenfeld and Philippa Back (Pan Books); *The Book of Herbs* by Dorothy Hall (Pan Books); *A Pattern of Herbs* by Meg Rutherford (George Allen & Unwin); *Herbs with Everything* by Sheila Howarth (Sphere); *The Penguin Book of Herbs and Spices* by R. Hemphill (Penguin); *Oriental Herbal Wisdom* by Masaru Toguchi (Pyramid); *Leaves From Gerard's Herball* arranged by Marcus Woodward (Thorsons); *Potter's New Cyclopaedia of Medicinal Herbs and Preparations* edited by R. W. Wren (Harper Colophon Books). C.L.

Herbs (medicinal)

Health food reform is sweeping the world. So many are realizing with horror the enormity of the pollution scare, and waking up to the devitalization, chemicalization, and adulteration of basic foodstuffs, that a substantial move has been made in the direction of wholefoods, naturally produced.

In this specific context we find nature's own remedies for sickness and disease as a natural and integral part of healthful living. After all, to insist on health foods and wholefood reform, to repudiate the use of insecticides, chemical fertilizers and the adulteration of water supplies with fluoride, etc., and yet to rely, when ill, on chemically manufactured and man-made drugs, with dangerous side effects, is a contradiction in terms.

Herbal treatment is safe, non-toxic and long-term in action. There must, however, be qualified help in this field, if the desired results are to be achieved. Self, and domestic, administration is not recommended and, as the potency of various herbal remedies varies, it is wise to seek the professional advice of a member of the National Institute of Medical Herbalists (Mrs E. G. Merritt, General Secretary, 19 Cavendish Gardens, Barking, Essex.)

The National Institute of Medical Herbalists was founded in 1864, following a charter by King Henry VIII in 1542. In 1957 the National Institute of Medical Herbalists was granted its own Armorial Bearing by the College of Arms, and the Medicines Act of 1968 contains provisions for the supply of herbal remedies by a consulting medical herbalist. The Medicines Act provision is to safeguard the public and consultant herbalist alike and states that the patient must be personally consulted and that herbal remedies may be prescribed according to the judgement of the herbal practitioner.

The consulting medical herbalist is trained by the National Institute of Medical Herbalists to a high standard in the following main subjects: (1) antomy and physiology; (2) pathology and diagnosis; (3) herbal *materia medica* and pharmacy. The entrance examinations include many hours of written, practical and oral work.

Accurate and exhaustive examination and diagnosis are the all-important prerequisites of successful treatment. Every patient is treated individually, as no two persons are alike – physically, physiologically, psychologically, temperamentally or environmentally. Every prescription is dispensed separately, uniquely and carefully – and no two prescriptions so dispensed are alike.

In this way the cause of the manifested trouble is eradicated and ultimately the symptom, or effect, is cleared as the body's own vital force defeats the enemy. Orthodox medicine today, apart from being inherently dangerous, seems to be using large chemical sledgehammers to crack comparatively small nuts. After all, to switch off the fire-alarm outside a burning building does not necessarily mean that the fire is out, or the danger past! The consulting herbalist is more concerned with extinguishing the internal conflagration that gave rise to the alarm, than tampering with the bell. (Pain is a simple warning device).

The range of diseases and ailments which may be successfully treated in this way is vast and includes the digestive and circulatory systems, the nervous and urino-genital system, the glandular organs, skin infections, rheumatism and arthritic complaints, respiratory troubles including chronic bronchitis and asthma, varicose ulceration, vertigo, hypertension, insomnia, depression and many other conditions.

Whereas modern drugs rather dictate to the system a course of action, and, in the case of antibiotics, leave the consequent resistance lower after treatment, herbal remedies, correctly administered, raise the whole general level of health and resistance so that the latter state is far better than the former, as the body itself has fought the fight and gained immunity to further attack (see also *Garlic, Herbs (culinary), Tisanes*). J.D.H.

Hesperedin
(See *Vitamin P*.)

Hiziki
This is a dark stringy seaweed from the ocean shore around Japan. It looks like dark spaghetti after it has been soaked for fifteen minutes. Hiziki seaweed is not used in large quantities because it is so rich in vital minerals and trace elements, particularly in iron, calcium, vitamin A and the B-complex vitamins. It is used as a sea vegetable, encouraging the digestive flora and acting to restore healthy condition to the intestines. The minerals and enzymes contained in hiziki aid the body in eliminating the effects of eating animal food and help it to adapt to vegetable quality food (see also *Calcium, Iron, Macrobiotics*). J.L.

Honey
Before this age of technological progress and the refinement of nature's other sweetener, sugar cane, honey has always been a favourite delicacy as far back as the Stone Age. Honey is today the only sweetener which offers life-giving properties of which no trace is to be found in any other.

The body has mineral requirements which must be met to establish and maintain good health, and honey contains iron, copper, manganese, silica, chlorine, calcium, potassium, sodium, phosphorus, aluminium and magnesium. These minerals are derived from the soil in which plants grow and are passed through by the plant to the nectar, which is the base substance used by bees to make honey. It will vary, therefore, in mineral content, according to the mineral resources in the soil where its evolution began. The darker honeys seem to have larger quantities of minerals than the light.

Honey also contains small amounts of the vitamins which nutritionalists consider to be necessary to health. Whereas cane sugar and starches must undergo a process of inversion in the gastro-intestinal tract by the action of enzymes to convert them into simple sugars, this has already been done for honey by the bees.

The advantages of honey over other sweeteners are that it is nonirritating to the lining of the digestive tract; it is quickly and easily assimilated from the stomach (alcohol is the only other substance which can also be assimilated from the stomach, and it is interesting to note that the cure for a hangover or excess alcoholism is frequent spoonsfuls of honey); it quickly, therefore, furnishes the demand for energy; it enables athletes and others who expend energy heavily to recuperate rapidly from exertion; it has a gentle natural laxative effect; it has sedative value, quietening the body. One teaspoonful in half a glass of hot water sipped before retiring will ensure a wonderful night's sleep and will also relieve coughing and the pain of arthritis. Honeycomb chewed and swallowed has remarkably good effects on sinusitis and catarrhal conditions of the nose and the entire gastro-intestinal tract, and as a healing agent it has been used since Hippocrates' day for burns, external ulcers, wounds and so on (see also *Sugar*). M.M.

Ice-cream

There are two groups of ice-cream manufactured commercially in this country, dairy ice-cream and non-dairy ice-cream.

The term 'dairy' ice-cream indicates that by law the product must contain no other fat but milk fat, the only exception allowed being small quantities of egg yolk. The milk fat should be at least 5 per cent, and the milk solids-not-fat at least $7\frac{1}{2}$ per cent.

Dairy ice-cream is a combination of cream, milk, and milk products to which sugar, a stabilizer, an emulsifier and flavouring materials have been added. The ingredients are mixed together and by law have to be heated to a specific temperature for a specific time, before being rapidly cooled and frozen. Air is beaten into the ice-cream during freezing; the quantity of air incorporated will give variations in texture, weight, and colour of the ice-cream.

The fat in the dairy ice-cream can be made up of cream, butter, or full-cream condensed milk. The sugar is usually sucrose or dextrose (it is illegal to use an artificial sweetener). The stabilizer is usually gelatin or a pure vegetable gum and is incorporated into the ice-cream to maintain an even texture and prevent the formation of ice crystals; the emulsifier can take the form of egg yolk or a synthetic material and is incorporated to give a stable, smooth ice-cream.

Dairy ice-cream, because of its derivation from milk, contains appreciable amounts of protein, calcium, phosphorus and vitamin B, and will provide calories from its content of sugar and fat. Its inclusion in the diet can make a useful contribution to the daily nutritional requirements, and is especially useful as an invalid food.

Non-dairy ice-cream has by law to contain 5 per cent fat, but this may be derived from any animal or vegetable source. It must also contain not less than $7\frac{1}{2}$ per cent milk solids-not-fat.

Various types of soft ice-cream are available; they are nearly always non-dairy ice-cream and incorporate a very large volume of air. They are made up from cold powdered mixes, frozen, and sold immediately.

It is now a legal requirement for vendors of ice-cream to display clearly which type of ice-cream is sold on their premises (see also *Butter*, *Cream*, *Milk*). O.A.H.

Iodine

Iodine is an essential element in that it is part of the hormone thyroxin manufactured in the thyroid gland which substance regulates the rate of metabolism of the human body. Deficiencies in iodine result in myxodoema, goitre, and cretinism. Traces of iodine are present in most plants, but the concentration will largely depend on the iodine content of the soil in which it is grown. Thus there are soils in certain areas, particularly chalky soils, where the iodine content of the plants grown thereon is exceptionally low.

Seaweed is a very rich source of iodine, and a small quantity each day will supply the body requirements. Seawater fish contain high quantities and shellfish most of all foodstuffs. Fish liver oils are particularly rich in this element (see also *Hiziki*, *Kelp*, *Kombu*, *Trace elements*, *Wakame*). R.F.M.

Iron

Iron is an essential trace element inasmuch as it is the keystone of the molecule haemoglobin, the respiratory pigment in the blood. Haemoglobin carries oxygen round the body, and removes carbon dioxide produced in the tissues as a result of the burning up of proteins, fats and carbohydrates. Iron is also present in certain oxidase systems which regulate the energy utilization in individual cells. A deficiency of iron will result in anaemia, as a result of the failure of the body to manufacture haemoglobin.

Meat proteins, particularly liver and kidneys, are rich sources of iron, as are egg yolk, poultry, game, and peas and beans. White bread, and dairy products, are poor sources, but wholemeal flour contains three times as much iron as white flour and this applies to the whole grain and refined flour of other cereals. Dried fruits, spinach, cress, etc., also contain useful quantities of iron. Vegetables may lose up to 20 per cent of their iron when boiled in water, and broccoli as much as 50 per cent (see also *Hiziki, Trace elements*). R.F.M.

Kelp

The kelps (order *Laminarales*) are one of the best-known varieties of brown algae. One of the commonest of seaweeds, they can sometimes reach the giant size of 60 m in a mere two years' growth. Most kelps form hollow air bladders which bring the top leaves near the surface where light is available for the vital photosynthesis process. Bladder wrack, a common kelp on most beaches, is readily recognizable by these air bladders.

Since ancient times kelp has been gathered in its locale and used as forage and fertilizer and until recently it was harvested commercially and used in soap making. The potash resulting from combustion of the kelp was used as the basis of the soap, but this has now, of course, been replaced by used or unusable animal fats. Some kelp is still harvested for its algin content. This colloidal substance is used in the food and adhesive industries of today.

With the growing acknowledgement that kelp is the richest of all foods in mineral content, it has been looked at more as a potential food source. Though appetizing recipes have yet to be discovered for kelp as a sea vegetable, it is sold by most health food stores in powder or granule form. Different brands and forms vary enormously in flavour. They can be taken either as a 'dose' of medicine or added to soups and vegetables in the cooking or on the table. Numerous preparations containing kelp are also available for use as footbaths, hair rinses, etc. These seaweed baths have long been used by nature cure practitioners in the treatment of tired or sore muscles.

Kelp is exceedingly rich in sodium, potassium and phosphorus. Its enormous calcium content is matched only by sesame seeds in the entire vegetable world (see also *Macrobiotics*). G.S.

Kombu

Kombu is a large seaweed from Japan, and, in common with other seaweed foods, it is rich in minerals and trace elements. It is sold dried, in long strips from 8 to 12 cm wide. As a flavouring agent it should be used in the ratio of one 10 by 5 cm strip to half a kilogram of beans, vegetables or soups. When soaked it becomes twice as thick and it can be spread on the bottom of the pan in which a casserole or pie is to be made. Kombu cuts the cooking time of hard vegetables and beans, and it can even be served instead of chips if it is cut into 2 cm strips and deep fried. It is a seaweed that is most famous for its ability to bring out the flavour in all foods. It is frequently hung up on a kitchen wall as a sculpture piece because its unusual shape is intriguing and its texture decorative (see also *Macrobiotics*). J.L.

Lentils

The lentil (*Lens culinaris*) is a small bean grown in the Near East, the Mediterranean regions, Germany, France, China, and North America. It is one of the earliest beans used by man and a staple part of the diet of the ancient Romans, Greeks and Egyptians. It was considered the 'poor man's meat' because of its nutritional value; it is high in protein.

Lentils were considered the best food to take on long journeys, and formed a regular part of the army diet of the Roman legions. They are easily digestible and cause less flatulence than any other known bean.

The lentil comes in different sizes and colours. The small, brownish lentil is considered the most flavourable, and is grown in China. The greenish coloured lentil is slightly larger, and found on the Continent. The greenish-brown lentils are grown in the Near East. The orange and reddish coloured lentils are dehusked and split, but are widely used commercially.

Lentils should be steeped overnight and then cooked in water for 30 to 45 minutes. In the Near East they are cooked with oil and vegetables in a hot skillet. Lentils are most frequently used in soups, casseroles, stews, and as a dish by themselves (see also *Macrobiotics*). J.L.

Macrobiotics

Macrobiotics is the art of choosing food that will make your life more adventurous, amusing, happy and healthy. Choosing food according to a set of principles is what distinguishes macrobiotics from other diets. These principles are based on common sense. To live and be active, man depends on food. To live naturally, man must eat natural foods. If he lives naturally, man can reach the above goals.

There is an order in the universe and it is important to understand the balance between man and nature. *Yin* and *yang*, representing the negative life forces, are symbols to help man to understand how the balance in nature works so that he can live in harmony with himself and his environment. When applied to food, the principle of *yin* and *yang* becomes effective as a means of guaranteeing that we get the most ideal proportions of foods at each meal.

Each food is predominantly *yin* or *yang*, and once this is understood it becomes very easy to achieve a balanced diet. There is no one macrobiotic diet for every man – the Eskimo may eat 85 per cent animal meat, the Lambaréné African may eat 90 per cent of a variety of sweet potato, yet both are eating according to the principle of macrobiotics. Eating the staple food of the region is the key factor in both cases.

To eat the food grown in one's own region is ideal, but in a complex society such as ours it is difficult to find all the necessary untreated, natural, nutritious foods in one region. Therefore it is important that the food we have to choose from other countries be perfectly balanced and contain all the nutritional qualities necessary. Grains meet these qualifications. In Britain, imported whole rice, buckwheat and millet, along with native wheat, barley, oats, corn and rye, make up the main part of the diet.

Grains are eaten at every meal. They are a 'live' food which is economical and plentiful and they have been used by man since the dawn of civilization. When used in their natural unrefined state, they do not have the stodgy characteristics of denatured, refined products such as white bread and white rice. Grains used within the principles of macrobiotics can actually be weight-reducing factors.

Secondary foods are cooked land vegetables, except members of the nightshade family – tomatoes and potatoes. Sea vegetables (wakame, hiziki), beans and seeds are essential supplementary foods. Other foods eaten occasionally are nuts, raw vegetables, fresh and dried fruits, fish and, very occasionally, animal products.

Seasoning is an area which combines heightened flavour with added nutritional value. Soya bean products (miso and tamari) are important sources of high-grade protein. Gomasio is made of sesame seeds and sea salt and is often used on the table in place of ordinary salt.

Sweet tastes are best obtained from natural food sources. No refined or brown sugars are used. Honey, naturally made, is used occasionally during warm periods.

Beverages are herbal (mu), or green teas. Twig tea, taken from the three-year-old bancha bush, is used primarily. Coffees are made from roasted grains, beans and the roots of the dandelion or burdock. All commercially grown teas and coffees are avoided. Liquids in general are severely curtailed. The kidneys are not clay pipes that have to be flushed constantly.

Before beginning the diet, observe your condition. Inactivity, day-dreaming, paranoia, timidity, fear, melancholy, exclusiveness, sexual impotence, hair loss, and poor circulation are some of the signs of an extreme *yin* condition. This is caused by taking too much sugar, liquid, alcohol, fruit, dairy produce, chemicals and drugs. Anger, hostility, impatience, rigidity, excessive masculinity are a result of eating too much meat, animal produce and salt. These are very *yang*. Either condition can be corrected by avoiding the extremes in food. A list of foods from extreme *yang* to extreme *yin* would go like this: meat – animal products – grains – vegetables – fruits – dairy foods – sugar – drugs. Grains and vegetables are at the centre and more easily balanced.

Climate plays an important role. In a cold climate the foods should be more *yang*, for instance fish, eggs, meat, buckwheat, rice and vegetables. In a hot climate the food should be more *yin*, rice, wheat, oats, salads, honey, raw vegetables and fruit. Way of life is also very important. People who do manual labour need more *yang* foods than people who do creative work.

Avoid dogmatism. Blindly following every thing that is written about macrobiotics is not understanding it. It is better to experiment, observe and make mistakes. Study the *yin-yang* theory; it will teach you to look on the world in a way that will bring you into close harmony with and appreciation of the simple things of life. The principles of macrobiotics are adaptable to a changing world. Nothing is permanent; everything changes. It is advisable to go into the diet slowly and not to follow any rule for ever.

Georges Ohsawa (1893–1966) introduced macrobiotics to the West and wrote extensively on the subject. His best-known works are *Macrobiotics: An Invitation to Health and Happiness; Zen Macrobiotics; The Guidebook for Living and The Book of Judgement*. For a practical application of macrobiotics to everyday living there is *About Macrobiotics* by Craig Sams. Mr Sams, with his brother Gregory, introduced Macrobiotics on a wide scale in Britain with their Harmony Foods and Ceres grain shop (269a Portobello Road, London W11)(see also *Aduki beans, Barley, Buckwheat, Chick peas, Hiziki, Kombu, Lentils, Maize, Millet, Miso, Mu tea, Rice, Sesame seeds, Soya beans, Tahini, Tamari, Twig Tea, Umeboshi plums, Wakame*). J.L.

Magnesium

Magnesium must really be considered as a macro element in that it is present in all animal tissues and all plant tissues at levels much greater than the substances previously mentioned. Quite high concentrations are found in some plants. It is fifth in order of concentration of all the elements in the human body. Besides being part of the skeletal tissue, magnesium is present in every cell of the human body to a measurable quantity. It is undoubtedly concerned with certain enzyme systems that are involved in the oxidation of sugar to produce energy. An adequate intake is ensured from the vegetable and fruit portion of a mixed diet (see also *Trace elements*). R.F.M.

Maize

Maize is unique among cereals because of its communal existence. The seeds are all crowded together on the central axis and the entire sheath is surrounded by leaf. Because of this unique format its seeds do not disperse and it is dependent upon man's care and attention for its continued survival. Evidence of man's maize cultivation in the New World dates back some 7,000 years and this is a fair measure of the respect that man has held for maize as a food. The ancient Incas and Aztecs based their entire diet on maize and, after the Spanish conquest of South America, maize spread around the world so rapidly that it was known in China as early as 1550.

In these excessively protein-conscious days, maize is often overlooked because it is the cereal least rich in protein. It is rich in vitamin A and other food elements, but possibly some of its food value has been lost through years of indiscriminate breeding for bigger and bigger ears.

In the health food shops maize is usually found in the form of a flour called maize meal. This is similar to the polenta which forms the basis of the diet of Northern Italy. Polenta is prepared by cooking the maize meal with 2 or 3 parts water to a thick cream and letting it cool to set. It is then cut into bars and served covered in stew or chunky gravy. The meal can also be used to add the corn flavour to pancakes, bread and biscuits, but be careful not to use too big a proportion of maize meal since it contains no gluten and will make things crumbly (see also *Macrobiotics*). G.S.

Manganese

Manganese is involved in a number of the enzyme systems concerned with cellular metabolism, and is also concerned with the action of vitamin B1. It has been established that in manganese deficiency states there are considerable upsets in reproduction and lactation. Manganese is found particularly concentrated in liver, and yeast is also a good source. The highest quantities are found in wholemeal cereals, nuts, certain berries, and crustacea. All vegetables, berry fruits and légumes contain appreciable quantities, but most fish, meat, eggs and dairy products are low in this element (see also *Trace elements*). R.F.M.

Meat

Formerly regarded, with fish, as the only source of first-class protein; this opinion has now been revised and a varied vegetarian diet recognized as equally nutritious. Scientists (Dr Hugh Sinclair, Dr Michael Crawford and others, see letter to the *Lancet*, 27 December 1969) have shown that free-ranging animals able to select their own food produce three times as much protein as fat; modern intensive rearing systems produce nearly three times as much fat as protein; the fat, infiltrated into the muscle, or lean, is low quality – in fact, obesity – induced by high-energy diet and lack of exercise. 'Factory-farmed' meat is therefore low in essential fatty acids needed for building and maintenance of cell tissue. Modern methods of indoor rearing also involve use of antibiotics and other growth-promoters, including arsenical compounds. The Swann Report revealed danger to human health by this practice, and some antibiotics are now coming under control. Factory-farmed meat includes almost all poultry (reared in their thousands in broiler-houses); most pig meat; and white veal (from anaemic calves kept in small crates). Animals grazing intensive grassland may also lack essential fatty acids which are obtained from seed-bearing plants growing under the grass and from wild herbage. The 'status-symbol' meat is now venison, coming from animals which really are free-ranging and able to choose their own food (see also *Factory farming*). J.B.

Milk (cows)

Milk is the liquid obtained from the mammary gland of the healthy and normally fed cow. The greater part of milk – about 87 per cent by weight – is water, the remainder is made up of solids which fall into five groups.

Butterfat

The fat forms about 4 per cent of the milk according to the breed of the animal and the season of the year. It is of importance because of its taste and its use in the manufacture of butter and cream. The fat is suspended in the milk serum in the form of minute globules. These globules, being lighter than water, rise to the top of the milk as cream.

Protein

The proteins in milk are of high nutritional value because they contain all the essential amino acids required by the human body.

Lactose

The milk sugar is the only carbohydrate present and is found only in milk.

Minerals

Milk contains some of all the mineral elements needed by the human body, and is exceptionally rich in calcium. Other minerals present are phosphorus, potassium, sodium, chloride, and traces of iron and magnesium.

Vitamins

Milk is an important source of riboflavin and contains appreciable quantities of vitamin A and thiamine, but is comparatively deficient in vitamins C and D. A and D are found in the milk fat, while those of the B and C group are found in the water and solid-not-fat portion.

Milk has a calorific value of 66 per 100 ml. It may be purchased in various grades defined by official regulations. All milk must contain a minimum of 3 per cent butterfat and 8.5 per cent solids-not-fat. Channel Island milk must contain at least 4 per cent butterfat. It is illegal to adulterate milk in any way, or to add preservatives or stabilizers. All milk must be produced from cows that have passed a test to certify them free from tuberculosis, known as 'attested'. It is recommended that where applicable *raw* milk products should also bear the word 'accredited', indicating that they are produced from cows certified free from the disease brucellosis or ungulant fever.

The main grades for milk are:

Pasteurized
The milk is heated to a certain temperature for a time just sufficient to kill any pathogenic or spoilage organisms without impairing the flavour or nutritive value of the milk (small losses of thiamine, vitamin B12 and vitamin C do occur during pasteurization).

Sterilized
The milk is bottled and heat treated to boiling point, thus giving the milk a much longer shelf-life, but the nutritive value is adversely affected.

UHT – ultra heat treated
Milk is heated to a very high temperature for a very short time and immediately packed into completely sterile disposable containers. This renders the whole product sterile and gives it several months of shelf-life so long as it remains unopened.

Untreated
Milk which is not subjected to any form of heat treatment and is bottled or cartoned on the farm or other premises, provided a licence is held by the farmer (see also *Butter*, *Buttermilk*,*Cheese*). O.A.H.

Millet
Millet (*Panicum milliaceum*) has been cultivated since prehistoric times. This small, round, yellowish grain is the staple food of many of the people of the East, Near East and Russia, and there are references to the use of millet flour in the Old Testament.

It has only recently been used as a food in Britain although the large British population of domestic birds have been fed on it for many years. In Britain today, however, one can buy millet wholegrain, millet flakes (which can be added to muesli) and millet flour. In Dartmoor and other parts of southern England, barnyard millet grows wild, while in America barnyard millet is cultivated and made into a flour that is noted for its concentrated nutrional value.

It is a very nutritious grain, high in carbohydrates, protein and fat in that order. It contains all eight of the essential amino acids and a high degree of minerals. It is also high in both thiamine and riboflavin.

Millet can be used as a grain by itself, mixed with vegetables, in soups, casseroles and stews; as a flour it lends a sweet taste to bread. It should be cooked in water (in a 3-to-1 ratio) for 20 to 30 minutes (see also *Macrobiotics*). J.L.

Miso

Miso is a soya bean product that is used as a basis for soups, casseroles, pies, spreads, and stews. It is in the form of a purée or paste. The whole soya bean, with the addition of barley, water, and sea salt, is put through a fermentation for a period of not less than eighteen months. This slow, natural fermentation method produces a high-grade vegetable protein that is used by many people as a daily source of digestible protein. Soup is perhaps the best-known use for miso and, in the East, miso soup is used every morning in the way that cornflakes are eaten in the West. Miso heightens flavour wherever it is used. It is a living food, containing live enzymes that are beneficial to the digestive process. Miso should be added to food towards the end of the cooking time to preserve its high nutrient qualities (see also *Macrobiotics, Soya beans*). J.L.

Molasses

According to Dr S. C. Roy of the Indian Central Sugarcane Committee, the thick black molasses sold in health food stores is 'both a food and a medicine'. It acts as a mild laxative, warms the body more effectively than sugar and is fed to mothers during lactation. It is also prescribed by *kavirajis* who practise *ayurveda* (an ancient system of medicine), for stomach and blood ailments, bile disorders and rheumatism.

It is rich in the minerals iron, calcium, magnesium, potassium, sodium, sulphur, chlorine and phosphorus; the vitamins aneurin, riboflavin nicotinic acid and pantothenic acid; and inositol and choline.

Molasses, the product of the first process of refining the juice of the sugar cane, is a rich food. Further stages of refinement produce first Barbados sugar, secondly Demerara sugar, and finally white sugar, which is a dead product and an irritant.

Inositol is a substance about which we know very little, although it appears to be essential to fat metabolism and helps to prevent deposits of fat forming in the liver and other organs. Choline helps to regulate the cholesterol level of the blood and prevent the hardening of the arteries.

Unlike refined sugar or treacle, molasses does not harm the teeth; according to Dr F. A. Sterling of New York, it actually strengthens them. After observing peoples in Africa and the West Indies, he wrote: 'The coloured race, particularly those living on whole corn meal and the unrefined sugar[molasses]diet of the southern plantations,have good teeth.'

Molasses can be taken in many ways: in warm water or milk; over porridge or wheat germ; with yoghurt; spread on bread like jam; or in cakes and puddings instead of sugar.

It is one of the best heart strengtheners known. Molasses is not a cathartic. It has a gentle laxative effect on the bowels and if a dessert spoonful of molasses is added to a tablespoon of unsweetened Prewitt's or Allinson's bran every day, with warm milk, it will quickly relieve all but the most stubborn cases of constipation (see also *Honey, Sugar*). H.D.

Molybdenum
Whilst it is certain that molybdenum is an essential element in the growth of plants, and certain bacteria, there appears to be no definite information relating to its uses in the human body. With plants, molybdenum is an essential growth factor in minute quantities of the order of 0.1 part per million in the soil. All animal tissues contain molybdenum of the order of 0.5 ppm and there is some evidence that the ratio of copper to molybdenum found in animal tissues has some significance in nutrition.

Molybdenum is present in the molecule of the enzyme xanthine exidase which is concerned with uric acid metabolism, and in animals the absence of this element prevents formation of the oxidase and so the final conversion of xanthine to the excretion product uric acid. In consequence of this xanthine stones tend to develop in the kidneys. It is not certain whether this condition can apply also in humans. Molybdenum is present to a small extent in all plants. It is found in higher concentrations in the livers and intestines of many animals. Some soils contain very high quantities of molybdenum and this results in high levels being transferred to grasses grown therein, producing extremely toxic symptoms to animals fed on these grasses (see also *Trace elements*) R.F.M.

Mu tea

Mu tea is a blend of ginseng and fifteen other roots and herbs: ligusticum, paonia root, cypress orange peel, ginger, rehmannia, cinnamon, cloves, peach kernel, coptis, liquorice root, cnicus, atractylis, moutan and hoelen. It is blended in Japan, although the ginsing is cultivated in Korea. Emperor Shen-ung listed ginseng as the most potent herb in his medical book. Orientals have been claiming for 5,000 years that ginseng retains youthful vitality and prolongs life. The name Mu tea is taken from the name of the Chinese god of the immortals. Mu tea has been used in ancient Japanese and Mongolian procreation rites.

Mu tea comes in individual sachets, each one sufficient to make 1,500 ml of tea. Put one sachet into 900 ml of water, and bring it to the boil and simmer for ten minutes. Repeat with same bag with 600 ml of water. It is advisable to use an enamelled tea pot, as metal creates a slightly metallic taste. No milk or sweetening is required. Mu tea is beneficial to people with high or low blood pressure. Its tonic effect upon the central nervous system regulates blood circulation (see also *Macrobiotics*). J.L.

Muesli

Muesli is the name given to a mixture of cereals, nuts, dried fruits and so on by the late Dr Bircher Benner who first introduced it as part of the diet at his nature cure clinic near Zürich, Switzerland.

When made at home it usually consists of rolled breakfast oats soaked in milk overnight if desired) honey and grated raw apple. To this can be added any other fruits, dried or fresh, and nuts.

There are many ready-mixed proprietary brands available in the UK which may contain some or all of the following ingredients: oats, wheat, barley, rye, millet, hazel, nuts, almonds, dried apple, sultanas, raisins, honey and dried skim milk powder. It is also possible to buy the mixed cereals alone as a base to which nuts, fruit and so on can be added.

Strictly, the ingredients should be raw, but in certain brands the cereals and nuts are toasted to give added flavour. It is desirable that the ingredients should be organically grown as far as possible. S.M.

Mung beans

Mung beans are the Chinese beans whose sprouts are served in Chinese restaurants all over the world. The Chinese have been sprouting beans for centuries, for in their wisdom and poverty they knew of the life-giving, youth-maintaining properties of young sprouted beans. Linda Clark, the nutritionist, says, 'Young growing substances from new growing sprouts will induce cells to grow younger.'

The food value in the sprouting bean is increased up to 600 per cent – especially the vitamin C and the B-complex group; vitamin E is also present. At about 72 to 90 hours' germination the quota of these vitamins and their attendant enzymes and mineral salts are at their highest and are therefore considered wonder foods.

It is important to check, when buying these seeds, that they are fresh, unsprayed and packaged as food. Seeds that are packaged for planting purposes may contain mercury compounds.

A continuous supply can be obtained by setting a few beans in containers each day. Cover the bottom of a flat container with the tiny, dried-out, green beans (which look quite unappetizing at this stage), cover with water for twenty-four hours, strain off, rinse and repeat the operation daily. When the sprouts are 3 to 5 cm high they are at the peak of their goodness and should be used.

They may be eaten as they are, mixed in a tossed salad, conservatively cooked as a vegetable or sprinkled over soups, cereals, stews and casseroles. Some of their value is, of course, destroyed in cooking. In many diet reform cook books there are atrractive recipes for the use of sprouted seeds. (Not all sprouts are beneficial – never experiment with potato sprouts, for instance, because they are poisonous.) M.M.

Niacin
(See *Nicotinamide*.)

Nicotinamide (niacin)
One of the B-group vitamins. Gross deficiency of this vitamin is responsible for the disease pellagra (whose symptoms include skin rashes, weakness and mental depression) which is found in cases of nutritional deficiency, particularly in those parts of the world where rice is the staple diet and polished rice the traditional dish.

Nicotinamide is found in wholegrain products, flesh foods, particularly offals and, surprisingly, in mushrooms. White flour, fruit, vegetables and milk are poor sources. Meat extracts contain high quantities.

Nicotinamides are heat stable and resist oxidation, and very little is lost from food during cooking and processing, other than that which is leached in cooking water. Even after the roasting of meat, not more than 20 per cent of this vitamin is lost. As with the other members of the B-group vitamins mentioned, nicotinamide is an essential feature of the complex enzyme systems involved with hydrogen transport in the cells, known as codehydrogenases.

Nicotinamide is required not only by human beings, but also by bacteria, and it is considered that the action of chemical therapeutic agents such as sulphanilamides in preventing the growth of bacteria is directly due to the ability of these substances to combine with the nicotinic acid which otherwise would be used for the growth of the bacteria.

As with other B-group vitamins, necessary daily intake is associated with the calorific intake, and on a diet of 2,500 calories per day, it is thought that the intake should be of the order of 12 mg per day (see also Vitamin B). R.F.M.

Nut butters

Nut butters provide a good alternative to dairy butter and margarine and are higher than either in protein content. They can be used both as a spread and in the making of pastry, puddings and cakes. There are also savoury nut butters on the market made from a combination of nuts with the addition of herbs and other flavourings.

The most popular nut butters are made from cashews and coconuts. Both can be converted into a cream or a milk if water is added and the mixture put through a blender for a few seconds. The cream can be used with fruit and sweets and as a salad dressing (the butter being only slightly diluted for the latter). The milk is an excellent alternative to cow's milk for children, especially with the addition of a little honey. Peanut butter, which is rich in vitamin B, is excellent as a spread but less useful as a cream or milk because of its strong flavour.

Nut butters are less acid-forming and richer in unsaturated fats than conventional dairy butter, and their mineral content is higher, although the vitamin content is probably lower (see also Nuts). J.D.L.

Nuts

Nuts are an excellent source both of protein and of unsaturated fat, and also contain small amounts of carbohydrate (with the exception of chestnuts which have a low fat and a high carbohydrate content). Most nuts, and especially almonds, have an alkaline effect on the body and are therefore useful for keeping the acid-alkaline balance.

Among the well-known varieties, peanuts are the richest in oil, although they are not rated as highly dietetically as the other varieties because their protein is of a poorer quality. Brazil nuts are also high in fat content and next come walnuts (probably the most easily digested of all), almonds, hazel nuts, barcelonas and lastly cashews. All of these have a good protein content and some proportion of calcium, iron, potassium and phosphorus. Most nuts contain almost negligible quantities of vitamins A and D, but only walnuts and almonds have a trace of vitamin C (although green walnuts are very rich in the C vitamin). All the commonly used nuts contain vitamin B, and peanuts are a particularly rich source.

Nuts can be used very satisfactorily as a substitute for meat and have the added advantage of not being acid-forming, as mentioned above. If the nuts are milled and combined with herbs and other vegetables, an almost infinite variety of savoury dishes, roasts, rissoles and so on can be achieved. Milled nuts can also be used instead of flour in the baking of pastry and puddings, greatly enhancing the protein value of the dish as well as the flavour (see also *Nut butters*). J.B.L.

Oats
Oats are one of the more recent additions to the list of cereals cultivated by man. They tolerate a wide variety of soils and growing conditions, and thrive on poor soils which cannot sustain a wheat crop. Historically, commodity prices have reflected oats' poor-sister status: the prices for bread wheats and malting barleys are traditionally higher than the prices for oats. Oats are rich in magnesium salts and are associated with the element fire. Rudolf Haushka, a student of Rudolf Steiner, describes them as the most typically warm and radiant of our cereals, with a similarity to rice. The spreading panicles of the oat plant are reminiscent of rice in appearance. They are high in protein content (thirteen per cent). In general, the high oil content of oats means that they are more susceptible to rancidity than other cereals. So it is better to use fresh oats when using groats or meals. The preparation of oatflakes nowadays ensures that the oil is stablized so that the bitterness that would otherwise develop is avoided. Oat flour is widely used in the 'ready-mixes' for cakes and some brands of margarine, because a small amount of oats in the preparations delays the onset of staleness. R.R.

Organic farming

There is no recognized definition of 'organic farming', but an organic farmer can well be described as one who conserves and fosters the biological activity of his soil, relying on the indirect feeding of crops through the agency of the soil life, feeding his livestock on the crops so grown and returning to the soil all so-called waste products, both animal and vegetable.

The basis of this approach is the recognition that a fertile soil is one that abounds in living creatures which in the course of their normal existence break down all animal and vegetable residues into plant food, increase the humus in the soil and convert minerals which are in a form unavailable to plants into one in which plants can use them.

The right conditions for soil life are produced by ensuring good drainage and aeration of the soil, maintaining good soil structure by building up organic matter and refraining from deep cultivations which disturb natural structure. In addition, as much animal and vegetable matter (especially the latter) as is possible must be returned to the soil. This is fundamental. Chemical sprays and seed dressings are undesirable; insecticides and fungicides should never be used, indeed they should be unnecessary. Occasional herbicides may be acceptable. Husbandry should be as mixed as economics permit.

The basic difference between organic and modern agriculture is that the latter ignores the biological aspect and depends for nutrition of crops on the application of soluble minerals, the emphasis being on feeding the crop and not the soil. Chemical sprays are widely relied on and there is high concentration on only a few enterprises on any one farm.

Perhaps the most important thing to remember is that, in farming, everything is interdependent and that each individual part affects and is affected by the whole environment. That is a vital part of the organic farming creed. S.M.

Pantothenic acid

Sources of this vitamin are wholegrains and most leafy vegetables, particularly broccoli. High concentrations are also found in liver and animal muscle. Cheese is a good source, and egg yolk contains quite large quantities, as does concentrated vegetable protein extract such as Marmite.

In the body, pantothenic acid is concerned in the enzyme systems involving oxidization. In experimental animals, deficiency results in many symptoms including loss of hair, dermatitis and general degenerative symptoms. It is estimated that a 2,500 calorie per day diet requires about 4.5 mg of pantothenic acid (see also *Vitamin B*). R.F.M.

Proteins

Proteins are essential features of all living organisms and play an important part in the formation of the structure of all cellular systems. Our information about protein is derived from a study of the products of decomposition by the process of breakdown by chemical or enzymic means.

It is found that all proteins are built up of amino acids of which some thirty are present in commonly occuring living things.

Besides being essential for maintaining the structure of cells, proteins may also be used as energy forces, because the constituent amino acids can lose the amino group and be oxidized in the tissues in the same way as sugars and fats.

The number of possible proteins is multitudinous and it is because of this that there is such diversity in the natural species, e.g. the cellular proteins of an elephant are quite different from those of a bacterium, although the amino acids present may be exactly the same. This is because in the building up of proteins from the simple amino acid units, the number of permutations and combinations is multitudinous.

Higher organisms react violently to the introduction of proteins from a foreign source. It is for this reason that there is frequent rejection of transplanted organs; this is also the reason for the shock that occurs when people are stung by bees and wasps. Asthma, hayfever and most allergic reactions are the result of the reaction of the body to foreign proteins. This is not the case, usually, with proteins taken as food, since they are broken down in the digestive system to the simple amino acids. With such, the body can build up its own proteins in the prescribed manner. Diets high in protein and low in sugars and fats are 'slimming' due to the increase in metabolic rate induced by the ingestion of certain specific amino acids (see also *Amino acids*, *Enzymes*). R.F.M.

Pyridoxin
(See *Vitamin B6*.)

Red kidney beans
Red kidney beans – they are kidney shaped – are mainly grown in South America. Black and white kidney beans are also available, although the strong-tasting red ones are the most popular, especially for making chilli con carne. They can also be mashed, for loaves, soups or salads (see also *Beans*). R.R.

Riboflavin
(See *Vitamin B2*.)

Rice
Wholegrain (unpolished) rice is an important cereal and a basic one for virtually half the world's population. It also forms a major part of the macrobiotic diet. It contains all the original nutrients of the grain and is a very good source, in particular, of vitamin B1 and other B vitamins. It also contains vitamin E, calcium, phosphorus and iron.

White (or polished) rice is not recommended. The milling removes all the nutrients. When the Chinese adopted the fashion of eating white rice a new disease appeared in the form of beriberi and caused untold suffering and death to millions of Chinese. Beriberi is the recognized diagnosis of a definite deficiency of vitamin B1; it is a form of paralysis and affects the nervous system.

That we do not suffer from beriberi from eating white rice is due to the fact that our western standard of living enables us to have a great choice of foodstuffs, and the lack of B1 and other nutrients in white rice is partly made up by vitamins in other foods – but only partly, because so many of our foods today are also nutrient-deficient due to processing and overrefining. It is therefore significant that a large number of the nervous diseases from which we suffer are similar to the early stages of beriberi, but seldom recognized as such. White rice, therefore, should be avoided by all who value their health. So-called 'enriched' white rice should also be shunned.

While there is no one perfect food, whole rice comes close to being the ideal food for man. It releases energy in the body over a long period of time. It is a balanced, live, versatile and economic food.

Wholegrain rice should be well washed before it is cooked in cold water. For each cup of rice use two cups of water and one teaspoonful of salt. Place in a stainless-steel (or heavy enamel) pan, bring slowly to a full boil, and then turn the heat to the lowest possible setting. The rice will take about 40 minutes to cook. When cooked (a grain rubbed between finger and thumb will be soft) turn off the heat and leave for a few minutes. The water should be completely absorbed and the grains separated. If the heat is difficult to regulate to very low, cook in the top of a double boiler. Takes very little longer to prepare than white rice, taking into account the more complicated procedure which the latter requires to produce well-separated grains (see also *Carbohydrates*, *Macrobiotics*, *Starches*, *Vitamin B*). D.G.

Rye

Rye or secale cereal, is grown mainly as a food crop but also for animal feeds, whisky manufacture (the distinctively sweet and smooth Canadian whiskies owe their flavour to malted rye grains), and brewing. In Northern Europe rye is still the main bread ingredient as it is prized for its sweetish-sour almost spicy taste. Scandinavian crispbreads owe much of their popular appeal to the use of rye. Much of the bread sold in Britain as rye bread contains little or no rye as such, and as white flour comprises the main ingredient, caramel colour or government-permitted brown food dyes are also added to produce the characteristic 'black bread' appearance. Pumpernickel, on the other hand, is one of the few breads that is made exclusively with rye flour. Rye can be used as a soup ingredient, and rye grains harmonize with many vegetable flavours and with maize. Rye flour can be used in sauces that call for wheat flour, and rye flakes can be added to muesli and to rissoles. In baking naturally leavened breads, it is much easier to obtain an effective 'starter' if the original mixture contains a good proportion of rye flour.

Rye's easy susceptibility to fermentation processes may also be related to the frequent occurrence of the fungus infection ergot in rye crops. Ergot is identified by the presence of black grains among the normally greenish-grey grains of the harvested rye. Ergot is a powerful drug, causing hallucinations in some people. A derivative, ergotamine is the basic ingredient in the manufacture of lysergic acid diethylamide (LSD). Today, stringent controls ensure that ergot-ridden rye grains are not milled into flour, but occasionally a black grain may be found among the healthy ones in a bag of rye. These can be easily removed and discarded. R.R.

Seaweeds
(See *Hiziki*.)

Sesame seeds

One of the oldest oil seeds known to man, sesame seeds are still a major food throughout the world. Whether used as a condiment in Japan or a bread dip in Greece, sesame seeds provide nourishment in the form of calcium and amino acids for millions of people. In the oldest legend of mankind, the Epic of Gilgamesh, we find bread made from sesame seed, barley flour and onions. For Queen Nefertiti sesame oil was the finest cosmetic. Today, in the health food shop, one can obtain these seeds in their various forms and benefit from their rich store of nutrients whilst revelling in their gourmet qualities.

Long known as a brain food, sesame is rich in calcium. No vegetable food known to man, even seaweed, rivals sesame in calcium content. Milk, the common calcium standard, contains exactly a tenth as much as the sesame seed. But for the high calcium count and the full store of essential amino acids it is important to find the whole, unhulled sesame seed. The hulled seeds are easily spotted by their white, waxy look. The natural seeds vary in colour, sometimes with cream, tan, brown and black seeds in one batch!

Sesame is available in many forms. Most favoured is the sesame cream or tahini that is used in sauces, spreads, dips and salad dressings. It is also the vital ingredient of a popular Middle Eastern candy called halvah. When mixed slowly with water, tahini first becomes thicker and then thinner and thinner until it has the consistency of cream of milk. In this form it can be added to 'cream' soups and salad dressings. When mixed with soya sauce and water or honey it forms an unbeatable sweet or savoury spread for bread. Sauces become creamier and richer by its addition.

As an oil, sesame is rivalled only by olive oil as the key to successful cooking. Both oils are among the few that can be cold pressed without becoming rancid in time. Be sure that sesame oil is cold pressed, as only by this method does it retain its full sweet aroma of sesame and its precious cosmetic properties. Simply use the unrefined oil as you would a fancy skin lotion for best results.

As a definite though delicate bread 'spice', sesame seeds are sprinkled on to loaves, rolls and biscuits before baking. In the baking they become toasted and lend a nutty fragrance and flavour to the baking. The flavour of toasted seeds is used to advantage in a condiment known as gomasio. To prepare this the seeds are toasted till lightly brown in an oven or dry frying pan and ground with one-tenth part sea salt.

As a spread, in a soup, or on your skin, this remarkable plant food yields its healthful properties in the most delightful ways (see also *Calcium, Macrobiotics, Tahini*). G.B.

Soya beans

The soya bean is renowned for its protein content. Weight for weight, this bean contains approximately twice the protein of meat, four times that of eggs, and twelve times that of milk. Nutritionists who recommend daily doses of meat, eggs and dairy products have only recently begun to realize that plant protein is easier to assimilate, less expensive, less toxic, and at least equal in nutritional value to protein from animal sources. Soya beans contain 40 per cent protein which, when broken down, contains all the essential amino acids, in ratios rendering them highly assimilable by the body.

As a crop, the soya bean can only be called the miracle food of the twentieth century. From a meagre harvest of 5,000 bushels of this crop in 1924, the United States production of soya beans had skyrocketed to 5,000,000 by 1956. The bean is extensively pressed for oil and is used for everything for 'stretching' sausage meat to making paints. It is only recently that it has been looked on as the future source of protein for mankind, and now greater and greater percentages of the world crop are used directly for food.

A member of the pea family (*Leguminosae*), there are over a thousand varieties of soya bean in colours of green, brown, black and golden yellow. It is usually the golden yellow bean which is found in the health food trade, the best beans being of a rich yellow colour with no black areas or broken beans.

Much work is now being carried out to process the protein of soya beans so that it resembles meat in taste and texture. This involves various highly questionable processes and the addition of dubious flavouring agents and conditioners. As we know, a food in its natural state is far preferable to the processed variety, and with the easy availability and delicious flavour of soya beans there is every reason to include them in any diet.

Before cooking, the beans must be soaked overnight (longer if you care). Then place in a pot with 3 parts water and slowly bring to simmering point. Cover and continue to simmer for $1\frac{1}{2}$ to 2 hours until tender. They will never become soft and mushy. Add a little sea salt to flavour and they are ready to use. To produce a more tender soya bean, you may add a small square of kombu seaweed to the water, as this will soften the beans. Although they can be served on their own, they are also very suitable with other foods, such as soups or stews or casseroles. They may also be sautéed for 10 to 15 minutes with onions or other vegetables.

Natural lactic fermentation processes have been developed which make soya beans available as a ready-to-use food (see *Miso, Tamari*). These processes 'digest' the soya bean over a long period of time and render its vital ingredients immediately available. Soya beans may also be purchased in the form of soya flour which makes a pleasant addition to most baking. Though unusable on its own due to lack of gluten content, this flour can be added to breads (up to 20 per cent) and cakes or biscuits. The result is usually a little richer flavour, a slightly crumbly texture and, of course, higher protein content (see also *Macrobiotics, Miso, Tamari*). G.S.

Soya milk
Soya milk can be taken as ordinary milk for drinking or cooking, and can be given to babies as it is as nutritious as cow's milk. Commercially it is available in tins and has added preservatives such as sugar and flavourings (see also *Soya beans*). R.R.

Soya oil
Soya oil is mainly used as a cooking oil but is also used in the manufacture of paints, plastics and soap. The oil contains a high proportion of linolectic acid, an essential unsaturated fatty acid, and lecithin, a fatty compound particularly needed for brain and nerve tissues. Both linolectic acid and lecithin are known to break up cholesterol deposits in the blood, thereby reducing the risk of illnesses caused by high levels of cholesterol (see also *Soya beans, Cooking oil*). R.R.

Sprouting
Sprouting beans is great fun and can be done at any time of the year, thus ensuring a healthy supply of protein, vitamins and minerals up to five times greater than in the unsprouted bean. All you need is a jam jar: either pierce the lid with lots of small holes or cover the open jar with a piece of butter-muslin secure with an elastic band. Then you will need a handful of any bean, seed, lentils, aduki beans, mung beans or wheat grain that you want to sprout. Also possible are chick peas, marrowfat peas, soya beans, buckwheat, barley and rye. Put the beans in the jar

and cover with the pierced lid or butter-muslin. Fill the jar with warm water and leave overnight. Drain the water off in the morning and thereafter simply rinse through with warm water and after 2 days put in sunlight. The sprouts are best eaten between 4 and 6 days later. Bean sprouts are good in salads and omelettes and added to bread, stir-fried vegetables, pies, sandwiches or fruit salads. R.R.

Starches

These foods come from grains. If derived from wholegrains they are better than sugar as sources of energy; they are assimilated more slowly than sugar and do not overload the insulin-producing function of the pancreas. It is not generally realized that about 60 per cent of the food we eat is converted into sugar.

It is of the greatest importance, especially for adults, not to overeat the concentrated starches such as bread and all other foods made with flour. It is therefore an excellent rule for health to include these at one meal only each day, unless you are very active.

Another good rule for health is not to include bread, flour-thickened sauces, biscuits, pastries or stodgy puddings at a meal containing fish, flesh or fowl, or at one containing acid fruits such as grapefruit, oranges, apples, berries and so on. Concentrated starches require a different medium for digestion to that required for animal protein and acid fruits. Experiments by the great Russian scientist Pavlov proved that a meal containing mixtures of concentrated starches and animal proteins was improperly digested and promoted a fermentative type of intestinal flora.

Since Pavlov's time the philosophy of correct food combinations has been preached by a number of physicians including the internationally famous William Howard Hay, Robert Shelton, Philip Norman, Daniel Munro, and Lionel Picton of the 'Cheshire Testament'. It would seem that medical interest in the importance of correct food combinations has recently been gaining ground. In the letter columns of the *Lancet* (3 July 1971) Dr P. D. Newberry suggests that combinations of food habitually eaten together may be even more important with regard to the development of atherosclerosis than the total quantity of any given food in the diet.

All refined starches should be shunned. The conception put forward by Surgeon Captain Cleave in *Diabetes, Coronary Thrombosis and the Saccharine Disease,* of the close relationship between the refined carbohydrates (white flour and refined sugar) and a number of serious degenerative diseases, is now receiving strong support from leading members of the medical profession. This conception may yet rank, according to some experts, among the most important medical discoveries of this century (see also *Bread, Carbohydrates, Flour, Rice, Sugar*). D.G.

Sugar

Sugar is a refined carbohydrate. The raw juice of unrefined sugar cane contains a considerable number of nutrients, including vitamins A and D and fractions of the B complex, and essential minerals such as calcium, manganese, magnesium, zinc, copper and iron.

Since the latter part of the nineteenth century the refining of sugar has removed all these nutrients. All that remains is fitly described as 'empty calories', and these cannot be properly metabolized in the body because of the removal of the vitamin-and-mineral catalysts. As a result, toxic substances are created, with eventual damage to the body cells. Mr Denis Burkitt, internationally known discoverer of Burkitt's lymphoma (a malignant tumour), warns in the *Lancet* (12 December 1970) that refined carbohydrates can alter the bacterial flora of the faeces and that this alteration produces injurious substances which could be the cause of benign and malignant tumours of the bowel.

Due to the increasing use of sugar in countless processed foods, plus the considerable amount of sugar added at table to food and drinks, the consumption of sugar has rocketed to well over one hundred pounds per annum. Leading members of the medical profession are now linking this increasing consumption (and the consumption of the other refined carbohydrates) to the concurrent increasing incidence of diabetes, dental decay, diverticulitis, obesity, varicose veins, appendicitis, and atherosclerosis.

The much-publicized 'quick-energy' factor in refined sugar constitutes one of its greatest dangers. Sugar floods the bloodstream so quickly that it plays havoc with the level of the sugar in the blood, which must remain constant for health, and pushes it up too high for safety. Nature did not equip the body to burn up avalanches of sugar, and in dealing with these the overworked pancreas pours out too much insulin and burns up more sugar than is necessary. High consumption of sugar can therefore result, paradoxically, in creating the condition known as 'low blood sugar', when discomfort and depression are experienced with an accompanying craving for 'something sweet'. And so more sugar is consumed and up goes the sugar level again, only to fall too low a little later – and a vicious circle is begun. An American researcher, Dr Benjamin Sandler, has warned that low blood sugar is a factor of susceptibility to polio and to other infections.

Over-consumption of sugar also has serious, and generally un-recognized social implications. A number of medical researchers have produced well-documented evidence linking low blood sugar states to juvenile delinquency, irrational behaviour, sadism and even homicide.

Apart from a real Barbados muscovado cane sugar, most brown sugars are little better than white. Honey is by far the best sweetening agent, but even this should be used sparingly. The best source of sugar in the diet is that supplied by ripe fruits (see also *Carbohydrates, Honey, Molasses, Starches*). D.G.

Supplements

Just as proper nutrition is the basis of good health, so it is equally true that virtually every disease known to man, from defective teeth to cancer, is due to poor nutrition. In the western world this is not due to lack of food itself but to the eating of food which is lacking in the essential elements which the body must have or die. Even food grown on un-contaminated soil is nowadays adulterated by the addition of chemical substances. It is also over-refined from its natural state and robbed of its vitamins and minerals by processing.

If our foods were available in their whole state, grown on naturally composted soil, untouched by fall out from a polluted atmosphere, chemical fertilizers and poisonous sprays – the milk products and eggs from healthy animals who eat uncontaminated grass and whose health is not exploited by drugs and hormones, there would be no need for supplements.

Whole, natural foods like this are obtainable in many parts of the country, but they are produced in comparatively small quantities and are therefore not available to the vast majority of people. This is why vitamin and mineral supplements, derived from natural sources, have value. They take the place of the elements which are lost when food is stored, peeled, left standing in water or cooked to death (especially if bicarbonate of soda is added to retain the green colour of the vegetable). Factory-produced foods which are chemically bleached or coloured emulsified, pasteurized, flavoured with synthetic chemicals, preserved and so on are lacking in these vital elements. Further, our polluted rivers and reservoirs must be chemicalized by chlorine processing before we can drink the water, and it is known that chlorine destroys vitamin E. Sodium fluoride, which is also added to water, causes further destruction both of vitamins and of precious enzymes. B-complex vitamins are destroyed by drugs used in sleeping pills and for the treatment of infectious conditions, and also by the arsenic and sulphur compounds, DDT and other hydrocarbons used in spraying fruits and vegetables. Atmospheric pollution (chimney smoke, exhaust fumes, tobacco smoke) destroys vitamin C complex. Our only means of self-defence is to make up for the losses and deprivations by taking vitamin and mineral substances which have been obtained from natural sources; that is to say, have been extracted from naturally grown foods. These will be balanced as they are in nature. Beware of chemically manufactured vitamins and multivitamins. These preparations can be ill-balanced – there is always something missing and always the 'intrinsic factor'. If the balance of nature is upset in the soil, the plant, animal or man, ill health will ensue (see also *Trace elements, Vitamins*). M.M.

Tahini

Tahini is made from natural sesame seeds. The seeds are finely ground into a smooth cream, with nothing else added. Tahini contains the full range of amino acids and is very high in calcium. It has been used for centuries in Greece and the Near East as a traditional sauce and spread. To use as a sauce, dilute with water and heat slowly. As a spread, it can be used on its own or mixed with miso. Chopped onions, herbs or orange peel add interesting flavours. Tahini is best used in moderate amounts to retain its mysterious flavour (see also *Miso, Sesame seeds*). J.L.)

Tamari

Tamari is a liquid sauce made from the soya bean. The bean, with the addition of wheat (or barley), sea salt and water, goes through a lactic fermentation process lasting eighteen months that results in an easily digestible, high-grade vegetable protein. (Tamari is widely used as a basic protein supplement, as well as a sauce, in the wholefood diet.)

The transformation of the whole soya bean under this natural fermentation and ageing process gives tamari the rich flour and nutritional qualities that commercial soya sauces are only able to achieve with the addition of sugar, caramel, monosodium glutamate or other additives. Commercial soya sauces use the soya bean after it has been pressed for oil – thus losing the nutritional value common to the whole soya bean.

Tamari is used in small quantities during the last few minutes of cooking time with all vegetables, beans, casseroles, stews, soups, fish and fowl. As a condiment, tamari often replaces salt. Tamari has been used for centuries in the East and is still made there by the traditional method (see also *Macrobiotics, Soya beans*). J.L.

Thiamine
(See *Vitamin B1*.)

Tisanes
The *Oxford Dictionary* calls them 'ptisan, a nourishing concoction made mostly of barley', but the French word *tisane* is now used in English with its French meaning, namely a herb tea. This can be made from leaves, flowers, berries or roots and usually has both refreshing and medicinal qualities.

The herbs for tisanes should be gathered on a dry day, preferably in the early morning when the dew has dried off. Herbalists and health food stores stock loose dried herbs, fresh herbs occasionally, and also sachets which should be steeped in boiling water for at least five minutes. When prepared from fresh ingredients, the majority of tisanes are made by pouring boiling water over the herbs and allowing it to stand, covered, for 5 or 10 minutes. The exceptions are tisanes made from roots, which have to be soaked and simmered for lengths of time which vary according to the particular plant. The following are some of the best-known herbal drinks.

Root
Marshmallow root (*Athea officinalis*) is used chiefly for the relief of chest complaints. This root contains a great deal of mucilage which is freed by long and slow simmering one hour at least) and which relieves a persistent cough. The strained liquid can be taken with honey.

Leaves
Sage (*Salvia officinalis*) has many virtues. The Chinese say, 'Where there is sage growing in the garden, no doctor will be needed.' Sage tea or tisane can be drunk as a general strengthener between or after meals. As a mouthwash or gargle, it is a most efficient cure for sore gums, blisters in the mouth or a sore throat. In the first two cases the liquid

should be held in the mouth for some time before it is spat out.

Mint (*Mentha viridis* or *Mentha piperita*) makes an excellent tisane to be taken after meals, as this herb helps to cure indigestion and flatulence.

Flowers
Lime blossom (*Tilia europaea*) promotes perspiration in the case of colds or fevers. It is a very pleasant drink and can also be served cold in the summer with the addition of lemon and honey.

Elder flower (*Sambucus nigra*) serves the same purpose.

Verbena (*Verbena officinalis*) contains a particular kind of tannic acid which acts as a nerve tonic and antispasmodic remedy, especially in the case of epilepsy.

Camomile (*Matricaria chamomilla*) helps to soothe and calm the nerves and is highly recommended by French doctors as a digestive tonic to be taken as a tisane after meals.

Berries
Rosehip (*Rosa canina* or *Rosa rugosa*) is highly appreciated in Switzerland and Germany. The dried hips, with their pips, are used. This tisane has a high vitamin content and a very pleasant flavour (see also *Herbs* (*culinary*), *Herbs* (*medicinal*)). L.D.

Tocopherols
(See *Vitamin E*.)

Trace elements
Besides the macro elements which make up the bones and the tissues of the body, e.g. carbon, hydrogen, nitrogen, oxygen, phosphorus, potassium, calcium and sulphur, there are certain minor or so-called trace elements which are essential for maintenance of health in the human body. These trace elements are necessary either in the production of the hormones which regulate body metabolism, or are involved (sometimes together with the vitamins) in the enzyme systems which regulate all chemical processes which go on in the cells. In some instances the quantities required are extremely small, but nevertheless, without these traces, ill-health in one form or another can result. In most instances the metals concerned are extremely toxic if taken in larger amounts and the margin between the necessary traces and the toxic dose is in most instances quite narrow. Thus the useful daily intake of flourine is put at

2 mg, but toxic symptoms may occur if 10 mg per day is consumed.

The term trace element does not mean that the metals concerned do not play an important role, and these substances are anything but insignificant in their activities. In most cases a balanced diet will automatically ensure an intake of an adequacy of trace elements, but there are circumstances when deficiencies may result and augmentation becomes essential (see also *Aluminium, Cobalt, Copper, Fluorine, Magnesium, Manganese, Molybdenum, Zinc*). R.F.M.

TVP

Textured vegetable protein is made from soya flour and water. It is used as a meat substitute in many developing countries, as a high source of protein, and in other countries as an alternative to meat. It is available in various shapes and flavours and is soaked in hot water before cooking (see also *Soya beans*). R.R.

Twig tea

Japanese twig tea is gathered from the lower part of the bancha bush. The leaves and stems are picked after three years, at the time when the sap is at its highest, thus obtaining maximum tonic effect. A generous pinch per cup is used, but roasted in a dry pan first to bring out the flavour. It is put in water, brought to the boil and simmered for ten minutes. The leaves and twigs can be used again. It is interesting to note that twig tea has a refreshing uplifting effect, but can be drunk shortly before retiring with no disturbance to sleep. It is ideal for pregnant women, and has been used for centuries as a soothing beverage for women in this condition (see also *Macrobiotics*) J.L.

Umeboshi plums

The umeboshi is a pickled plum from Japan. It is picked while still young and placed on straw mats in the early summer where the moisture of the night and the heat during the day combine to prepare the plum for its unique salting process. With the addition of beefsteak grass (red grass) the plums are sprinkled with sea salt and put under pressure in large jars where they remain for two or three years before they are ready.

The umeboshi is used both as a culinary delight and a medicinal remedy. As a food, it is sliced and used in salads, vegetables or cooked with grain dishes. Only a small amount is needed – two plums at the most for a dish which is to serve two people. As a relief from stomach acidity, eat one umeboshi. A pinch of pulped umeboshi will refresh the mouth, quench the thirst and stimulate a flagging appetite (see also *Macrobiotics*). J.L.

Veganism

Veganism is a way of living which excludes all forms of exploitation of, and cruelty to, animals, and includes a reverence and compassion for all life.

It applies to the practice of living on plant products such as salads, vegetables, fruits, nuts, grain, pulses, seed oil, to the exclusion not only of flesh, fish and fowl, but also of eggs, honey, animal milk and its derivatives, and encourages the use of alternatives for all commodities derived wholly, or in part, from animals.

Veganism seeks to remember man's responsibilities to the earth and its resources and to bring about a healthy soil and plant ecosystem and a proper use of the materials of the earth.

It is thus not just a simple abstention from those foods which make up most of the diet of our countrymen, nor attachment to this or that particular reform, whether it be concerned with the alleviation of the human condition, the abolition of cruelty to various kinds of creature, or some misuse of the plant and mineral resources, it is also the acceptance of stewardship and guardianship for that part of the earth's life which comes under the person's power or influence. Becoming a vegan involves a whole series of changes of thought, habits and customs concerning food, dress, amusement, medicine, science and sport, etc., and this step should be undertaken with care and forethought.

Whilst many of the principles of veganism have been and are today an accepted part of the tenets of many religious faiths and groups in various parts of the world, it was only in the nineteenth century that vegetarianism gradually coalesced into a movement in the West and as recently as 1944 that the Vegan Society, with its stricter principles, was born in London, out of the Vegetarian Society, into the difficult conditions of food and clothes rationing. The new vegan today need not feel himself to be an isolated unit struggling with difficulties of all kinds, for the Vegan Society is there to help him during the adjustment period, and its journal, other literature and counsel will guide him to self-reliance when buying food, clothing and other commodities. (Secretary: Mrs K Jannaway, 47 Highlands Road, Leatherhead, Surrey.)

The Vegetarian Society's Nutritional Research Centre states that in plants are to be found all the nutritional factors that man must obtain from his food, and the study of anatomy reveals that man was designed to live on an all-plant diet. Many farming methods used today work against nature, sickening the soil and producing man-made defects, grossly exploiting animals, and serving the flesh food 'needs' of the few at the expenses of the nutritional needs of the many – it is still not generally realized that three-quarters of the earth's fertile land is used in feeding food animals, and that a change-over to a vegan economy would feed at least three times as many people.

Vegans believe that as man ceases his parasitical relationship with the animal kingdom and adopts the instruction given in Genesis i, 29, his physical health and the quality of his emotional, mental and spiritual life will improve along with his greater perception and awareness, and the growth of the spirit of compassion within him will reveal avenues of service all around him (see also *Vegetarianism*). J.S.

Vegetables

This term includes the tomato (truly a fruit) and sweet corn (a cereal). Vegetables, whether raw or cooked, constitute a major part of the balanced diet. Until recent years the health-conscious bought shop vegetables by their face value. This is not safe today. Shop vegetables are suspect. They may appear to be in prime condition, but in fact they have probably been culled from unhealthy plants, palliated by man-made poisons. Most commercially-grown vegetables are produced in this manner. The soil in which the plants grow is polluted with chemical fertilisers by which the plants are sustained directly – instead of by a natural indirect feeding. Weaknesses of plants sustained in this manner lead to major attacks by natural scavengers, pests and diseases. Farmers and growers combat these attacks by spraying and powdering with more products of the chemical industry. Gas masks and protective clothing are recommended for those applying these highly dangerous potions. Apart from invisible, poisonous residues which are present on the harvested vegetables produced by this unnatural procedure, the sap of the vegetables may also be highly contaminated. Certain chemical products are liable to change the flavour of some vegetables and can thereby be detected. On the whole, the result of chemicalisation in agriculture leads to (a) a lack of the expected flavour, (b) an offensive smell when chemicalized vegetables are cooked, and (c) the intake by the consumer of doses of highly dangerous products from the chemical industry.

How the human system, physical and mental, is affected by the regular ingestion of minute quantities of factory-made chemical residues and by a diet known to include a high proportion of vegetables produced by chemicalized methods, is not known. The human metabolism is not geared to deal with a number of these residues which may be stored in various parts of the body and not ejected from it. Organic gardeners claim that only unchemicalized (whole) food is health-giving, and that chemicalized vegetables and other foods are not only unhealthy but dangerous to the consumer.

It is to be hoped that, with a better understanding of what constitutes wholefood as opposed to chemicalized food, there will shortly be a demand by those without gardens for land around our towns where organically grown vegetables may be raised in the sort of leisure garden already popular in Holland and Germany. The Report of the Departmental Committee into Allotments (HMSO, £2.10p) recommends that all local authorities should be permitted and encouraged to provide half an acre of leisure garden for each thousand of their population. Even were the present government to implement this recommendation, it is doubtful if sufficient land is available to many large authorities. Will this mean that wholefood vegetables will be denied to future generations? It seems possible that with an increased demand for healthy, unchemicalized vegetables, commercial growers will be forced to revert to traditional, unchemicalized vegetable production which, if humanity can control other forms of pollution, will lead to a healthier, cleaner environment and to superior commercially grown vegetables.

At the present time unchemicalized vegetables are in very short supply commercially. The fortunate owner of a garden may grow his or her own. Where books are not readily available to guide the home vegetable grower intent on raising crops by the organic method, the Soil Association, Haughley, Suffolk IP14 3RS, will be pleased to send a list of helpful publications. Please enclose a stamped, addressed envelope with your request. B.F.

Vegetarianism

The basic definition of a vegetarian diet is one which does not contain fish, flesh or fowl, nor any of the derivatives of slaughter. In contrast to vegans, vegetarians can and do include in their diet eggs and dairy produce. It is important (particularly in the early stages of the change-over from meat-eating) to be certain of having adequate protein, and the use of dairy produce ensures this. Most vegetarians are advocates of the 'middle way' in diet, avoiding extremes of asceticism in addition to flesh-eating. Both have an adverse effect on their health and useful functioning in society, while the latter also goes against their ideals.

In considering a vegetarian diet, there is no need to have any worries regarding deficiency or lack of variety. Using eggs (preferably free-range), cheese, milk, nuts, pulses (lentils and beans) and whole cereals, with herbs, spices and vegetable extracts for additional flavouring, it is possible to produce a wide range of tasty and wholesome vegetarian savouries – meals which are in fact far more varied than the traditional 'meat and two veg'. Add to these a full range of vegetables and fruit, which can be prepared in many different ways, and it will be seen that this diet is both nutritious and varied.

Some vegetarian protein dishes are designed specifically to replace meat, and even try to imitate its flavour and texture. Examples include the various forms of textured vegetable protein (TVP) and the different kinds of 'meatless steaks'. There is, however, no need for the would-be vegetarian to continue to be a slave, even in word and thought, to meat. Many savouries have a character all their own, and call to mind no association with flesh foods.

The majority of vegetarians are also advocates of wholefoods, although this does not always follow. However, it would seem logical that devitalized and over-refined foods, such as white flour and sugar, lack certain vital elements which are essential to a fully balanced diet. A basic vegetarian diet has often been shown to contain all the ingredients necessary for health, so it is clearly a good thing to take the foods in as whole a state as possible.

A vegetarian diet has an important contribution to make to the world food shortage. The breeding and raising of animals for protein is very wasteful of land resources, the production of plant protein having been estimated to be between three to five times as efficient. Experiments are now going on into the various forms of algae and single-cell protein which will produce over twenty times as much protein per acre as the most modern of factory farms, yet involve neither the cruelty nor the pollution caused by intensive farming.

Detailed information and recipes can be obtained free of charge from the Vegetarian Society at Parkdale, Dunham Road, Altrincham, Cheshire (061-928 0793), or 53 Marloes Road, London W8 6LD (01-

937 7739). They will also send a free specimen copy of the monthly newspaper, the *Vegetarian* which deals with all aspects of the subject (see also *Veganism*). J.P.

Vitamins

Whilst food is necessary in order to supply the body's energy requirements, and this can be measured in the number of calories derived from proteins, fats and carbohydrates, it is also necessary to have an adequate intake of ancilliary substances, for instance minerals and trace elements, and also vitamins.

The discovery of vitamins was made in the very early part of this century by nutritional workers. Originally the factors found were divided into two groups, water-soluble and fat-soluble. Subsequently the component members of the two groups have been isolated and named, so that today the number of vitamins known is about twenty. It is quite possible that there are still vitamin substances to be identified.

A good, sensible diet, made up of whole food and including a proportion of fresh, uncooked fruits and vegetables, should supply the necessary vitamin requirements. Certain of the water-soluble vitamins are not stable to heat and consequently tend to be destroyed by cooking and become leached out by the cooking water. This is one reason for advocating some uncooked foodstuffs each day. During infection and illness there is evidence that the vitamin requirements are increased or the body's reserves may be depleted. Under these circumstances it can be useful to augment the diet with vitamin pills, preferably obtained from vitamins, which have come from natural sources in view of the interdependence of one vitamin upon another in the body physiology, and the fact that naturally occurring vitamins would include other compoments (see separate entries). R.F.M.)

Vitamin A

This is known as the anti-infective vitamin, and it seems to protect the body against invasion by low-grade infections. It is found particularly in fish oils and in extremely high concentration in certain fish liver oils (halibut and tuna as well as cod). The livers of other animals which are used for food also contain high levels of vitamin A.

There is virtually no vitamin A in vegetable products, but these do contain carotene which can be converted within the body to vitamin A.

This reaction, however, is not complete, and large quantities of carotene are therefore necessary if there is no augmentation from animal or fish sources.

The highest concentrations of carotene are found in green leaves such as kale, parsley, lettuce, cabbage and, of course, carrots.

Because vitamin A is fat soluble, it tends to migrate to fatty substances and is stored therein in the body. The daily requirements are about 5,000 international units.

Deficiency produces symptoms of night blindness and follicular skin eruptions (see also *Vitamins*). R.F.M.

Vitamin B

This comprises a group of water-soluble vitamins, the bulk of which are present in the seeds of cereals. Because of this it is essential that the whole cereal should be eaten and not merely that portion which remains after the bran endosperm and germ have been removed by the milling process. Although white flour as presently sold in this country contains an augmentation with certain members of the B vitamin group, this is not sufficient to ensure that there is an adequate intake of other members of the B vitamin group which are not present in white flour, and which are present in the wheat germ.

The main components of the B vitamin complex are vitamin B1 (thiamine), vitamin B2 (riboflavin), nicotinamide (niacin), vitamin B6 (Pyridoxin), pantothenic acid, biotin, folic acid, vitamin B12 and folinic acid. In addition there are certain other substances which have been given doubtful vitamin status by the nutritionists. There are inositol, choline, paraminobenzoic acid and vitamins B13 and B14. Vitamin B13 is an identified growth factor isolated from dried distillers' yeast, and vitamin B14 is a crystalline compound which is extremely active in stimulating new cells in the bone marrow (see also *B1 B2, Nicotinamide, B6, Pantothenic acid, B12, Flour, Rice, Vitamins, Wheat germ, Yeast*). R.F.M.

Vitamin B1 (thiamine)

This vitamin was found to be essential for the prevention of the disease beriberi which was very prevalent among populations feeding largely on polished rice. It is water-soluble and is fairly resistant to heat provided the cooking temperature is not much above 100°C. Considerable inactivation can occur with pressure cooking, and it is estimated that about 25 per cent of the value may be lost during baking. If soda is used, then the cooking losses are very considerably increased.

Fruits which have been preserved by the use of sulphite are generally low in thiamine and because of its water-solubility considerable quantities may be leached out in cooking.

The best sources of the vitamin are the germs of cereals such as oats, wheat, barley and rice. White flour contains about one-tenth the thiamine of the wholegrain. There are substantial quantities of thiamine in most fruits and vegetables, although if these are cooked then there are the inevitable losses. Fish roe is a useful source of this vitamin, as is the liver of fishes such as the halibut, herring and haddock. Higher quantities are found in the offal of animals than in the flesh.

Yeast is a particularly useful source, in particular brewers' yeast which contains as much as 20 mg per 100 gm.

The intake of vitamins of the B group should be closely associated with the starch intake of the diet. This is true with a number of the members of the vitamin B group, but particularly so with thiamine. In consequence, diets which contain a high proportion of refined carbohydrates tend to deplete the body's reserves of thiamine if they are not adequately made up from other sources. This is because thiamine is a necessity in the enzyme system whereby carbohydrates are oxidized in the tissues to give energy.

Based on a 2,500 per day calorie intake, the thiamine requirements have been estimated as approximately 1 mg, but if the diet is predominantly carbohydrate, then proportionately larger quantities are required (see also *Vitamins, Vitamin B*). R.F.M.

Vitamin B2 (riboflavin)

Daily requirements have been estimated as between 1.5 and 2.0 mg. It is widely distributed in plants and in animal tissue, but the best sources are yeast, milk, eggs, fish roe, offal and growing, leafy vegetables. In green vegetables it is the leafy portions and growing parts which contain most riboflavin, and as the leaves get older and dryer so the content diminishes. Thus cows fed on fresh young grass give milk containing more riboflavin than those fed on dry grass or roots. Fresh raw peas and beans are good sources, but white bread is poor. Riboflavin is particularly stable to cooking unless soda is present, but considerable quantities may be leached out into the cooking water.

In cereals, riboflavin is particularly concentrated in the germ. Most nuts are good sources, and there are appreciable quantities in most fruits. As with thiamine, so riboflavin is concerned with the enzyme systems in the body, in particular those which are involved with oxidation processes in every cell of the body, which is the reason for the importance of this substance (see also *Vitamins, Vitamin B*). R.F.M.

Vitamin B6 (pyridoxin)

Wholewheat, légumes, nuts, yeast and yeast extracts are rich sources of this vitamin. Appreciable quantities are found in vegetables, including potatoes, tomatoes and turnips. Of the fruits, strawberries and citrus fruits contain the highest quantities. As with other members of the B group, B6 is concerned with the enzyme systems in the cells. Deficiencies result in multiple symptoms, including muscle weakness. It is probable that daily requirements are of the order 1.5 to 2 mg, but where the metabolism is increased then these requirements are likewise raised (see also *Vitamins, Vitamin B*). R.F.M.

Vitamin B12 (cobalamine)

The history of vitamin B12 began in 1926 when it was discovered that taking liver by mouth cured pernicious anaemia. It is now known that the active substance is a complex organic compound containing the metal cobalt. It is definitely concerned with the regeneration of red blood cells, but needs to be activated by a so-called intrinsic factor which is normally present in the gastric juices.

Rich sources of vitamin B12 are liver extracts of all kinds and other offal. A good source is fish, particularly herring, and also egg yolk. The muscle of meats contains a fair proportion, but fruit and vegetables are poor sources.

It is thought that the reason why vegans do not suffer as much from pernicious anaemia as might be expected is because there is evidence that this vitamin can be produced in adequate quantities by the micro-organisms present in the intestinal tract (see also *Vitamins, Vitamin B*). R.F.M.

Vitamin C (ascorbic acid)

This vitamin is of importance in that a lack produces the disease scurvy, which used to be prevalent amongst sailors, travellers and others who lived on salted and preserved foods. It is abundantly present in fresh fruits and vegetables of all kinds. Particularly rich sources are rose hips, black and red currants, citrus fruits, strawberries and green vegetables. Reasonable quantities are found in potatoes, spinach, watercress, turnips and sprouts, but légumes, particularly dried, and cereals, are rather poor sources.

Ascorbic acid is very sensitive to heat and becomes rapidly inactivated, especially if soda is present. Large quantities of ascorbic acid are leached out if fruits and vegetables are cooked, and the loss of the vitamin is increased in the presence of oxygen. Long boiling of potatoes and green vegetables destroys most of the vitamin.

Vitamin C is essential for wound healing, bone formation and repair and in maintaining the strength of the capillary walls. It is essential for the enzymes concerning cellular oxidation and amino acid metabolism. Deficiencies result in low resistance to infections, especially respiratory infections, skin manifestations and, finally, scurvy. The recommended daily intake is about 75 mg for an adult, which could be contained in about 100 grams (4 ounces) of orange juice (see also *Vitamins*). R.F.M.

Vitamin D

This vitamin is essential for the correct development and maintenance of bone material, and a deficiency in intake results in the condition of rickets which even today is widespread in undernourished children, particularly those living in the northern hemisphere.

It is fat soluble and is found in abundance in the liver oils of cod, halibut and tunny. Fish generally have a high content, but fruit and vegetables contain virtually none. Neither do the whites of eggs, although the yolk is a good source, particularly in the summer months.

Vitamin D is a sterol called calciferol and can be formed in the skin on exposure to ultraviolet light. This is why it is called 'sunshine vitamin' and why rickets is more prevalent in sunless climes. Human milk contains practically no vitamin D unless the mother is taking supplements. This is probably the best way to get the vitamin to the baby.

Vitamin D is quite stable to heat and in consequence there are no losses in cooking. The maximum dosage for children would appear to be about 1,500 international units per day, but the dosage for adults is probably not more than 500 international units. There is some evidence that vitamin D deficiency is associated with dental decay and certain forms of tuberculosis (see also *Vitamins*). R.F.M.

Vitamin E (tocopherols)

Sometimes called the anti-sterility vitamin, deficiency will undoubtedly cause delay in the development of the foetus in experimental animals. It is largely found in the germs of seeds and in green leaves. Wheat germ oils is the best source of vitamin E. During the modern milling process for flour the wheat germ oil is removed to a great extent and the vitamin destroyed by the heat produced. This vitamin, which is oil soluble, has anti-oxidative properties. It is recommended, in the form of wheat germ oil, for the treatment and prevention of coronary heart disease, and antioxidants of the type vitamin E are known to slow down ageing in experimental animals.

A diet which is largely composed of white bread and refined carbohydrates would be lacking in vitamin E. Apart from the germ oil from many plants, particularly cereals, vitamin E is found in most fish and in leafy vegetables, but not in fruits. Of the vegetable oils which contain high quantities can be mentioned maize, soya bean and sassaras. The daily requirements of this vitamin are not accurately known, but it would seem that on a mixed diet and intake of about 5 mg is usually achieved. In some spheres vitamin E has been considered to be of no importance in human nutrition, but in view of the profound effect deficiency has on monkeys, ducks, dogs and beetles, a very wide variety of species, it is difficult to believe that it is not essential for man. Moreover, there is abundant evidence that muscular dystrophy is associated with a deficiency in this vitamin (see also *Vitamins*). R.F.B.

Vitamin F

This vitamin is found in certain fatty acids but its precise function has not yet been properly established. It is credited with assisting to maintain the normal resilience and lubrication of all cells. It would appear to be beneficial to the skin, hair and nails. Vitamin F is found in butter, lard, linseed oil and some vegetable oils (see also *Vitamins*). R.R.

Vitamin H

This is a recently discovered vitamin, about which very little is known as as yet. It is an acid substance and is credited to have value in a type of baldness, alopecia areata, and also in a form of eczema. The richest sources of it are liver and kidney, dried yeast and watercress (see also *Vitamins*). R.R.

Vitamin K

This vitamin is concerned with the clotting of blood, and a deficiency will produce a low prothrombin level and consequently a tendency for continuous bleeding.

The best sources are deep green leaves, especially cabbage, cauliflower and spinach. High quantities are found in animal livers and eggs. Tomatoes are also high in this vitamin. Measurable quantities are found in fresh peas, rose hips, strawberries and carrots. Fruit appears to be particularly lacking in this factor.

Natural vitamin K is fat soluble, but synthetic substances have been produced, with vitamin K activity, which are water-soluble. The daily requirements are not known, but it is generally considered that on a good mixed diet the intake is adequate, particularly since this vitamin is synthetized to some extent by the microorganisms of the intestine (see also *Vitamins*). R.F.M.

Vitamin P (hesperedin)

Fruits are the richest source of vitamin P, followed by green leaves. There is very little in roots or seeds, although it increases in the latter on germination. It is a fairly stable substance, so very little is lost in the processing of commercial fruit concentrates and syrups, but over a period of time it becomes sensitive to light and temperature. Thus blackcurrant concentrates kept under normal storage conditions can deteriorate considerably over a period of six months. Apart from blackcurrant, good sources are other berries, grapes, rose hips, oranges and other citrus fruits, cherries, prunes, walnuts and apricots. The daily requirements are not known, but are certainly not less than 33 international units. Deficiency of the vitamin results in a tendency for petechial haemorrhage due to a loss of strength in the capillary walls, giving rise to bleeding from the gums and a tendency to bruise very readily (see also *Vitamins*). R.F.M.

Wakame

A seaweed from the rich waters of the ocean currents of the islands of Japan. It is cut into short strips of about 45 cm and when soaked in water for about 20 minutes it enlarges to five times its size. About 12 cm of dried wakame is sufficient for a pot of soup, or 8 cm per person when used as a sea vegetable. The minerals and enzymes contained in this seaweed aid the body in eliminating the effects of eating animal food and help it to adapt to vegetable quality food. Wakame is rich in protein, carbohydrates, calcium, sodium, phosphorus, iron and vitamins B1 and B12. It is helpful to people with hair problems and circulatory blockage (see also *Macrobiotics*). J.L.

Wheat germ

Wheat germ includes the wheat embryo and its associated cells the scutellum, and it is the growing point of the wheat. That is to say, it is the part that actually produces the new tiny rootlet, and the new shoot, when a wheat grain starts to germinate or grow. Nourishment for this first growth is drawn from the rest of the grain, but without the vital wheat germ there could not be any growth from any other part of the grain.

It is only a tiny part of the grain, about $1\frac{1}{2}$ to 2 per cent, but nutritionally it is concentrated goodness. It is our best source of vitamin E, and an excellent source of the B vitamin complex and of essential unsaturated fatty acids. As well as this, it contains some very good protein and is a rich source of phosphorus, so it is clear that the nutritional losses suffered by the consumer when this small fraction is removed are serious.

As the percentage of wheat germ in a grain of wheat is just about 2 per cent, and as about 5,000,000 tons of wheat are milled annually to provide us with about 3,600,000 tons of white flour, this means that every year about 100,000 tons of wheat germ, which is the most nutritious part of the wheat, are removed from the consumers' bread. A small amount of this removed wheat germ is packaged up for human consumption, and can be bought as a valuable, but rather expensive, food adjunct, which is said to be particularly good for sufferers from arthritis and for expectant mothers; but most of the tonnage removed, along with the bran, is bagged up and used for feeding farm animals and racehorses.

Unfortunately, when the cells of the wheat grain are broken up by milling, the fatty acids become exposed to the air, and so in due course become oxidized and start to go rancid. For this reason, before the wheat germ is packaged up for selling in the shops, it is usually 'stabilized', which means that it is treated, most often by heat, to prevent it from going rancid, so that it can have a long 'shelf life' in the shops. However, this stabilization reduces the nutritive value of the wheat germ. When buying packaged wheat germ it is therefore helpful, whenever possible, to buy a packet of wheat germ which has not been stabilized.

Wheat germ comes in the form of tiny dry flakes, and is excellent if added to muesli for breakfast, which a little milk. It can also be added to soups, or stirred into a glass of hot milk as a bedtime drink, and it is very good for children.

The fact that the fatty acids of the wheat germ, once released from their protective cell coverings, all too quickly deteriorate is one of the reasons why there is a movement, notably on the Continent, to promote 100 per cent wholewheat bread, baked from freshly milled wheat. Certainly such freshly milled wholewheat flour does make particularly delicious bread, and it is to be hoped that the time will come when bread of this quality is available to everyone (see also *Bread, Flour, Vitamin B*.) M.Y.B.

Whole cereal grains

Whole cereals are the original dried instant food. Plant one single grain and it will grow into a whole head of grain ready to be harvested. It will store almost indefinitely and yet contain within itself the seed of life. Grains have been known to sprout and grow even centuries after being first picked. Each wholegrain seed contains many elements sealed in its protective outer layers – minerals, vitamins, oils, protein and carbohydrates – that make it the perfect basis of human diet.

Since the beginning of civilization, man has centred his diet on cereals as his principal food – rice and millet in Asia, millet and corn in Africa, buckwheat in Russia, wheat, rye, barley and oats in Europe. Grain comes in many forms – mainly whole, as flour, as meal (coarse flour) and as flakes. All these can be incorporated into cooking, but it is best to use wholegrains as much as possible. When the delicate structure of the sealed grain is broken open and exposed to the air, it soon oxidizes and loses its vitality. Refined grains and their products (white rice, white bread) should be avoided wherever possible. When valuable parts of the grain are discarded, the result is an unbalanced food of little virtue. It is these refined grain products that contribute to obesity, since they lack the bran and other nutrients that help us to digest and use the starch properly. Always try to obtain grains that are organically grown. Grain must be chewed thoroughly to obtain maximum benefit. The saliva in the mouth begins the digestion of the starch. The muscles of the jaw are exercised and the teeth and gums strengthened (see also *Barley, Buckwheat, Maize, Millet, Oats, Rice, Rye, Wheat*). R.R.

Wholewheat

Wheat is normally ground into flour but is a deliciously chewy grain when cooked whole. There are many wholewheat products which can be used in a variety of ways. For instance, bulghur is a Middle Eastern preparation made by cooking wheat, then drying and cracking it. It comes in varying degrees of coarseness and is easy to cook. Couscous is an Arab dish, made from the inner layers of the wheat grain and should ideally be steamed. Also, wholewheat flakes can be used for porridge, stews and muesli, and wholewheat macaroni, spaghetti and noodles are available in health food shops (see also *Whole cereal grains*, *Wheat germ*, *Bread*). R.R.

Yeasts

The group fungi consists of moulds and yeasts, and yeasts are distinguished from moulds in that they are usually unicellular, although many of them form mycelium under special conditions.

Yeasts are a particular form of microorganisms belonging to the fungi group, and are usually associated with the process of fermentation. Yeasts are widely distributed in nature and are intimately concerned with the spoilage of foodstuffs, particularly fruit.

We are commonly made aware of yeasts in the production of alcohol, wines, beer, etc., and in the making of bread. In the manufacture of wine the yeasts react with the sugars in the grape juice to produce alcohol. A further stage in the fermentation process is the production of acetic acid from alcohol when the wine becomes transformed into wine vinegar.

In brewing, the yeasts act upon the maltose of the beer to produce the desired alcohol content.

In the making of bread, the yeast is incorporated to cause the dough to rise. The mechanism is that the starch becomes partly broken down to simple sugars, which then become fermented to produce small quantities of acetic acid and alcohol. At the same time, carbon dioxide gas is produced which causes the dough to become aerated.

There are an enormous number of different kinds of yeasts, and as many as twenty to thirty may be found during the fermentation of grape juice.

Besides the production of alcohol during the fermentation process, the yeast produces many by-products, and it is these by-products which give the different wines their characteristic flavours. It is because of certain by-products in fermentation that yeasts are of value for nutritional purposes. Among the by-products of yeast growth are produced a whole series of vitamins of the B group, in particular thiamine, riboflavin, nicotinamide, pyridoxin, pantothenic acid and biotin. Many of these vitamins are concerned with enzymic processes involving sugar metabolism, and therefore should be fed at the same time as carbohydrates, as is indeed the case when such are in the form of whole grain cereals. Supplementary addition to the diet of yeast and yeast extracts is very useful, particularly when refined cereals are being eaten (see also *Carbohydrates, Sugar, Vitamin B*). R.F.M.

Yoghurt

Yoghurt is made from milk which has been allowed to ferment under controlled conditions by the action of two bacteria, *Lactobacillus bulgaricus*, and *Streptococcus thermophilus*. (Occaionally a third group, *Lactobacillus acidophilus* may also be included.)

Cow's or goat's milk yoghurt is the most common. The usual manufacturing procedure is to break down and disperse the fat globules by homogenization and then to heat the milk to just under boiling point which renders it nearly sterile. The milk is then cooled to a temperature at which optimum growth of the bacteria is encouraged, the bacteria added, and the milk held at this temperature until it has set into a firm curd. The curd is the result of the production of acid by the bacteria which clots the milk and creates the well-known yoghurt consistency. The yoghurt should then be transferred to a cool store to retard further growth of the bacteria and firm and strengthen the curd. Live yoghurt is available plain or in a variety of sweet or savoury flavours, achieved by the addition of fresh fruit or herbs which contain no additives or preservatives. Plain yoghurt can be eaten as it is or sweetened with honey, molasses or muscavado sugar.

Yoghurt has immense value as a food because not only does it contain all the nutrients of the milk from which it was made (with the exception of lactose which is used by the bacteria), it also provides the body with vigorous health-protecting bacteria which speed up the breakdown of the milk with which they are incorporated and of the other foods we eat as well. It is recognized that these bacteria help the body to build up resistance to illnesses by fighting off pathogenic bacteria, and are of inestimable value in establishing a vigorous digestive system.

True yoghurt is a living culture and the milk in which it lives must be free from inhibitors, preservatives and colourings. The sharp, tangy, clean taste of the yoghurt is lost if these are added, and the culture is forced to change in its struggle to survive. It is then no longer a health food.

Yoghurt should be stored carefully. The bacteria are still active, hence the term 'live' yoghurt, and in order to prevent it from becoming over-acid and unpalatable it should be kept at all times in a refrigerator. If the container is shaken, the curd will break and whey will be released. This affects the appearance only and does not result in any loss of flavour or food value. A.H. O.A.H.

Zinc

As with some other of the trace elements, zinc is essential for many of the enzyme reactions which take place in the body. It is known to be essential in order to promote healthy growth in all young animals. It is part of the enzyme carbonic anhydrase, which is essential for the respiratory function of every cell of the body. This enzyme is important for the function of red blood corpuscles in that it allows the interchange of carbon dioxide and oxygen with the haemoglobin part of the cell.

Wheat germ and the germ of other cereals contain high quantities of zinc. It is also present in high concentrations in liver, eggs, nuts, meat and légumes. Low concentrations are found in fruits and vegetables, other than berries (see also *Trace elements*). R.F.M.

Part three

A Guide to Places

1 Organizations, Clubs and Societies

Parts One and Two of *The Health Food Guide* should by now have shown you how much you need Part Three. The shops, restaurants and health farms which are listed from this point onwards are there for those who have become personally committed to health food or wholefood; campaigns for change must, however, involve more than individual efforts, and this chapter gives details of how you can join others in bringing about the changes you want to see occur. Be sure to include a large stamped addressed envelope whenever you write for details, because many of the organizations listed here are run with entirely voluntary help, and every pound of postage saved will be a pound well spent in your own interests.

The organizations mentioned are open to all and are directly concerned with food or its manufacture in some way; they all need active members. Other organizations which are not included because they do not precisely accord with this definition include the following: *The National Institute of Medical Herbalists,* Honorary Secretary, Mr E. Robson, 22 Osborne Avenue, Jesmond, Newcastle-upon-Tyne, Tyne and Wear (telephone: 0632 813922); *The Herb Society*, The Secretary, 34 Boscobel Place, London SW1 ('Herbalism is the science which deals with those plants, parts of which are used for the benefit of man as food or medicine, domestically, or for their scent or flavour'); *The British Homoeopathic Association*, Basildon Court, 27A Devonshire Street, London W1N 1RJ (01–935 2163).

Beauty Without Cruelty 1 Calverley Park, Tunbridge Wells, Kent TN1 2SG (0892 25587); journal: *Compassion* (twice yearly) 30p, postage extra.
Beauty Without Cruelty was formed in 1959 by Lady Dowding and a handful of friends with the help and guidance of the late Air Chief Marshal Lord Dowding, whose work for animals is as well known as his victory in the Battle of Britain.

The idea came quite suddenly to Lady Dowding when she was attending a lecture and became unbearably aware of the thousands of animals which must have been killed to provide the fur coats worn by women in the audience. The movement snowballed and soon encompassed all aspects of animal exploitation in the interests of beauty preparations, clothing and household goods.

The organization, founded in the UK, has grown into a worldwide federation dedicated to the total abolition of abuses and to propagation of the truth about the cruelties they entail. Qualified observers are sent on fact-finding missions to all parts of the world where man in his greed has forgotten that animals too have rights.

Beauty Without Cruelty is an international educational charitable trust and is associated with Compassion in World Farming, the International Society for the Protection of Animals, and the World Federation for the Protection of Animals. Its commitment is *total*, and it offers viable alternatives to animal products whose manufacture or testing involves cruelty.

A range of first-rate cosmetics, soaps, shampoos, scents and household cleaners is manufactured by Beauty Without Cruelty Ltd, and in addition, simulated fur and leather goods are obtainable from Beauty Without Cruelty boutiques.

Lists are available of their approved complexion creams, lotions, powders, deodorants and antiperspirants, foods and natural supplements, hair preparations, household cleaning preparations, perfumes and toilet waters, soaps, sun preparations, talcum and bath preparations, toothpastes and veterinary preparations. The organization also has branches in Australia, British Columbia, Canada, Finland, India, Iceland, Japan, Kenya, Netherlands, New Zealand, Zimbabwe, Scotland, South Africa, United States of America, and Wales.

Beauty Without Cruelty averages between seven and nine lectures a week in the UK, and these in some instances involve full-scale fashion or film shows – not intended for the converted but for women's institute members, business and professional women, luncheon clubs, youth clubs, religious groups, parliamentary groups, schools, etc. Details on application with the usual stamped addressed envelope, or you may call personally; office opening hours 9.00–16.30.

Biodynamic Agricultural Association Woodman Lane, Clent, Stourbridge, West Midlands DY9 9PX (0562 884933); journal: *Star and Furrow* (biannual) £1.50.
The association exists to futher the study and practice of agriculture and

horticulture along lines indicated by the late Dr Rudolf Steiner, and also to promote research.

In biodynamic practice, the emphasis is on the highest standards of health and quality in soil, produce and livestock, rather than on maximum production by the intensive use of specialized methods. Such methods have raised yields and lowered costs, often spectacularly, but the price is a variety of ecological side effects and declining nutritional value of produce. Biodynamic methods are based on a comprehensive study of the interrelationships of the living organisms and processes involved in farming and gardening; this means adapting methods to suit local conditions, cropping-systems which emphasize lasting producivity, and the use of farm-produced manures and composts as the basis of fertility, supported by special preparations for treating composts, soil and plants. In animal husbandry, the emphasis is on optimum production and good health, rather than on excessively high yields. These measures can eliminate a number of diseases and pests; commercial orchards under intensive biodynamic management can dispense with chemical sprays.

The quality of biodynamic produce has been demonstrated in practice and confirmed by scientific tests, and manifests in the balance of proteins, carbohydrates, minerals and vitamins, as well as in colour, flavour, keeping qualities, etc. In Britain, these products are marketed as Biodynamic Produce. On the Continent they are marketed under the registered trademark Demeter. Demeter products are often recommended by doctors for patients, convalescents and babies.

Biodynamic farms and gardens exist in several European countries, in the United States, Central America, Africa, Australia and New Zealand, as well as in the British Isles.

The association's headquarters are at Woodman Lane, where an advisory service is available, as well as supplies of special preparations for compost-making and sprays, and further literature, including Steiner's fundamental course, *Eight Lectures on Agriculture*, and copies of the biannual journal *Star and Furrow*.

Research is conducted at Emerson College, Forest Row, Sussex, a centre for adult education based on Steiner's work. A school of biodynamic agriculture and earth sciences has recently been established at the college. Students normally follow a two-year course, which includes a broad foundation year, followed by more specialized studies. There are opportunities for practical work on the college's 250 acre farm, and for participation in research projects. Further details are available from the Secretary, Emerson College, Forest Row, Sussex. (There are also research centres at Darmstadt in West Germany, Järna in Sweden, and Spring Valley, New York State, USA.)

Community Health Foundation (East West Centre) 188 Old Street, London EC1V 9BP (01–251 4076)
The Community Health Foundation (or East West Centre, as it is also known for some of its activities) has grown out of the work of the East West Centre of London, a group of concerned individuals who sought to establish a teaching unit in London that could provide practical information for the public on ways in which good health can be established and maintained. For over two years the group planned and prepared their aims, methods to be used and the facilities needed. In July 1976 the CHF was registered with the Charities Commission to perform educational work in the field of preventive medicine, and in October 1976 the foundation signed a lease on the St Luke's School building in order to provide a permanent home for its facilities.

The CHF's leaflet for inquiries includes this comment on what they are about:

The relationship between humanity and the environment has now been strained to the point where we are no longer aware of the profound relationship between the food we eat, the water we drink, the air we breathe and our own well-being. Yet health is a creative and dynamic balance between the individual, society and the environment. Good health cannot be established by dependence on medicine or drugs, nor can it be gained by legislation or even research. Good health is a direct result of our own lifestyle. Many of the tools that are effective in establishing good health lie buried within the traditions of both eastern and western culture. This wealth of practical information is being ignored. The main goal of the CHF is to serve as a bridge between the traditional wisdom of our ancestors and our present scientific knowledge.

The CHF, among its many activities, runs courses on nutrition, macrobiotic food, and the traditional healing arts which include holistic and macrobiotic medicine, Iyengar, Oki Do, T'ai Chi Ch'uan, Aikido and Shiatsu (including Do-in, self-massage). Michio Kushi, who is associated with these valuable courses, is the founder and president of the East West Foundation of America.

Full membership costs £12 a year, which entitles you to a 10 per cent discount on all class fees, the booklets published twice yearly, one free introductory daily workshop, and advance notice of events. Family (husband and wife) membership costs £20 a year, and patron membership £50. The Natural Snack is within the CHF (see Chapter 9, *Restaurants*).

Compassion in World Farming 20 Lavant Street, Petersfield, Hampshire GU32 3EW (0730 4208); subscription UK and overseas, £3 (which includes the journal); journal: *Ag*.

This is a vigorous organization with its own newsletter, *Ag*. It is concerned with the full range of horrors in factory farming, and holds regular meetings, campaigns and demonstrations. *Ag* lists area contacts to assist and coordinate effective activity, and the journal also makes use of illustrations to bring home the points it makes. Chapter 1 of *The Health Food Guide* has emphasized the necessity for organizations such as this.

The Free Range Egg Association (FREGG) 39 Maresfield Gardens, London NW3 2UA (01 435 6688); annual subscription £1 (minimum).

FREGG consists of a group of volunteers who promote the sale of free-range eggs by inspecting apparently genuine free-range chicken farms which, if approved, are then put in touch with suitable retailers. The shops receive the FREGG sign to look out for – a hen's head in a yellow triangle. The farms are inspected every eighteen months or so; if they are too far from a FREGG inspector, a Weights and Measures inspector supplies a report, which is specially verified later by FREGG if it contains doubts. Following a successful inspection, FREGG invites the chicken farmer to complete a questionnaire about his method of husbandry and to sign a contract confirming that he does not mutilate his birds, feed routine antibiotics or hormone supplements, and that his birds have access to outside runs; the farmer is also asked to inform FREGG of any changes in his methods and to agree to a forfeit of £50 if he violates his contract. Such contracts are renewable each year, so FREGG does indeed offer an invaluable service. The association has available a list of FREGG-approved farmers and egg stockists (not necessarily health food shops) around Britain.

The Soil Association Walnut Tree Manor, Haughley, Stowmarket, Suffolk IP14 3RS (044970 235/6); journal: *Quarterly Review*.

This well-known organization, founded in 1946, has members in over seventy countries; it is thus well placed to achieve its aims, which include collecting information on farming systems and their effects on the environment, on the long-term fertility of the soil, and on the health of those who consume the produce. It believes that there is a relationship between soil treatment and the nutritional quality of crops grown in that soil, and that farming methods affect human health. Research at Walnut Tree Manor has been going on for very many years, and the resulting information about organic farming methods finds its way into the Soil Association's extensive list of books, booklets and leaflets on agriculture, gardening, food, health and the environment; some are technical and some are designed for general use. Annual membership is £8, which includes the journal.

The Vegan Society 47 Highlands Road, Leatherhead, Surrey; journal: *The Vegan* (quarterly, free).

The society was formed in 1944 by former vegetarians who believed that it was wrong to tolerate the use of animal products such as eggs and milk. The literature that the society has available makes it quite clear why the beliefs of the founders are now widely shared. Living the vegan way, Britain could in fact feed itself and have large land areas left for trees, wildlife, and for 'energy farming' which would make the use of nuclear power unnecessary; at present 90 per cent of Britain's 46 million acres of agricultural land is devoted directly or indirectly to livestock.

The Vegetarian Society of the United Kingdom Parkdale, Dunham Road, Altrincham, Cheshire WA14 4QG (061 928 0793); London centre and Bookshop: 53 Marloes Road, London W8 6LA (01 937 7739); journal: *Vegetarian* (bimonthly 35p; £3 p.a., post free).

The Vegetarian Society was founded in 1847 and the London Vegetarian Society started in 1888; the two merged in 1969 and are affiliated to the International Vegetarian Union which was founded in 1908. Today, the society is extremely vigorous in all its activities, and membership will be approaching 10,000 fairly soon. In return for a large stamped addressed envelope you will receive free literature, but membership brings you their journal and also, every two years, the *Vegetarian Handbook*, which otherwise costs £1.50.

Vegetarians believe that the slaughter of animals for food and the consumption of fish, flesh and fowl are bad for man, individually and socially, from a health point of view, ethically and spiritually. Unlike vegans, vegetarians may choose whether or not to include dairy products and eggs in their diet. Through its food and cookery sections the society organizes cookery courses, lectures and demonstrations and, naturally, the scientific aspects of nutrition are stressed. A research section was started in 1974 with the aim of establishing the scientific basis of vegetarianism, and to show economists, agriculturalists and scientists how vegetarianism can provide a permanent solution to the worsening world food situation.

2 Shops

The expansion of the health food retail business continues apace, as the number of entries in this chapter demonstrates. The search for health food has been made easier since the last edition of this guide by a great number of dedicated individuals who have been 'filling the gaps' in towns and cities throughout Britain. Indeed, saturation point has surely been reached in some places, such as Chichester, Sussex where there are three health food shops within 150 yards of each other in the very centre of the city. As I have pointed out before, it is often well worth patronziing old-fashioned or 'high-class' grocers, because there you may find really fresh cheeses, yoghurt, free-range eggs, etc. Family-owned butchers' shops in country districts are also worth exploring. The market forces of supply and demand, where health food shops are concerned, require that you demand, constantly, what you need; nowadays the sources of supply for most basic health foods exist (goat's milk, buttermilk, fresh yoghurt, halva, free-range eggs, fresh butter, fresh cheeses, wholemeal bread, etc.).

The following abbreviations have been used in this chapter:

BHMA   British Herbal Medicine Association
CHF   Community Health Foundation
FNWC   Federation of Northern Wholefood Collectives
FOE   Friends of the Earth
HFTA   Health Food Trade Association
HSW   Health Food Stores (Wholesale) Ltd
ICOM   Industrial Common Ownership Movement
NAH   National Association for Health
NAHS   National Association of Healthfood Stores
NWFWC   North West Federation of Wholefood Cooperatives

Aberdare

Natural Selection 27 Whitcombe Street, Aberdare, Glamorgan, Wales (0685 877066); open Monday to Saturday 9.00–18.00
Newly opened store with horizons which extend beyond profits. Excellent stock of grains, pulses, flours, cereals, dried fruits, and shelled nuts sold loose from sacks. Honey, oils, fruit juices, vegetarian specialities, goat's milk and fresh yeast. Herbs and spices for culinary and medicinal use also sold loose. Good selection of herb teas. Also sell local craft pottery, cooking utensils, and books on cookery, gardening, yoga, natural health and similar subjects. Owners now trying to extend goods on offer by finding regular supplies of free-range eggs, wholemeal bread and fresh yoghurt. It may soon be necessary to move to larger premises! Very willing to order goods at customer's request.

Aberdeen

Aberdeen Grain and Herb Store and **Health and Dietary Food Stores** 81/81a Rosemount Place, Aberdeen, Grampian, Scotland (0224 630570/ 630845); buses 13, 22; open Monday to Saturday 9.00–17.00
The owner of the Health and Dietary Food Stores, which have been open for twelve years, has now launched the Herb and Grain stores next door. Enormous variety of wholefoods on offer, in bulk or sizes from 1 lb upwards – seventeen kinds of flour, twenty kinds of bean, huge choice of nuts and dried fruits, raw sugars, soup mixes, cooking and baking mixes, pastas, oils, fruit and vegetable juices, beverages including real and grain coffees, sweets such as boilings and rock from brown sugar, good range of honey, and wide selection of Indian, Chinese and Asian foods. Dazzling choice of herbs, including several of which we have never heard. Food for your pets, natural remedies and vitamin supplements for you and them. The original shop also sells vegetarian special foods, natural cosmetics, slimming supplements, and books on health; all you need for home beer and wine-making. Goods sold retail and wholesale; please give twenty-four hours' notice for bulk supplies. Deliveries arranged for a small charge, large orders free. Mail order service. Old age pensioners given discount. Stock list available. (NAHS)

Health Food Centre 14 Waverley Place, Aberdeen, Grampian, Scotland (0224 55208); buses (to door) 18, 19, (300 yards) 4, 5; open Monday to Friday 9.00–12.30, 14.00–17.30, Saturday 9.00–12.30
Manager here very helpful and prides himself on the range of his satisfied customers as well as on his stock; this includes good range of wholefoods (which can be bought in bulk with prior warning) including flours and other cereal produce, grains, nuts and seeds, sea salt, dried fruits, preserves, speciality teas, and vegetarian special foods. You can also buy natural cosmetics and herbal remedies, as well as the usual vitamin and mineral supplements. Bread and other wholewheat bakery goods delivered from local baker. No delivery service, but if customers unable to come into shop because of illness, special arrangements will be made; goods sent by post with minimum of delay. (NAHFT)

Abergavenny

Cornucopia 13 Market Street, Abergavenny, Gwent, Wales (0873 5346); open Monday to Saturday 9.00–17.30
One regular customer maintains that the best bread in Wales is found here and she is referring to the wholewheat bread made fresh on the premises each day. All the other fresh items, such as cakes, sell very quickly too. Full range of groceries and dried goods.

Aberystwyth

Maeth Y Meysydd 11 Princess Street, Aberystwyth, Dyfed, Wales (0970 612946); open Monday to Saturday 9.30–17.00, Wednesday to 13.00
At present this shop stocks the usual cereals and grains, beans, lentils dried fruits, juices, spreads and honeys, herbs and spices but hopes to sell bread in the future. Muesli is prepared on the premises and a takeaway service for such items as pasties, scones, flans, flapjacks is available. Cookery books on sale together with environmental magazines and papers. Two large noticeboards display small advertisements and events.

Windsor Health Centre 28 Chalybeate Street, Aberystwyth, Dyfed, Wales (0970 612915); buses apparently unnumbered, but stop 25 yards

from shop; open Monday to Saturday 9.30–17.30, Wednesday to 13.00
Shop recently changed hands and about to undergo a complete refit,
when range of stock will be increased. At present, however, sells all
normal items to be expected in a well-managed store plus spring waters,
live yoghurt, goat's milk and vegetarian products. Fresh wholemeal
bread, baked locally, available daily. No van deliveries, but an excellent
postal service, also omnibus delivery service covering twenty-mile
radius. (NAHS)

Abingdon

Frugal Food 17 West Saint Helen Street, Abingdon, Oxfordshire (0235
22239); open Monday to Saturday 9.00–17.30
The shop, in a pretty and well-restored sixteenth century building in
town centre, has only been open a year. Combines good health food
stock with high-class grocery provision, and some Greek, Chinese and
Indian foods. Staff friendly and will give advice on the use of anything
unfamiliar. Stock includes dried wholefoods weighed out for you –
thirty-two pulses and fourteen kinds of flour; herbs and spices; coffee
ground as you wait; vitamins and dietary supplements; some homoeopathic supplies. Most goods selected for their lack of colouring and
preservatives. Also a take-away snack service, offering food cooked
daily by local people, including quiches, vegetable curry, pasties and
cakes, e.g. banana loaf.

Airdrie

Airdrie Health Foods 174 Clark Street, Airdrie, Strathclyde, Scotland
(023 64 63379); open Monday to Saturday 9.00–17.30, early closing
Wednesday
Vegetarian and vegan special foods, wholefood dried goods and wholemeal bread, honey and preserves, some Deves products, fruit juices.
mineral waters; very good stock of vitamin and mineral supplements,
homoeopathic herbal and biochemical remedies. Also free-range eggs,
and books on health. Deliveries to Airdrie and Coatbridge.

Aldeburgh

The Health Food Shop 183 High Street, Aldeburgh, Suffolk (072 885 2234); open Monday to Saturday 9.30–13.00, 14.30–17.00, early closing Wednesday, closed Mondays in winter
Delicatessen and health food shop popular with local music luminaries, stocking packaged health foods, dairy goods, fruit juices, vegetarian specialities, honey and preserves, herbal cosmetics and remedies, vitamins. Fresh bread three days a week.

Aldershot

Southern Health Foods 8 Wellington Centre, Victoria Road, Aldershot, Hampshire (0252 315738); open Monday to Saturday 9.00–17.30
One of the well-established Southern Health Foods chain, stocking good range of packaged wholefood and vegetarian groceries including slimming, diabetic and diet foods; yoghurt and other dairy goods; a range of wholemeal flours; wholemeal bread delivered daily by a local baker; free-range eggs; compost-grown vegetables in season; and usual range of vitamin and mineral supplements, natural cosmetics, books on health subjects. (NAHS)

Alfreton

Herbs and Health Foods 55a Mansfield Road, Alfreton, Derbyshire (0773 811 840); open Monday to Saturday 9.00–17.30, Wednesday to 13.00
This shop, opened in 1973, is looking for larger premises closer to town centre. Stocks usual range of dried health food goods, herbs, herbal and vitamin products, natural fruit juices, preserves.

Altrincham

Health Food Centre and Nutcracker Restaurant 41/43 Oxford Road, Altrincham, Greater Manchester (061 928 4399); bus 364 to Lloyd Street; open Monday to Saturday 9.00–17.30, Wednesday to 13.00

This well-established shop and restaurant maintains high standard of provisions. Wide range of products, and organically grown vegetables on sale in season. Stocks natural cosmetics and vitamins. Deliveries in the area. Restaurant serves wide range of vegetarian savouries, salads, wholewheat cakes and snacks. All dishes prepared fresh each day.

Naturally Yours 154 Ashley Road, Hale, Altrincham, Greater Manchester (061 928 5828); open Monday to Friday 9.00–13.00, 14.00–17.30, Wednesday half day
Stocks fresh wholewheat bread (delivered daily), cereals, grains, nuts, fruit preserves, honey, slimming aids, Beauty Without Cruelty cosmetics, herbal remedies and herbal cigarettes, vitamin and mineral supplements, fresh farm eggs.

Amersham

Amersham Whole Foods 156 Upper Station Road, Amersham, Buckinghamshire (024 03 6752); open Monday to Saturday 9.00–17.30
Good shop, well stocked, with friendly service. Aims to expand still further. They sell their own stoneground flour – mill is in shop – and bake their own 100 per cent wholemeal bread and cakes. Muesli mixed on premises. Other goods include over 200 herbs and spices, grains and flakes, pulses, cereals, macrobiotic foods, honey and preserves, free-range eggs, dairy goods such as vegetarian cheese, some fresh fruit and vegetables, and natural cosmetics. Also the usual vitamin and mineral supplements, herbal remedies, books on health and whole foods. Deliveries in South Buckinghamshire and Hertfordshire. Mail order service. (NAHS)

Holland and Barrett 100 Sycamore Road, Amersham, Buckinghamshire (02403 7273); buses stop 25 yards away; open Monday to Saturday 9.00–17.30
One of well-known Holland and Barrett chain of health food shops, which is the largest in the country. In bright, well-organized surroundings they stock their normal range of cereals, soya products, yeast, unsweetened juices, natural cosmetics, herbal remedies and teas, vitamins, natural food supplements, dried fruit, nuts, pulses, etc. Some shops have fresh food daily to take away, which might include wholemeal sand-

wiches with a variety of fillings, salads, flans, vegetarian quiches, pies, slices, cakes, biscuits, rolls and loaves of wholemeal bread. A few Holland and Barrett outlets also sell hot snacks and soups over the counter, and others prepare fresh muesli on premises. None of their shops undertakes local deliveries or maintains credit accounts. The latest issue of the magazine *Here's Health* (published by another company in the group) will generally be on sale, to complement the useful range of books in stock.

Andover

Carlyns 36 Chantry Way, Andover, Hampshire (0624 2003); open Monday to Saturday 9.00–17.30, Wednesday 9.30–13.00
Sells packaged health foods, dairy goods, vitamin and mineral supplements, natural cosmetics, home beer- and wine-making supplies.

Ashford

Country Life Wholefoods 15 Bank Street, Ashford, Kent (0233 35724); open Monday to Saturday 9.00–17.30
This is a shop in the expanding Country Life chain of health food retail outlets. They all stock usual range of vitamins, remedies, natural cosmetics, tissue salts, loose muesli, bran, pulses, honey, flour, dried fruit, nuts, herbs, unsweetened fruit juices, cereal products, books, etc. Yoghurt and cheeses also available. Country Life have their own bakery; all their Sussex shops receive bread and cakes daily for sale, but at present the Kent branches have cakes only. (NAHS)

Ashton-under-Lyne

Harry Harrop & Son Ltd 133 Stamford Street, Ashton-under-Lyne, Greater Manchester (061 330 1064); buses 216, 218, 219 (75 yards); open Monday to Saturday 9.00–17.30, Tuesday half day
A herbalist since 1917. Keeps a good range of herbal and natural remedies, together with prepacked health and vegetarian foods. (NAHS)

S. Harrop, Herbalist 24 Old Street, Ashton-under-Lyne, Greater Manchester (061 330 9853); buses to town centre; open Monday to Friday 9.00–17.15, closed Tuesday
Established in 1848 as a herbalist. Has excellent range of herbs, roots, barks, and other herbal remedies; advice given on their use. Also stocks dietary supplements and some health foods. (BHMA, BHU)

Axminster

Healthwise Lyme Street, Axminster, Devon (0297 32397); bus 231 stops 200 yards from shop; open Monday to Saturday 9.15–13.00, 14.15–17.00, early closing Wednesday
Run by young vegetarians, tries to stock as much compost-grown fruit and vegetables from local suppliers as possible. Good stock of health and wholefoods includes herbs and spices sold loose from jars, fresh yeast and other home-baking supplies, muesli and muesli bars made on premises, herbal remedies and nutritional supplements. Section selling local pottery. Deliveries to Seaton and Axe Valley.

Aylesbury

Delikatesserie and Health Food Store 56 Kingsbury, Aylesbury, Buckinghamshire (0296 23487); buses to Kingsbury; open Monday to Saturday 8.30–17.30, Thursday to 13.00
Shop catering for all delicatessen, health foods and supplements, together with restaurant offering salad lunches, sandwiches, coleslaw, etc.

Ayr

McWhirter 8 Cathcart Street, Ayr, Strathclyde, Scotland; open Monday to Saturday 9.00–13.00, 14.00–17.00, closed Wednesday
Range of mostly packeted general health food, together with herbal remedies. Muesli mixed on premises. (HSW, NAHS)

Nature's Way 17/19 Carrick Street, Ayr, Strathclyde, Scotland (0292 61099); open Monday to Saturday 9.00–17.30
Recently opened wholefood shop – no one of chain by same name. Stocks very full range of dried goods, dairy products, beverages, herbs and spices, as well as bread and bakery products made fresh on premises. Deliveries within Ayr. (NAHS)

Babbacombe

Lynnell's Health Food Store 120 Reddenhill Road, Babbacombe, Torquay, S. Devon (0803 37102); buses 130, 151; open winter Monday to Saturday 8.45–8.00, summer every day 8.45–21.15
Florist and fruiterer are combined with this health food store where wholemeal bread and rolls are baked on premises. Also stock both loose and prepacked grains, dried fruits, vitamins, minerals, and range of country wines, cider, mead. Owners will endeavour to obtain any item not normally in stock.

Ballymena

Jack McLelland Naturally 16 Ballymoney Street, Ballymena, County Antrim, Northern Ireland (0266 2922); open Monday to Saturday 9.00–17.30, closed Wednesday
Stock is completely vegetarian and preferably vegan; includes wholefoods both loose and packaged (muesli mixed in shop), dairy goods like vegetarian cheese, free-range eggs, fresh compost-grown fruit and vegetables, bread from main shop in Belfast, natural medicines, nutritional supplements. A naturopathic consulting room, for which you need an appointment. (HSW, NAHS).

Banbury

Ann Elisabeth Health Foods 48 Parsons Street, Banbury, Oxfordshire (0295 2681); open Monday to Saturday 9.00–17.30, Tuesday to 13.00
Store housed in oldest building in central Banbury, dating from fifteenth century. Comprises shop, take-away snack bar and beauty *salon*.

Full stock of honey, beverages, cooking aids, herbs, dried fruit, nuts, grains, dairy produce, wholewheat in every form, vegetarian savouries, vitamins, herbal preparations and cosmetics. Snack bar specializes in vegetarian food and cakes baked on premises. They hope in the future to extend to coffee-roasting and home coffee-making equipment. Also bulk supplies of about 100 culinary and medicinal herbs.

Bangor (N. Ireland)

Little & McQuoid 20 Central Avenue, Bangor, Northern Ireland; buses to Bangor station; open Monday to Saturday 9.00–17.00
Mainly packaged health foods, organically grown where possible, with good selection of herbs, vitamin supplements, books on diet and health.

Bangor (Wales)

Harvest Wholefoods 23 Wellfield Court, Bangor, Gwynedd, Wales (0248 4518); opposite bus station; open Monday to Saturday 9.30–17.00, Wednesday to 13.00
An excellent range of dried goods packed in the shop, including grains, flours, pulses, nuts and fruits, herbs and spices, muesli, pasta; also vegetarian cheese, yoghurt, juices, oils, Asian foods and spices, preserves, fresh wholemeal bread, organically grown vegetables in season, books on health and wholefood cooking. A café and restaurant with a take-away counter is planned.

Barking

Competitive Remedies 39 Ripple Road, Barking, Essex (01 594 3016); open Monday to Saturday 9.00–18.00
This small shop has been run by the same man for fifty-two years, and he is an excellent advertisement for his own shop, being eighty-two and fit as a fiddle. He is helped by his daughter, now as knowledgeable as her father. They specialize in herbal remedies and vitamin and mineral supplements; stock health foods – branded dried goods, dairy products including goat's milk, wholemeal bread; also nicotine-free tobacco products, slimming aids, natural cosmetics. (BHMA)

Barnstaple

Barnstaple Health Store 25 High Street, Barnstaple, Devon (0271 5624); buses to The Strand; open Monday to Saturday 9.00–17.30, early closing Wednesday
The shop, which may move along the street, leaving this to house the cosmetics part of the business, stocks a wide range of wholefood groceries and speciality vegetarian foods, yeast, Twinings teas, Brewhurst products, vitamins and remedies.

Barrow-in-Furness

Furness Health Food Store 58 Crellin Street, Barrow-in-Furness, Cumbria (0229 25834); buses to town hall; open Monday to Saturday 9.00–17.30, closed Thursday
The owners of this bright modern shop have recently opened another branch in Ulverston. They stock full range of wholefoods – pulses, vegan and vegetarian foods, yoghurt, goat's milk, dried fruit and nuts, vitamins and mineral supplements, honey, juices, herbs and spices, herbal remedies; good range of mueslis and breakfast cereals; natural cosmetics; home beer- and wine-making equipment.

Basingstoke

Holland & Barrett 15 Wote Street, Basingstoke, Hampshire (0256 58324); open Monday to Saturday 9.00–17.30, Tuesday 9.30–17.30
For further information see entry on Holland & Barrett chain in Amersham (page 180).

Trivedi Health Foods 4 Winton Square, Basingstoke Hampshire (0256 28680); open Monday to Saturday 9.00–17.30, Thursday half-day
Shop opened in 1972. Good stock of dried wholefoods, herbs and spices, free-range eggs, wholemeal bread (including French loaves), vitamins and herbal remedies natural cosmetics. Deliveries within Basingstoke. When new premises can be found, a snack and juice bar will be opened.

Bath

Culpeper 28 Milsom Street, Bath, Avon (0225 66321); open Monday to Saturday 9.30–17.30

During the 1970s the famous herbalists opened several country branches like this one. Live herbs in pots sold, as well as huge selection of dried herbs and spices. There are other condiments: mustards and sauces, natural curry powders, vinegars, flavourings like natural vanilla; natural foods including raw sugars, honey, organically grown wholewheat flour; drinks like tea and apple juice; and preserves – the marmalade is quite special. Also herbal remedies and topics, herbal beer extracts, potpourri and pomanders, herbal cushions and sachets, seeds; huge cosmetic range includes hair and bath preparations, toilet waters and colognes, essential oils, real sponges. Additional odd delights such as real liquorice, prunes, crystallized seeds and leaves from France, dried flowers and flower cards, and good range of books on herbs. No deliveries but goods valued over £3 will be sent by post anywhere.

Harvest Wholefoods 37 Belvedere, Lansdown Road, Bath, Avon (0225 23407); open Monday to Saturday 9.30–18.00

Bath and its outlying districts must be full of wholefood eaters – the owner of the very successful Harvest in Lansdown Road has opened another, bigger, shop in the city at Walcot Street; which must make a record number for a town of this size. Original shop was one of the earlist new-style shops selling from bulk, and has helped many younger businesses. Shops stock very full range of dried goods, organically grown where possible: grains, flours, flakes, mueslis, dried fruit, shelled nuts, herbs, spices; honeys and preserves; beverages; Japanese traditional foods; fresh wholemeal bread and fresh dairy produce; juices; a range of useful books and kitchenware. Service is from bins – the customer helps himself; but advice is readily available, and orders are taken for some items in bulk. Juice bar is planned for the new shop in Walcot Street.

Harvest Wholefoods 37 Walcot Street, Bath, Avon (0225 23407); open Monday to Saturday 9.30–18.00

Recently moved to larger premises. Good stock of wholefoods, grains, pulses, nuts, and herbs. Fruit, fresh vegetables and baked goods sold; bulk items sold at discount. Books and cooking-ware also stocked.

Orders worth £30 and over delivered within immediate neighbourhood of city.

Health Foods 6 Green Street, Bath, Avon (0225–62449); open Monday to Friday 9.00–17.00, Saturday to 13.00
Attractive Georgian shop established in 1905; proprietors interested in promoting food reform. Good range of vegetarian foods, wholefoods, herbal remedies and natural cosmetics, herbs and spices, yoghurt, free-range eggs. Deliveries in Bath and immediate district. (NAHS)

Realfoods (Holland & Barrett) 6 Cheap Street, Bath, Avon (0225 62280); buses to High Street; open Monday to Saturday 9.00–17.30
For further information on the Holland & Barrett chain see entry for Amersham (page 180).

Beckenham

Herb of Grace 276 High Street, Beckenham, Kent (01 650 4753); buses 54, 276, 289; open Monday to Saturday 9.00–17.30, Wednesday to 13.00
Good range of usual dried wholefood and packaged brands of vegetarian provisions, with dairy goods, fruit juices, vitamin and mineral supplements.

Nature's Way Ltd 256 High Street, Beckenham, Kent; open Monday to Saturday 9.00–17.30
One of the chain of Nature's Way health shops stocking wide range of all types of health foods, dairy goods, vitamin supplements, herbal remedies. Shop is particularly bright and attractively laid out. Prices of own-packed goods very cheap.

Bedford

Bedford Health Food Centre Ltd 104 Bromham Road, Bedford (0234 50947); near main bus station; open Monday to Wednesday 9.00–17.30, Thursday to 14.30, Friday and Saturday to 18.00
Owner prides himself on keeping an old-style shop where stock and

information are nevertheless up to date; he is now looking for premises in the shopping centre. Although small, the shop sells almost every brand of health foods, natural remedies and nutritional supplements. Fresh local bread available on Fridays and Saturdays; free-range eggs sold when obtainable, as are some compost-grown vegetables. Dairy products include vegetarian cheese and local goat's yoghurt. Helpful staff well able to give advice on dietary problems; deliveries can be arranged in case of customers' illness. (HSW)

Holland & Barrett 10 Horne Lane, Harpur Centre, Bedford (0234 52866); buses to nearby bus station; open Monday to Saturday 9.00–17.30
For further information on the Holland & Barrett chain see entry for Amersham (page 180).

Sunflower Wholefoods 103 Castle Road, Bedford, buses 104, 105 stop outside; open Monday, Tuesday, Thursday 18.00–21.00, Wednesday 9.30–13.00, 14.00–21.00, Friday 9.15–21.00, Saturday 9.15–17.30, Sunday 9.15–13.00
A well-stocked collectively run shop selling from bulk, so prices are low. Goods on sale include juices, seeds, flours, flakes,grains, muesli, pastas, sea salt, raw cane sugar, vinegar, oils, dried fruits, nuts, pulses, honey, spreads, fresh yeast, sauces, miso, free-range eggs, yoghurt, cheese, goat's milk. Besides food, you can buy seeds for planting; recycled paper toilet rolls; T-shirts; badges; books on health and environmental issues. The collective acts as a centre for the collection of paper for recycling, and as an information service on bottle exchange and gardensharing schemes. Displays on ecological subjects, and a library of books on the environment. Information service is to be extended as far as resources will allow.

Bedworth

Health Food and Homebrew 59 Mill Street, Bedworth, Warwickshire (0203 315627); open Monday to Saturday 9.00–17.30
Sell complete range of packaged health foods, herbal remedies, vitamin supplements, home beer- and wine-making supplies. Hope to expand the herbal medicine service. Some deliveries by arrangement.

Belfast

S. G. Donaldson 22 North Street Arcade, Belfast, Northern Ireland (0232 26967); open Monday to Saturday 9.30–17.00, Wednesday to 13.00
Long-established store selling complete range of vegetarian foods, wholegrains, wholemeal flours, natural fruit juices, herbal preparations, vitamins, and books on health subjects. On second floor, resident consultant herbalist. (HSW)

Jack McClelland Naturally 327 Antrim Road, Belfast, Northern Ireland (0232 744590); open Monday to Saturday 9.00–17.30, closed Wednesday
Stock here is completely vegetarian and preferably vegan; includes popular brands of packaged wholefood groceries; yoghurt, vegetarian cheese, free-range eggs; fresh compost-grown fruit and vegetables when available; bread baked on premises; natural medicines and nutritional supplements; a naturopathic clinic you can visit by appointment. (HSW, NAHS)

Sassafras Wholefoods 102 Great Victoria Street, Belfast, Northern Ireland (0232 24549); open Monday to Saturday 9.30–17.30
As much organically grown food as possible is stocked – grains, beans, cereals, flour, nuts, dried fruit, honey, sea vegetables, large selection of herbs and spices. Also goat's milk and cheese, free-range eggs, cold-pressed oils. Wholemeal bread available daily. Deliveries to any area, but on wholesale basis only.

Scrabo Health Products Ltd 267 Ormeau Road, Belfast, Northern Ireland (0232 646721); open Monday to Saturday 9.00–17.15
Opened in 1977, with a sister store in Newtownards; further shops will be opened in County Down. Accent here is on health and homoeopathy rather than just provisions; consultant is available for private advice, and the business holds the licence for Scleriso, a remedy against arteriosclerosis, which it markets throughout the country. Packaged wholefoods on sale, as well as homoeopathic medicines, vitamin and mineral supplements, herbal products. Goods mailed anywhere in the world. (BHMA)

Beltinge

Beltinge Health Stores 146 Reculver Road, Beltinge, Herne Bay, Kent (022 73 62910); open Tuesday, Wednesday, Friday, Saturday 9.00–17.30, Thursday to 13.00, closed Monday
The usual range of branded health foods, vitamin and mineral supplements, baked goods such as date and walnut cakes, fruit cake, flapjacks fresh daily. Deliveries within locality.

Benllech

E. & A. Huddart The Delicatessen, Benllech, Anglesey, Gwynedd, Wales (024 874 2756); open Monday to Friday 9.00–13.00, 14.00–17.30, Thursday half day
Village grocer stocking packaged health foods; willing to order items for local residents. Stock free-range eggs, local cheeses, and fresh wholemeal bread made near by.

Beverley

G: Jack & Son Ltd 6 Wednesday Market, Beverley, Yorkshire (0482 882437); open Tuesday to Saturday 8.30–17.30
Family grocery business which launched into health foods a few years ago. They stock vegetarian and wholefoods, free-range eggs, some compost-grown vegetables when available from local suppliers, local dairy produce, natural remedies, good stock of mineral and vitamin supplements. Local deliveries.

Bexhill-on-Sea

Health Food Store 11 London Road, Bexhill-on-Sea, Sussex (0424 212203); open Monday, Tuesday, Thursday, Friday 8.45–13.00, 14.15–17.45, Wednesday to 13.00, Saturday to 17.00
Established business with muesli, bran, cereals and pulses packed on the premises. Also vegetarian section and wide range of honey. Medicinal herbs, culinary herbs and spices sold loose; locally baked bread available daily. Deliveries in and around Bexhill. (NAHS)

Nature's Way Ltd 10 Devonshire Road, Bexhill-on-Sea, Sussex (0424 216860); open Monday to Saturday 9.00–17.30
One of the chain of Nature's Way shops. For further information on this chain see entry on Nature's Way Ltd in Beckenham (page 187).

Bexley Heath

Better Health 208a Broadway, Bexley Heath, Kent (01 303 6792); open Monday to Saturday 9.15–17.30, Wednesday 9.00–13.00
In addition to the normal range of wholefoods you will find fresh yeast, dairy goods including yoghurt and goat's milk, herbs and spices, slimming and diabetic speciality foods. A section selling books and magazines on health topics. Deliveries locally. (HFTA)

Bicester

S. Savins & Son 92 Sheep Street, Bicester, Oxfordshire (086 92 41405); buses to Sheep Street; open Monday to Friday 8.30–17.30, Thursday to 13.00
General store selling large variety of goods, stock some health food items such as wholemeal flour and cereals, vitamins, free-range eggs, locally grown fruit and vegetables including owner's own.

Bideford

Health Food Stores 14a Bridgeland Street, Bideford, Devon (023 72 2715); open Monday to Saturday 9.00–13.00, 14.00–17.00, Wednesday to 13.00
Wide range of herbal remedies, herbal supplements, diabetic foods and juices, fruit and nuts, spring waters, muesli, bran, herbs spices. (NAHS)

Bingley

Fodder 2 Norfolk Street, Bingley, Yorkshire (097 66 5187); bus 691 to Mornington Road; open Monday and Wednesday 10.00–17.00,

Tuesday to 13.00, Thursday 9.00–17.00, Friday 9.00–18.00, Saturday 9.30–16.30
Small shop selling wide range of goods direct from supplies bought in bulk, thus, generally, enabling them to reduce prices. Huge stocks include various flours and pastas, honey, pulses, cereals, fruit, nuts and seeds, fruit juices, over eighty varieties of herbs and spices. Also hold requirements for home beer-making. Should you see one of the local emergency services at the door, this is not necessarily a cause for alarm; both fire brigade and police have been known to stop for home-mixed muesli for which this store us justly renowned. (FNWC)

Holland & Barrett 6 Myrtle Walk, Bingley, Yorkshire (09766 7399); open Monday to Saturday 9.00–17.30
For further information on the Holland & Barrett chain see entry for Amersham (page 180).

Birchington

Spice of Life 87 Station Road, Birchington, Kent (0843 43610); buses 49, 50; to station or Birchington Square; open Tuesday to Saturday 9.00, closes Tuesday and Thursday 17.30, Wednesday 13.00, Friday 18.00, Saturday 17.00
Usual choice of packaged health foods, which the owner is planning to extend; also good range of high-class groceries – tea and coffee, preserves, cheeses. Range of herbs and spices is imaginative; full selection of vitamins and remedies. Deliveries within Thanet area.

Birkenhead

The Happy Nut House 183 Princes Pavement, Birkenhead, Merseyside (051 647 6361); open Monday to Saturday 9.00–17.30
A branch of Hillstart Ltd's chain of health food shops in the Midlands and the North of England. Stock a full range of dried fruit, nuts and pulses, a particularly wide range of honeys from all over the world, a big choice of culinary herbs, also medicinal herbs. They will order any herbs you need that are obtainable, provided you negotiate a reasonable minimum order with the manager. The branches also cater for the needs

of sportsmen who want vitamin supplements and biochemical remedies. Although nothing is made or prepared on the premises, the shops sell wholemeal bread and free-range eggs; from their dairy cabinets, yoghurt and (when they can get it, not easy in some areas) goat's milk. Herbal shampoos and natural beauty products stocked; so are books and booklets to help you understand and use your purchases. (NAHS)

Birmingham

City Health Centre 10 Union Street, Birmingham, West Midlands (021 236 7175); on most bus routes; open Monday to Saturday 9.00–17.30, Wednesday to 13.30
There has been a health and herb shop on this site since 1862, and owner tries to preserve traditional atmosphere. He claims to run the only real herbal and vegetarian shop in Birmingham, and will give personal diagnoses and prescribe the appropriate herb treatment. Compost-grown fruit sold in season, also free-range eggs, vegetarian foods, and usual nutritional supplements. Shop acts as agent for many suppliers. Deliveries locally. (NAHS, NFDA)

The Healthy Brewer 18 Newton Road, Great Barr, Birmingham, West Midlands (021 357 2255); open Monday to Saturday 9.00–17.30, closed Wednesday
Wide range of cereals, pulses and dried good, all packed on premises. Vegetarian and kosher foods available. Stock large supplies of spices, herbs, nuts, vitamins and minerals; also have natural cosmetics and biochemical remedies on sale. Bread and goat's milk available fresh daily. (NAHS)

Holland & Barrett 20 Priory Queensway, Birmingham, West Midlands (021 236 2513); buses 3, 6, 9, 12, 23, 31, 92 to Corporation Street and Priory Queensway; open Monday to Saturday 8.30–17.30
For further information on the Holland & Barrett chain see entry for Amersham (page 180).

Prana Wholefoods Ltd (Holland & Barrett) 33 Alcester Road South, Kings Heath, Birmingham, West Midlands (021 444 7000); buses 48, 50 stop outside shop; 99a High Street, Harborne, Birmingham (021 427

6494); buses 3, 12, 22, 23 stop outside shop; both open Monday to Saturday 9.00–17.30
For further information on the Holland & Barrett chain see entry for Amersham (page 180).

Prana Wholefoods Ltd (Holland & Barrett) 22 Grosvenor Shopping Centre, Northfield, Birmingham, West Midlands (021 475 4932); buses 4, 18, 27, 61, 62, 63, 144 Midland Red – Worcester; open Monday to Saturday 9.00–17.30
For further information on the Holland & Barrett chain see entry for Amersham (page 180).

Redbeans Wholefood Collective 7a High Street, King's Heath, Birmingham, West Midlands (021 443 4414); bus 50 from city centre; open Monday ,Thursday, Friday 10.00–18.00, Saturday to 17.00
Opened in 1979, the collective sells at competitive prices wholefood grains, cereals, pulses, etc. from bulk; also honey and preserves, oils, lots of herbs and spices. A good choice of books on health and wholefoods and related matters, and a section selling local crafts.

Sunrise World Shop 247 Dudley Road, Summerfield, Birmingham, West Midlands, (021 454 0435); buses 11, 82, 85, 86, 87 to City Road junction; open Monday, Tuesday, Thursday, Friday 10.00–17.30, Saturday to 15.00
A cooperative staffed mainly by volunteers, dedicated to increasing the area's awareness of the exploitation of Third World countries. Acts as an information centre and as local depot for recyclable goods such as paper and aluminium – proceeds go to Third World charities. A very good range of wholefoods sold from bulk in any quantity; includes all the usual dried goods like grains and pulses; honey and spreads; beverages like herb teas and grain coffee; nuts; oils; cheeses and other dairy goods; free-range eggs; specialities such as miso and molasses. Also a good variety of Third World and local crafts; recycled paper goods; books on cookery and gardening, and numerous publications on the Third World and on the environment. (MWC)

Bishop Auckland

Black's Health Food Centre 4a Bondgate, Bishop Auckland, County Durham; open Monday to Saturday 9.00–17.30
One of the four Black's shops owned by the athlete G. B. Black – you will be encouraged to exercise as well as improve your diet. Sells packaged wholefood groceries, vegetarian and special diet foods, natural juices, honey and preserves, mineral water, dairy goods, vitamins, herbal remedies, natural cosmetics, body-building supplies.

Bishop's Stortford

Holland & Barrett 12a South Street, Bishop's Stortford, Hertfordshire (0279 51637); buses stop 50 yards from shop; open Monday to Saturday 9.00–17.30
For further information on the Holland & Barrett chain see entry for Amersham (page 180).

Blackburn

CNS Moorey Herbs and Health Foods 41 Salford Road, Blackburn, Lancashire (0254 53245); open Monday to Saturday 9.00–17.30
As well as wholemeal bread, dairy goods, vegetarian and wholefood groceries, vitamins, Royal Jelly and ginseng preparations, an enormous range of herbs and herbal medicines. One of very few shops in the North offering a fully comprehensive advisory service on nutrition and herbal biochemical and homoeopathic treatments. Over 200 herbs in stock, in crude form, extracts of tinctures. Herbal medicines will be made up for you, either as prescribed by the staff or as you wish, and sent to any part of the world. Also toiletries from natural sources which have involved no animal suffering in their production. (HSW, NAHS)

Holland & Barrett 26 Lord Square, Blackburn, Lancashire (0254 64909); buses to Boulevard bus station; open Monday to Satuday 9.00–17.30
For further information on the Holland & Barrett chain see entry for Amersham (page 180)

Lovin' Spoonful 42 Mincing Lane, Blackburn, Lancashire; open Tuesday to Saturday 9.00–17.30, Thursday to 13.00
The proprietors maintain that taking on this shop was the best thing they've ever done. They say it's hard work, but satisfied customers make it worth while. They stock an extensive range of groceries and dried goods, a variety of wholewheat flours, nuts, various herbal teas. Fresh wholewheat bread each day and fresh vegetables, some organically grown. Also natural cosmetics, soaps, shampoo, etc. A small restaurant where snacks are sold all day, and a full meal 12.00–15.00. All dishes served in restaurant prepared fresh daily on premises.

Blackpool

The Circle Health Food Centre and Vegetarian Restaurant 311 Dickson Road, Blackpool. Lancashire (0253 52865); buses 7a, 7b, 7c 15a, 25a to Glynn Square; open Tuesday to Saturday (9.00–18.00 (shop) 12.00–14.00 (restaurant)
Has now been open more than ten years; the friendly and knowledgeable proprietors will give help and advice, though the layout is self-service. A good range of wholefoods is stocked: dried goods, free-range eggs, local dairy produce including butter, vegetarian cheeses and goat's yoghurt, vitamin and mineral supplements, natural cosmetics, herbal remedies. Organically grown vegetables fresh daily from the shop's own market garden. Wholemeal bread and cakes baked on premises daily; salads and vegetarian specialities also prepared here for the recommended twenty-seven-seat restaurant. (SA)

Cleveleys Healthfood Store 44 Victoria Road West, Cleveleys, Blackpool, Lancashire (0253 853482); adjacent to bus station; open Monday to Saturday 9.00–19.30
Emphasis here is on health rather than provisions; the shop is attached to the nearby naturopathic clinic. The usual branded health food groceries sold, with good stock of vitamins, minerals, herbal preparations; some vegetarian goods including dairy produce.

Health Food Store 207 Lytham Road, Blackpool, Lancashire; buses 5, 12 at 50 yards; open Monday to Saturday 9.30–17.30, early closing Wednesday

Originally a herbalist, now stocks a good range of packaged health and wholefood groceries, vitamin and mineral supplements, herbal remedies, vegetarian foods, a large range of honeys, and Denes pet remedies. Deliveries within Blackpool.

Health Food Store 7 Newton Drive, Blackpool, Lancashire (0253 33959); open Monday to Saturday 9.00–18.00, Wednesday to 13.00
Sells the usual popular health food brands, dairy goods, herbs, herbal remedies, vitamin supplements.

The Health Food Stores 74 Talbot Road, Blackpool, Lancashire (0253 23285); open Monday to Friday 9.00–17.30
Well-established store selling the general range of groceries, herbs and herbal remedies, Royal Jelly, comprehensive choice of honeys, and Jersey dairy products, some imported directly by the owner. Deliveries locally.

Hellon's Health Food Store 84 Highfield Road, South Shore, Blackpool, Lancashire (0253 43427); open Monday to Saturday 9.15–13.00, 14.15–17.30, closed Wednesday
Opened in 1976, sells the general range of branded health foods and vegetarian specialities, honey, vitamins, herbal and biochemical remedies, natural cosmetics, books.

Blandford

Blandford Wholefoods 7 Georgian Passage. East Street, Blandford, Dorset (0258 54070); open Monday to Saturday 9.00–17.00, Wednesday to 13.00
Has only recently opened and stocks packaged rather than fresh food. Items include dried fruits, pulses, cereals, honey and preserves, dairy produce, vitamin and mineral supplements, herbal remedies, wholemeal bread.

Bodmin

Health Stores 5 Honey Street, Bodmin, Cornwall (0208 2378); buses to
Mount Folly; open Monday to Saturday 9.00–17.30
Personal advice always available in this pleasant shop; to aid your diet
they stock fresh fruit and vegetables in season, fresh and dried herbs,
lots of different honeys, fresh dairy produce, free-range eggs, and the
usual dried health food groceries. Also vitamin and mineral supplements
natural medicines, herbal remedies.

Withiel Valley Dairy The Cheesebarn, Withiel, Bodmin, Cornwall
(0208 83 485); open usually Monday to Saturday 10.00–16.00, but best
to ring first
A lovely dairy, the only one for many miles using traditional methods to
produce delicious fresh products with original old dairy equipment. They
supply health food shops all over Cornwall, and hope to expand their
mail order business. In the dairy they make traditional hard and soft
cheeses with herbs, garlic and black pepper; wholemilk yoghurt; fresh
butter; mouthwatering cream – only on sale when there is an excess
after supplying shops.

Bognor Regis

Bognor Health Store 29 Station Road, Bognor Regis, Sussex (024 33
21420); open Monday to Saturday 9.00–12.30, 13.30–17.15
Range comprises many wholefoods, vitamins, mineral and herbal
remedies, wholemeal bread, home-brew and wine-making requisities.

Nature's Way Ltd 3 York Road, Bognor Regis, Sussex (024 33 828324);
open Monday to Saturday 9.00–17.30
One of the chain of Nature's Way shops. For further information on
this chain see entry on Nature's Way Ltd in Beckenham (page 187).

Bolton

The Health Food Centre 89 Deansgate, Bolton, Lancashire (0204 25084);

main bus station nearby; open Monday to Saturday 9.00–17.30. closed Wednesday

Established in 1925, is now self-service, but trained assistants can advise on health and nutritional problems, and on how to use the goods on sale. Large range of dried wholefoods packed from bulk on the premises: pulses, grains, dried fruit, nuts, cereals. Also raw sugar preserves, honey, fruit juices, natural flavourings, herbs and spices, herbal remedies, natural cosmetics, low-tannin teas, grain coffees, free-range eggs, vitamins, minerals, protein supplements, vegetarian special foods, TVP, dairy goods including yoghurt and goat's milk. Local baker delivers fresh wholemeal bread and cakes daily. In keeping with policy of providing maximum information, free leaflets are displayed for customers to take. (NAHS)

Impulse Wholefoods 102 Newport Street, Bolton, Lancashire; open Monday to Saturday 10.00–17.00, Wednesday to 13.00

Substantial range of packaged groceries and dried goods; various other foods – cakes, pies, flans, soups, pizzas, wholewheat bread – made on premises. Snack bar open at lunchtime for vegetarian meals.

Bootle

Holland & Barrett 10 The Palatine, New Strand Shopping Centre, Bootle, Merseyside (051 922 9157); buses 28, 55, 58, 59, 92 to New Strand; open Monday to Friday 9.00–17.30, Saturday 08.30–17.00

For further information on the Holland & Barrett chain see entry for Amersham (page 180).

Boscombe

Holland & Barrett 539 Christchurch Road, Boscombe, Dorset (0202 37864); buses 1, 21, 22, 23 to nearby stop; open Monday to Saturday 9.00–17.30

For further information on the Holland & Barrett chain see entry for Amersham (page 180).

Bournemouth

Earthfoods 75 Southbourne Grove, Southbourne, Bournemouth, Dorset (0202 422465); 250 Old Christchurch Road, Bournemouth, Dorset (0202 20099); buses 21, 22 to both shops; open Monday to Saturday 9.00–17.30, Wednesday to 13.00

Both shops opened since 1976, run by young people with the aim of 'gradually changing customers' eating habits towards wholefoods and vegetariansim, and acting as an outlet for alternative and new-age ideas'. The owner wants to carry on selling 'down-to-earth foods at down-to-earth prices'; to this end all the food is dispensed from open tubs and vats, packaged at the point of sale. A very extensive range of dried wholefoods, oils, honeys, herbs and spices: staff will give advice on their use. Also books, local crafts, a few locally made natural cosmetics. Take-away vegetarian wholefoods served at Christchurch Road only, where everything is made fresh on premises. The shops collect clean jars and newspaper for recycling. Deliveries only if quantities cannot be carried, e.g. whole sacks.

Gerard House 736 Christchurch Road, Bournemouth, Dorset (0590 72931); buses 1, 21, 23; open Monday to Saturday 9.00–17.00

This shop, with its sister store in Lymington, was opened in 1974 and has since established one of the largest book sections in the South on wholefoods, health and ecology-related subjects. Claims to be the best shop in Bournemouth. A fine selection of vegetarian and vegan foods, herbs and herbal preparations, dried wholefoods, natural cosmetics, vitamin and mineral supplements. (BHMA, NAHS)

Harvest Health Foods 152 Seabourne Road, Southborne, Bournemouth, Dorset (0202 424957) buses 22, 24 stop 100 yards away; open Monday to Saturday 9.15–17.30, Wednesday to 13.00

A health food shop for twenty-five years, but taken over by a new couple in 1978; we hope this will lead to an improvement in the stock. The new owners hope to move to larger premises so that they can bake their own bread, a good sign. Meanwhile, the shop sells normal range of packaged health foods, loose herbs and spices, diabetic special foods, herbal teas and tisanes, sea salt, cereals, baking requisites, honey, preserves, fruit juices, dairy produce including goat's milk yoghurt. Also a wide choice of packaged vitamin and mineral supplements and natural remedies.

Holland & Barrett 19 Criterion Arcade, Bournemouth, Dorset (0202 24982); buses 3, 9, 34 to Old Christchurch Road; all corporation buses; open Monday to Saturday 9.00–17.30

For further information on the Holland & Barrett chain see entry for Amersham (page 180).

Just Natural 573 Wimbourne Road, Winton, Bournemouth, Dorset (0202 526562); buses 3, 31 to Winton library; open Monday to Saturday 9.15–13.00, 14.00–17.30, Wednesday half day

Has just undergone reorganization; stocks the general range of health and wholefoods, vitamin and mineral supplements, natural remedies, natural cosmetics. Deliveries within district.

Natural Foods 101 Southbourne Grove, Bournemouth, Dorset (0202 428160); buses 21, 22 stop 100 yards away; open Monday to Friday 9.00–17.30

Advisable to phone before visiting, as the shop was intending to move in early 1980. Owner places great emphasis on ethics, and will not stock anything of doubtful content or provenance. Carries very large stock of herbal remedies and vitamins, and all the usual proprietary health foods. Deliveries within ten-mile radius. (BHMA, NAHS).

Nourishment 8 Royal Arcade, Boscombe, Bournemouth, Dorset (0202 301947); buses to Boscombe Arcade; 5 Burlington Arcade, Bournemouth, Dorset (0202 301947); both open Monday to Saturday 9.00–17.30, Wednesday to 13.00

Most prices cheaper than usual in these new discount shops. Muesli made fresh on premises, all pulses and fruits packaged from bulk in the shops. Good stock of wheat germ, grains and beans, vitamins and food supplements, juices, oils, honey, confectionery. Soya products stocked. Deliveries for orders over £5 within Bournemouth area. New shop and juice bar planned for Southampton.

Scoltocks Natural Food and Beauty Shop 8 The Arcade, Westbourne, Bournemouth, Dorset (0202 764637); open Monday to Saturday 9.00–17.00

One of three shops of same name in Dorset and Hampshire, each with own specialization. This one not only a health food shop with a range of grains, pulses and traditional health foods, but also a natural beauty shop. You can have a facial in their natural beauty parlour and see the range of natural herbal creams and preparations made. As is common to all three outlets, they make and market their own-brand museli. Deliveries within five-mile radius. (NAHS)

Westbourne Health Foods 18 The Arcade, Westbourne, Bournemouth, Dorset (0202 764064); open Monday to Saturday 8.45–17.30
Has recently changed hands, and the owners are mulling over changes in the format. At present stocks a full range of proprietary health foods, vitamin and mineral supplements, bread, free-range eggs. Wholemeal salad rolls to take away. Deliveries locally.

Winton Health Foods 33 Withermoor Road, Winton, Bournemouth, Dorset (0202 527871); buses 7, 8 to Wimbourne Road; open Monday to Saturday 9.00–13.00, 14.00–17.00, Wednesday to 13.00
Yet another shop opened in Bournemouth, which have the best-fed residents in England. Friendly service, and staff who will take time when giving advice; stock comprises the usual packaged health foods, honey and preserves, beverages, dairy produce, free-range eggs, vitamin and mineral supplements, natural cosmetics. Deliveries locally. (HSW)

Bracknell

House of Natural Foods 22 High Street, Bracknell, Berkshire (0344 55313); open Monday to Thursday 9.00–18.00, Friday to 15.30 (winter), to 17.30 (summer), Sunday 10.00–15.00, closed Saturday
Owned and run by Seventh Day Adventists, who are keen on promoting health foods. They provide a wide range of take-away wholefoods popular in the area: wholemeal rolls, homemade cakes and savouries, salads of many kinds. Also stock wholemeal bread, honey, live yoghurt and yeast, free-range eggs, natural remedies and supplements.

Bradford

Noble's Health Food Stores Ltd 18/20 Sunbridge Road, Bradford, Yorkshire (0274 31601); 18 John Street, Bradford, Yorkshire; open Monday to Saturday 9.00–17.30, John Street shop closed Wednesday.

Long-established shops; Sunbridge Road shop recently made self-service. Stock Allinson's bread, live yoghurt and vegetarian cheese, packaged wholefood groceries, herbal remedies, nutritional supplements. (NAHS)

Braintree

James Bowtell 31 Bank Street, Braintree, Essex (0376 24046); open Monday to Saturday 8.30—17.30; Thursday to 13.00
Founded last century as a provision merchants. Owners have now extended into health foods which include stoneground flour and speciality cheeses. Muesli mixed to their own recipe, coffee roasted, hams cooked and prepared on premises. Deliveries within six-mile radius. (NFGA)

Braunston

Goodness Foods 78/80 High Street, Braunston, Northamptonshire (0788 890305); open Monday to Saturday 9.00–17.30, Friday to 1800
The same concern runs similar shops elsewhere in Northamptonshire Their own packs come in sizes from 125 grams to 25 kilos, and the stock is extensive; dried goods include grains, cereals, flours, pulses; oils and spreads; ginseng and vitamins; coffees and teas; herbs and spices; dairy goods and free-range eggs; very good loaves including granary, which they bake themselves; fresh produce from their own farms – organically grown fruit and vegetables; Indian foods; natural cosmetics. All goods offered at wholesale as well as retail prices. Goodness Foods will deliver within radius of ten miles, but intend to increase this as business expands; particularly want to enlarge the wholesale side, and may also open further shops, which will be welcome. This branch has its own bakery.

Brentwood

Holland & Barrett 5 High Street, Brentwood, Essex (0277 211123); buses 251, 262, 260, 339, 351; open Monday to Saturday 9.00–17.30

For further information on the Holland & Barrett chain see entry for Amersham (page 180).

Bridgend

Bridgend Health Foods 36 Market Hall, Bridgend, Glamorgan, Wales; buses to Bridgend bus station; open Monday to Saturday 9.00–17.00; Wednesday to 13.00

Expansion planned in near future for this store, which specializes in dietary, homoeopathic, herbal and vitamin therapy, Also stockists of wholefoods, honey, juices, dried fruit, nuts. (BHFTA)

Bridgwater

Bridgwater Health Foods 58 High Street, Bridgwater, Somerset (0278 56545); open Monday to Saturday 9.00–17.00

Excellent shop with very wide selection of wholefoods both packed and in bulk, 120 varieties of herbs and spices sold loose, complete range of Judges honeys, home-made bread, free-range eggs, vitamins, herbal remedies, natural cosmetics, equipment for home wine-making. Also stocks requirements for organic gardening and will give advice. The Thursday Cottage brand of preserves for the health food trade is made here. *TV Times* astrologer a regular customer. The shop may move to larger premises, but now opposite the central bus stop, where there is according to the owner, a 'non-existent service'.

Bridlington

Bridlington Health Food Store 93 Promenade, Bridlington, Humberside (0262 75973); near bus station; open Monday to Saturday 9.00–13.00, 14.00–17.30

An old-established shop recently been taken over by a new manager; she is very friendly and helpful and ready to order anything customers need which is not already stocked. She sells the full range of packaged health foods, wholemeal bread, free-range eggs, many kinds of honey, herbs for medicinal or culinary use, natural remedies and cosmetics, vitamin and mineral supplements, books on health.

Bridport

Health Foods 60 West Street, Bridport, Dorset (030 885 447); open Wednesday, Friday, Saturday 9.30–15.00
In spite of odd opening times, the shop is well established. Carries wide range of grains, cereals, pulses, herbs, spices, etc. sold from bulk; also honey and preserves, juices, raw sugars, dairy goods including cheeses, oils and fats, confectionery, full range of vitamins and natural medicines. Deliveries within a ten-mile radius.

Brighton

Country Life Wholefoods 9 Garnet House, St George's Road, Kemptown, Brighton, Sussex (0273 602515); open Monday to Saturday 9.00–17.30, Wednesday to 13.00
A shop in the expanding Country Life chain of health food retail outlets. For further information on this chain see entry on Country Life Wholefoods in Ashford (page 181).

Country Life Wholefoods 146 North Street, Brighton, Sussex (0273 24591); open Monday to Saturday 9.00–17.00
A shop in the expanding Country Life chain of health food retail outlets. For further information on this chain see entry on Country Life Wholefoods in Ashford (page 181).

Culpeper 120 Meeting House Lane, Brighton, Sussex (0273 27939); open Monday to Saturday 9.30–17.30
For further information on the Culpeper chain see entry for Bath (page 186).

Herb and Health Food Store 106 Trafalgar Street, Brighton, Sussex (0273 67704); 113 Dyke Road, Brighton, Sussex; both open Monday to Saturday 9.00–17.30
Pleasantly intimate and old-fashioned shop – the business was founded in the late nineteeenth century; customers feel very much at home. Owners do their best to provide fresh foodstuffs in season and do not try to pass off anything for what it's not. Packaged health foods, fresh

eggs, dairy goods, fresh fruit, herbs and vegetables in season, vitamin and mineral supplements. Also herbal remedies and toiletries. Deliveries within East Sussex. (HSW)

Holland & Barrett Churchill Square, Western Road, Brighton, Sussex (0273 25563); open Monday to Saturday 9.00–17.30
For further information on the Holland & Barrett chain see entry for Amersham (page 180).

Infinity Foods 25 North Road, Brighton, Sussex (0273 603563); open Monday to Saturday 9.30–17.30, Wednesday half day (shop and bakery), Tuesday to Saturday 10.00–13.00, 14.00–17.00 (wholesale)
Three separate sections in this shop run by workers' cooperative of eight years' experience. Geared to the understanding and advancement of better methods of agriculture for health and ecology reasons; only food – not vitamins or supplements – sold, since their belief is that organically-grown food dispenses with the need of these. Emphasis on self-help; practices such as meditation and yoga are encouraged. The collective aims to set up an alternative holistic health centre in the building. The retail shop is in sections; the shelves of goods packed from bulk by the shop; the herb section with loose and packed herbs, vegetables which are compost-grown unless marked; the bottle-refilling section where you can fill from bulk bottles and jars previously bought at the shop *only* with further supplies of oils, pastes, spreads, etc.; barrels for dried goods on which a discount is given for bulk buyers. Also sells books on all aspects of health and wholefoods, including gardening, cooking, alternative medicine. Warehouse next door sells only in bulk – purchasers spending a minimum £5 – and goods are very cheap. List available; orders must be phoned twenty-four hours in advance. Bakery, on other side of main shop, sells all wholemeal bakery products, and snacks like flans, to take away. Also sold are free-range eggs and some dairy produce mainly from goat's milk. Where bakery food is not wholly vegetarian, this is stated; no sugar is used and only cold-pressed unrefined oils and sea salt are employed in cooking. Deliveries free in Brighton area only for orders worth £50 and over; otherwise prearranged charge made.

Nature's Way Ltd 35 Duke Street, Brighton, Sussex (0273 21574); 89 Western Road, Brighton, Sussex (0273 26181); both open Monday to Saturday 9.00–17.30
For further information on this chain see entry for Beckenham (page 187).

Simple Supplies 11 George Street, Kemptown, Brighton, Sussex; open Monday to Saturday 9.30–17.30, Thursday to 13.30
One of the new wholefood cooperatives; acts as a centre for alternative and environmental issues. Actively promotes the idea of shared work and production for minimum profit; publishes *Whole Earth* magazine, and encourages recycling of materials by acting as the local campaign centre. Badges, books and magazines on environmental issues sold, also recycled paper products. Almost all the food products packed on premises: pulses, whole and flaked grains, cold-pressed and heat-extracted oils, Sussex-grown wholemeal flour, nut butters and spreads, honey from their own hives, muesli mixed on premises, wholemeal bread baked fresh daily, pickles and spices including a large range for curries. Take-away service for freshly baked wholemeal sweet and savoury snacks. Deliveries in Brighton area for goods worth £25 and over.

Bristol

Didcott Medical Herbalist 117 Landseer Avenue, Bristol, Avon (0272 697798); open by prior appointment
Not open yet on permanent base; postal and delivery service. S.a.e. produces price list, etc.

The Grain Stall 92/94 St Nicholas Market, St Nicholas Street, Bristol, Avon; open Monday to Saturday 8.00–17.00
Helpful stall in covered market in centre of Bristol. Stocks very wide range of pulses and grains, including several kinds of brown rice, the largest selection of flours in Bristol, dried fruit, nuts, cereals, large stock of loose herbs and spices, medicinal herbs, other culinary condiments, unyeasted bread, jams and preserves, beverages. Muesli and roasted sunflower seeds prepared fresh. Plans for take-away stall in vicinity of market.

The Health Food Store Northview, Westbury Park, Bristol, Avon (0272 36531); buses 28, 87 to White Tree; open Monday to Friday 9.00–18.00
Long-established shop specializing in vegetarian and diet items – e.g. for diabetics. Good specially blended muesli without sugar, goat's milk and several kinds of yoghurt, vegetarian and ordinary cheeses, compost-grown fruit and vegetables, local eggs, fresh wholemeal bread, brown sugar confectionery. Extensive range of vitamin and mineral nutritional supplements. (HWN)

Health Foods (Bristol) Ltd 16 North Street, Bedminster, Bristol, Avon (0272 634292); open Monday to Saturday 9.30–13.00, 14.00–17.30
A very new shop specializing in grains and other dried wholefoods; also sells herbal remedies, vitamin supplements, wholefood snacks (range to be extended) including sandwiches and cakes. Policy is to provide friendly service and reasonable prices.

Holland & Barrett 19 Broadweir, Bristol, Avon (0272 294195); buses 9, 10, 14, 30, 31, 32, 33, 34, 87; open Monday to Saturday 9.00–17.30
For further information on the Holland & Barrett chain see entry for Amersham (page 180).

Larder 11 Lower Redland Road, Britol, Avon (0272 311959); buses 21, 22, 23, 324, 87, 88; open Monday to Saturday 10.00–18.00
An extensive range including rice, flour, bran, grain, tea, preserves, fruit juices, herbs and spices, dried fruit and nuts; bread, cakes, savouries prepared on premises.

Manor Farm Health Store 259 Gloucester Road, Bishopston, Bristol, Avon (0272 47696); buses to Nevil Road; open Monday to Saturday 9.00–13.00, 14.00–18.00, closed Wednesday
Popular brands of packaged health foods, Allinson's bread and bakery produce, fresh eggs, dairy goods including clotted cream, vegetarian special foods, natural medicines and nutritional supplements, some local vegetables in season.

Moore's Health Foods 308 Lodge Causeway, Fishponds, Bristol, Avon (0272 651221); buses 14, 89; open Monday to Friday 9.00–17.30, Sunday 9.30–12.30
A very go-ahead enterprise, staffed entirely by vegetarians; free health

course with every purchase and a 'stop smoking' clinic, augmented by talks which can be arranged on all aspects of healthful living. Stocks wholemeal bread, goat's milk and cheese, honey, free-range eggs; food supplements, natural cosmetics, over 100 herbs and spices. Also health and Christian literature. Deliveries in Bristol area.

Poets Corner Limited 1 Passage Road, Westbury-on-Trym, Bristol, Avon (0272 507937); buses to Westbury village; open Monday to Saturday 9.00–13.00, 14.30–17.00, Wednesday to 13.00
Owner of shop a trainee naturapath; he practises what he preaches. He originated GHM parties, becoming very popular in the Bristol area. He also founded the Bristol Jogging Club. Stocks extensive supplies of wholefoods, vitamin and mineral supplements, vegetarian foods, coffee alternatives, herbal teas, herb cosmetics. Dairy products include goat's milk, yoghurt, Thayers ice-cream; organic vegetables sold in season, also free-range eggs. Various savoury foods prepared on premises for take-away; Poets Corner muesli, packed in the shop, contains no sugar. No reasonable request for assistance with delivery refused.

Redland Health Stores Limited 29 Zetland Road, Bristol, Avon (0272 46505); open Monday to Saturday 9.00–13.00, 14.00–17.30, Wednesday to 13.00
Groceries, dry goods, butter, yoghurt, fresh bread, natural medicines, nutritional supplements, natural cosmetics, medicinal herbs. (NAHS)

The Salad Kitchen 4a Charlton Road, Keynsham. Bristol, Avon (027 56 4631); open Monday to Saturday 9.00–17.30 (closed for lunch), Wednesday to 13.00
One of three branches in Bristol area. Each shop carries a good range of groceries, vitamins, herbs, dried fruits, nuts, grains, pulses, nutritional supplements, natural cosmetics. (NAHS)

The Salad Kitchen 18 Park Row, Bristol, Avon (0272 24539); open Monday to Friday 9.00–19.00, Saturday to 18.00
The main shop of three of same name in Bristol area. Carries a wide range of foods, vitamins, herbs, dried fruits, nuts, grains, pulses, nutritional supplements, natural toiletries, A fully vegetarian restaurant at this address where all dishes are prepared on premises (HSW, NAHS)

The Salad Kitchen 15 Straits Parade, Fishponds, Bristol, Avon (0272 652387); open Monday to Saturday 9.00–17.30, (closed for lunch)
See entry for Charlton Road Salad Kitchen (page 209).

Spa Health Foods 60 The Mall, Clifton, Bristol, Avon (0272 560103); buses to Clifton; open Monday to Saturday 9.30–17.30
Attractive shop in Georgian spa of Clifton, a stone's throw from Brunel's suspension bridge. You can see the goods on shelves through the big picture windows; stocks all manner of dried foods, wholemeal bread, cakes and baps baked locally, free-range eggs, locally grown fruit and vegetables when compost-grown available. Also natural cosmetics, vitamins, biochemical and herbal remedies. Expansion planned.

Broadway

The Relaxation Centre 71 High Street, Broadway, Worcestershire (038 681 2337); open shop Tuesday, Wednesday, Friday 9.30–17.30 (shop), treatments Monday to Saturday
Fifteenth-century Cotswold house where health food shop is run in conjunction with a physiotherapy and yoga centre, sometimes used by the Royal Shakespeare Company, Stratford-on-Avon. Centre soon to be extended to include slendertone massage and steam-bath facilities. A general stock of wholefoods, vegetarian and food s upplements; specializes in vitamins, herbal products, tissue salts.

Bromley

Bromley Health Centre 54 Widmore Street, Bromley, Kent (01 460 3894); bus 227; open Monday to Friday 8.30–17.00, Saturday to 13.00
Strictly vegetarian. Has been functioning since 1932, and you are welcome to call on the vast experience of the owner. Sells good range of wholemeal flours and bread, muesli mixed daily on premises, unpolished rice, grains and pulses. many kinds of honey, free-range eggs, goat's milk, natural fruit and vegetable juices. Wide range of herbal remedies, vitamin/mineral supplements. (HSW, NAHS)

Buckhurst Hill

Buckhurst Hill Health Foods 24 Queens Road, Buckhurst Hill, Essex (01 504 0913); open Monday to Friday 9.00–17.30
Friendly store frequented by local television personalities, stocking usual range of packed health foods and food supplements.

Bude

Wonnacotts Wholefoods 2 Lansdown Road, Bude, Cornwall (0288 3105); open Monday to Saturday, 8.30–17.30 (winter), 8.30–21.00 (summer)
A dairy and wholefoods shop; the owners have their own farm which supplies fresh vegetables, dairy produce, free-range eggs. Also sell wholemeal bread, home-made fudge, clotted cream (can be sent by post), home-made pies, fresh cheese, goat's yoghurt, vitamins, minerals, natural remedies, dried wholefoods, local cider. Products can be sent by post. Deliveries within fifteen-mile radius.

Budleigh Salterton

Cornucopia 30 Fore Street, Budleigh Salterton, Devon (03 954 3003); open Monday to Saturday 9.00–13.00, 14.00–17.00, early closing Thursday, hours extended July, August
Recently opened shop selling bulk and prepacked grains, cereals, pulses, nuts, dried fruits, herbs, spices; also cheese, yoghurt, eggs; vitamin supplements, herbal remedies, natural cosmetics; fresh bread; herb and hop pillows. Also a comprehensive selection of books on related topics. Deliveries within eight-mile radius.

Burgess Hill

Holland & Barrett 71 Church Road, The Martlett, Burgess Hill, Sussex (0446 42724); buses 135, 170, 171, 172; open Monday to Saturday 9.00–17.30
For further information on the Holland & Barrett chain see entry for Amersham (page 180).

Burnham-on-Sea

Burnham Health Food Centre 4/6 Victoria Street, Burnham-on-Sea, Somerset (0278 3866); open Monday to Saturday 9.00–13.00, 14.15–17.00
Good stock including large selection of cereals and whole cheeses. Also flours, vegetable fats, beans, pulses, pasta, unpreserved dried fruit, twenty-four different honeys. Comprehensive range of herbal and homoeopathic remedies and cosmetics. Home-baking on premises – wholemeal flour/low-fat cakes, quiches, their well-known flapjacks. Plans for opening juice bar.

Burnley

Health Food Centre 3 Parker Lane, Burnley, Lancashire (0282 35910); open Monday to Saturday 9.30–17.30, closed Tuesday
Situated on pedestrian route from bus station to shopping precinct. A bright modern shop with staff who can and will advise, although basically a self-service store. Dried wholefoods, such as grains, pulses, shelled nuts, and dried fruits bought in bulk and packed in shop for the shelves. Free-range eggs, wholemeal bread, oils, yeasts, diabetic foods, gluten-free products, high-quality teas and coffee, honey and preserves, slimming products, body-building requisities, fruit juices, TVP products. Sells very large range of health aids: vitamins, nutritional supplements, herbal remedies, ginseng products, biochemical tissue salts, books on health. (NAHS)

Burton on Trent

The Happy Nut House 3 St Modwen's Walk, Burton on Trent, Staffordshire (0283 67734); open Monday to Saturday 9.00–17.30
A branch of Hillstart Ltd's chain of health food shops in the Midlands and the North of England. For further information on this chain see entry on The Happy Nut House in Birkenhead (page 192).

Bury

Bury Health Food Centre 14 Union Arcade, Bury, Greater Manchester (764 2499), open Monday to Saturday 9.00–17.30, closed Tuesday
Here you may meet stars from *Coronation Street* or *Look North*, and you can buy a good selection of fried wholefoods packed at the shop's own warehouse; herbs, dried fruit, nuts, cane sugars, flours, pulses, grains, cereals.

Bury St Edmunds

Health Food and Herbs 9 The Traverse, Bury St Edmunds, Suffolk (0284 4986); adjacent to bus station; open Monday to Saturday 9.00–13.00, 14.15–17.30, Thursday to 12.30
Blue- and maize-coloured shop; a herbalist since 1840 and still fitted with old-fashioned herbalist's fixtures including cut-glass knobs on the individual drawers. Health foods sold since 1955: groceries, dried wholefoods, fresh wholemeal bread three times a week, dairy goods like yoghurt, natural medicines, nutritional supplements. No objection to dogs (unlike most health food shops), on the grounds that they are better behaved than most children. (BHMA, HSW, NAHS)

Calne

New Age Health Shop 27a Church Street, Calne, Wiltshire (0249 812675); open Monday to Saturday 9.00–17.30.
Normal range of wholefoods together with free-range eggs, wholemeal bread, yoghurt, vegetarian cheese, goat's milk. Also wide range of herbs and spices, herbal remedies, shampoos, cosmetics.

Camberley

Holland & Barrett 19 Grace Reynolds Walk, Camberley, Surrey (0276 64043); buses to station; open Monday to Saturday 9.00–17.30
For further information on the Holland & Barrett chain see entry for Amersham (page 180).

Camborne

Health Foods (Camborne) 62 Trelowarren Street, Camborne, Cornwall (0209 714242); buses to main shopping street; open Monday to Saturday 9.00–17.30, Thursday to 13.00
Extensive stock of packaged goods, some packed on premises. Dried fruit, nuts, cereals, pulses. Also vitamin supplements and remedies. In future may prepare fresh dishes, such as quiches, for take-away. Local deliveries, and will send items by post.

Cambridge

Arjuna 12 Mill Road, Cambridge (0223 64845); 5 minutes' walk from bus station; open Monday to Wednesday 9.30–18.00, Friday to 19.00, Saturday to 17.30, closed Thursday
A workers' cooperative since 1976. Emphasis again on keeping prices as low as possible and encouraging collective lifestyle. Very good choice of dried goods such as grains, flakes, cereals, pulses, nuts, flours, bran, wheatgerm, fruits, sold from bulk. Uncommon items stocked like Japanese fermented products, as well as vegetarian rennet cheeses, goat's milk, goat's milk yoghurt, cold-pressed oils, organically grown fruit and vegetables, free-range eggs, cosmetics, incense, cooking utensils. Books and magazines on wholefood cookery and alternative lifestyles sold. A good take-away section, with wholefood pizza, rissoles, cakes and biscuits made daily on premises. Yoghurt also made fresh each day. (FREG, FSWC, ICOM)

Culpeper 25 Lion Yard, Cambridge (0223 67370); open Monday to Saturday 9.30–17.30
For further information on the Culpeper chain see entry for Bath (page 186).

Health Food Stores (Cambridge) Ltd 3 Rose Crescent, Cambridge (0223 353305); open Monday to Friday 8.30–17.30, Thursday to 13.00
Specializes in a variety of dried fruits, shelled nuts, honeys, raw sugar jams and marmalades, and large range of vitamins and nutritional supplements, herbal remedies, cosmetics; 100 per cent wholemeal bread and flour sold. Will deliver to Cambridge and nearby villages. (HSW)

Cannock

The Herb Garden Queens Square, Cannock, Staffordshire (05435 2418); on all local bus routes; open Monday to Friday 9.00–17.30, closed Thursday
Stocks the normal range of health foods, with a good selection of herbs.

Canterbury

Gateways Wholefoods and Provisions 15 St Dunstan's Street, Canterbury, Kent (0227 69839); open Monday to Saturday 9.00–17.30
Very good stock of wholefoods, free-range eggs; large section of Indian and Middle Eastern speciality foods and spices. Shop first opened as vegetarian café; no longer has seating, but a justly renowned supply of well-filled wholewheat sandwiches to take away. Also sells items like natural shampoos. A selection of books on health and health foods.

Oasis Wholefoods 74 Northgate, Canterbury, Kent (0227 63941); open Monday to Saturday 9.00–17.30
Opened in 1977. Plans a move to larger premises in Palace Street, so check before you make a special trip. Sells wide range of naturally grown foods, especially grains; most dried goods can be bought in bulk. Good choice of honeys, take-away cooked snacks, natural and bio-chemical remedies, vitamins, books on health topics. A family concern where you are given personal service; deliveries in case of difficulty. (NAHS)

Southern Health Foods 47 Burgate, Canterbury, Kent (0227 63063); open Monday to Saturday 9.00–17.30
One of the well-established Southern Health Foods chain. For further information on this chain see entry on Southern Health Foods in Aldershot (page 179).

Stanards 20 Love Lane, Canterbury, Kent (0227 58069 shop, 52112 restaurant); near main bus station; open Monday to Saturday 9.00–17.00, restaurant 9.00–16.00
Well-stocked wholefood shop, opened in 1977, with the usual health food groceries, tea and coffee, wholewheat bread, herbs, juices. You can

also buy books on health and wholefoods, cooking utensils. Self-service restaurant seats 100; all food prepared fresh daily. The business is run by Oasis Wholefoods in the Northgate.

Canvey Island

C. P. Venables 76 High Street, Canvey Island, Essex (037 43 2695); buses 3, 151 stop outside shop; open Monday to Saturday 9.00–18.00
In existence for more than half a century. Stocks all flours, brans, honeys and muesli, together with Vegex mince, protoveg, ginseng products, biochemical tissue salts. Deliveries in Canvey/Benfleet area. (NPA)

Cardiff

Beans 'N'erbs 171a Kings Road, Canton, Cardiff, Wales; open Monday to Saturday 9.30–18.30, closed Thursday
Recently opened and well-stocked shop selling almost all goods loose from bulk. Large and small quantities sold; you are encouraged to bring your own containers, jars and bottles for the shop. Full range of dried wholefoods – fifteen varieties of pulses, brown rice and other grains, muesli and muesli base, various nuts and seeds. All produce marked whether organically grown or not. Cold-pressed oils; Japanese products including tea; honey and spreads; large selection of herbs and spices both whole and ground. Immediate plan is to improve herb selection to include all manner of rare ones, with leaflets on their uses. The owner is making efforts to find a good supply of fresh organically grown vegetables. Shop also acts as a centre for 'alternative' activites in Cardiff; noticeboard gives information on everything going on in the area, and a book is kept with a register of community groups and anyone else who may be of interest to those building an alternative way of life.

Canton Health Foods 327 Cowbridge Road East, Cardiff, Wales (0222 397983); open Monday to Saturday 9.00–17.30
Mainly vegetarian foods, dried wholefoods, honey, several kinds of flour, fresh and dried yeast. Also vitamin supplements, natural cosmetics.

Holland & Barrett 11 High Street, Cardiff, Wales (0222 374985); buses stop just outside shop; open Monday to Saturday 9.00–17.30
For further information on the Holland & Barrett chain see entry for Amersham, (page 180).

Roath Health Foods 39 Wellfield Road, Cardiff, Wales (0222 36023); open Monday to Saturday 9.00–17.30
Under the same ownership as the Canton Health Foods store; this also sells vegetarian special foods, dried wholefood groceries, bread-making supplies including several kinds of flour, both fresh and dried yeast. Also vitamin supplements, natural cosmetics.

Wholefood Shop 1a Fitzroy Street, Cathays, Cardiff, Wales (0222 395388); open Tuesday to Saturday 10.00–13.00, 13.30–17.00 Friday to 18.00, closed Monday
An established concern with many plans for the future; in the words of the owners, they 'cater for the converted'. Comprehensive stocks of wholegrains, flakes, flours, pulses, nuts, dried fruit, honey, tea. Also wholewheat pasta, tamari, miso, cosmetics; books and cookware; oils, seeds, ginseng.

Cardigan

Picton's Delicatessen and Health Foods 55 Pendre, Cardigan, Dyfed, Wales (0239 2187); buses to Finch Square; open Monday to Saturday 9.00–17.30, Wednesday to 13.00
Primarily a delicatessen, nevertheless stocks a good range of packaged wholefoods, dairy produce, free-range eggs, vitamins, herbal remedies.

Carmarthen

Aardvark Whole Foods 2 Mansell Street, Carmarthen, Dyfed, Wales (0267 32497); open Monday to Saturday 10.00–17.00, closed Thursday
Fairly new shop, still expanding the scope of its stock. Dry prepacked wholefoods include muesli, grains, flakes, pulses; also honey and spreads; tahini, tamari and olives; dried fruit and shelled nuts; various varieties of rice. Natural cosmetics include soap and shampoo. Teas and bis-

cuits; herbs and spices; recycled paper goods. Fresh wholemeal snacks made on premises from Wednesday onwards. Owner's immediate plans include stocking fresh compost-grown fruit and vegetables, and providing bulk supplies.

Waverley Health Stores Lammas Street, Carmarthen, Dyfed, Wales (0267 6521); open Monday to Saturday 9.00–17.30
This shop, long established, has recently been taken over by new owners who plan to expand stock and service. They sell dried wholefoods, grains and pulses, packaged health foods; dairy goods including goat's milk; fresh organically grown fruit and vegetables; biochemical preparations; gardening supplies such as seed compost, beekeeping equipment; and useful aids to country living, e.g. wood-burning stoves. Fresh wholemeal bread delivered daily. Definitely worth a visit. Some deliveries in locality, by arrangement.

Castle Douglas

Health and Wholefood Store Castle Douglas, Dumfries and Galloway, Scotland (0556 3174); open Monday to Saturday 9.00–13.00, 14.00–17.00, Thursday to 13.00
Energetically run shop, opened in 1976. Has a thriving postal service as well as loyal local customers; sends health foods to the Isle of Arran, health magazines to Australia, herbal remedies to Austria, vitamins to New Zealand. Well stocked with wholemeal flours, twenty varieties of honey, thirty kinds of breakfast cereal, wheat germ, brewer's yeast, yoghurt, vitamins, minerals, herbal remedies, free-range eggs, wholemeal bread. Hope to further extend the stock when the partitions are moved to make more room. Home wine-making equipment also on sale. (NAHS)

Caterham

Caterham Health Foods 81 Croydon Road, Caterham, Surrey (22 44261); bus 197; open Monday to Friday 9.15–13.15, 14.15–17.15, Saturday to 16.00, closed Wednesday
Under new management; and changes being made to range of stock as

well as to appearance. Hoped to invest all profits for some time to increase goods on offer; at present a wide range of wholefoods, vitamin supplements and herbal remedies like homoeopathic salts; a large selection of herbs amd spices, free-range eggs, locally made goat's cheese.

Chard

Hayman's Bakery and Health Foods 43 Holyrood Street, Chard, Somerset (046 06 2941); bus 213 stops outside shop; open Monday to Saturday 9.00–17.00, Wednesday to 13.00
A wonderful bakery – will even bake to your own requirements. Many kinds of bread, cakes, scones on sale, freshly warm from the ovens. Health foods sold include most things you might need – their own mixed muesli is recommended – and if you can't find it, they will order for you. (NAMB)

Chatham

Country Life Wholefoods 116 High Street, Chatham, Kent (0634) 409291); open Monday to Saturday 9.00–17.30
This is a shop in the expanding Country Life chain of health food retail outlets. For further information on this chain see entry on Country Life Wholefoods in Ashford (page 181).

Cheam

Health Foods Cheam Village 60 The Broadway, Cheam, Surrey (01 643 5132); buses 213, 213a to Cheam village; tube to Morden; open Monday to Saturday 9.00–17.30, Wednesday half day
A big bright shop run by people experienced in the business and able to offer advice. Further expansion planned. Already stocked: all the branded health foods, herbal remedies, vitamin and mineral supplements; some dairy produce including goat's milk and natural ice-cream; local fruit in season; a very full range of natural cosmetics. Bulk supplies of health foods at a discount to order. (NAHS)

Chelmsford

Holland & Barrett 3 Exchange Way, Chelmsford, Essex (0245 353157); buses 40, 42, 44, 44a, 44b, 43, 55 to Market Road; open Monday to Saturday 9.00–17.30
For further information on the Holland & Barrett chain see entry for Amersham (page 180).

Cheltenham

Barley Corn 317 High Street, Cheltenham, Gloucestershire (0242 41070); open Monday to Saturday 9.30–17.30, Wednesday to 13.00
Groceries and dry goods, natural cosmetics, herbal shampoos, etc. Fresh bread and wholegrain baked items. Deliveries within ten-mile radius for orders of £25 or over.

Bath Road Wholefoods 133 Bath Road, Cheltenham, Gloucestershire (0242 514150); open Monday to Saturday 9.00–13.00 Wednesday to 13.00
Good range of mostly prepacked wholefoods, honey, live yoghurt, wholewheat bread, cheeses, preserves. Also natural vitamins and cosmetics, herbal remedies, vegetarian products. (NAHS)

Cavendish House The Promenade, Cheltenham, Gloucestershire (0242 21300); all local buses; open Monday to Friday 9.00–17.30
An attractive shop opened in 1818, now owned by a national chain of department stores. Stocks normal range of packaged dried goods; dietary specialities including vegetarian; herbal remedies, vitamins, juices, and a good range of honeys. Deliveries within fifteen-mile radius.

Southern Health Foods 20 Clarence Street, Cheltenham, Gloucestershire (0242 28749); open Monday to Saturday 9.00–17.30
One of the well-established Southern Health Food chain. For further information on this chain see entry on Southern Health Foods in Aldershot (page 179).

Chester

Dutton's Health Foods Godstall Lane, St Werburgh Street, Chester; open Monday to Saturday 9.00–17.30
One of three shops owned by the Dutton family who have many years' experience in the business. The manageress has been chosen for her similar knowledge and can give advice. Stock includes muesli mixed on premises, dried wholefoods, decaffeinated wholeberry coffee ground to your choice, wholewheat bread, dried fruit and nuts sold loose, herbs, vegetarian and slimming diet foods, free-range eggs, yoghurt and goat's milk, honey, juices, preserves, vitamin and mineral supplements, natural remedies, books on health. (NAHS)

The Granary 108 Northgate Street, Chester (0244 318553); open Monday to Saturday 9.00–17.00
The owner, who opened this shop in 1976, has since opened another two in Wrexham and Macclesfield. He hopes soon to open a wholefood café in the Chester shop. Stock is sold loose in any quantities required. Usual wide range of grains and flakes, pulses, cereals, flours; dairy produce including yoghurts, organically produced butter, cottage and vegetarian cheese, goat's milk; very good stock of herbs and spices, with over 100 medicinal herbs; vitamins and herbal oils; attractive range of general goods such as beansprouts, rock and sea salts, vinegars, honeys, raw sugar preserves, oils, spreads, pickles, herb teas. Book section stocks over 500 titles. Bulk supplies are the speciality, but they do not deliver.

Holland & Barrett In Owen & Owen, Bridge Street, Chester (0244 23112, ext. 254); open Monday to Saturday 9.00–17.30
For further information on the Holland & Barrett chain see entry for Amersham (page 180).

Chesterfield

The Happy Nut House 19 Glumangate, Chesterfield, Derbyshire (0246 75928); open Monday to Saturday 9.00–17.30
This is a branch of Hillstart Ltd's chain of health food shops in the Midlands and the North of England. For further information on this chain see entry on The Happy Nut House, in Birkenhead (page 192).

Chichester

Beanfeast 25b Southgate, Chichester, Sussex (0243 783823); buses to bus station; open Tuesday to Saturday 9.30–17.30, closed Monday
Located a mile or so from Chichester Theatre, this shop prepares and sells take-away food on premises. Specialities include pizza, rice salads, soups, flans, and freshly baked rolls with or without a variety of fillings. Stocks cereals, grains, pulses, dried fruits, nuts. Has a wide range of sugarless jams and marmalades, wholegrain mustards, pickles, apple and grape juices. Good supply of various honeys. Selection of helpful books on cookery and wholefood living; also cast-iron cookware. Owner is thinking of opening a wholefood snack bar – for hungry, healthy theatregoers no doubt.

G. L. Blagg 7 The Ridgeway, Sherbourne Road, Parklands, Chichester, Sussex (0243 782695); bus 249; open Monday to Saturday 9.00–13.00, 14.00–17.30, Wednesday to 13.00
A chemist selling a selection of branded health food groceries, fruit juices, etc., as well as vitamins, herbal and other natural remedies and supplements.

Chichester Health Foods 64 South Street, Chichester, Sussex (0243 784746); open Monday to Saturday 9.00–17.30
Opened in 1979, already has a good basic stock or packaged health foods, vegetarian speciliaties, dried fruit and nuts, flour, fresh and dried yeast, honey and preserves, fruit and vegetable juices, herb teas, herbal remedies, nutritional supplements, homoeopathic remedies, natural toiletries, wholefood confectionery. Wholemeal bread fresh daily.

Holland & Barrett 13 East Street, Chichester, Sussex (0243 78446); buses 249, 250, 251, 254, 260, 266, 270 to West Street; open Monday to Saturday 9.00–17.30
For further information on the Holland & Barrett chain see entry for Amersham (page 180).

Chorley

Chorley Health Food Store 23 Cleveland Street, Chorley, Lancashire (025 72 76146); open Monday to Saturday 9.00–17.15, closed Wednesday
Complete range of nuts, fruit, beans, flours, vegetarian foods, herbal remedies; also beer- and wine-making supplies. (HFTA)

Christchurch

The Healthy Way 360 Lymington Road, Highcliffe, Christchurch, Dorset (042 52 6929); open Monday to Friday 8.30–17.30, Wednesday half day
The owners opened the shop in 1978 after their conversion to wholefood way of life; before that they ran a well-known restaurant in Bournemouth. There is a snack bar and take-away service selling home-made date slices, wholemeal rolls and sandwiches, which they plan to extend as soon as possible. Extensive range of wholefoods, with bread fresh daily, vitamins and nutritional supplements, natural cosmetics, health aids for pets, books and magazines on health and health food subjects. Planning to stock home beer- and wine-making equipment. Deliveries in immediate area.

Natural Fayre The Fountain, 1 High Street, Christchurch, Dorset (0202 471152); open Monday to Saturday 9.00–17.30; Wednesday to 13.00
A well-stocked shop, with wholemeal bread and cakes, goat's milk, yoghurt and cheese, vegan margarine; large range of dried fruits and nuts; items like bran and rice sold in bulk. Sandwiches and flans made on premises. Deliveries within five-mile radius.

Clacton-on-Sea

Health Food Centre 18 Orwell Road, Clacton-on-Sea, Essex (0255 21256); near station and bus station; open Monday to Saturday 9.00–13.00, 14.15–17.30, Wednesday to 13.00
Dried health foods and groceries; dairy produce including goat's milk;

bread, herbal remedies, vitamins and minerals, natural cosmetics. A self-service shop where help is offered if you need it. (HSW)

Clay Cross

Health Foods 34 High Street, Clay Cross, Chesterfield, Derbyshire (0246 863477); buses 53, 54, 60, 61 to High Street; open Monday to Saturday 9.00–17.30, Wednesday 9.00–13.00
A range of packaged health foods, natural medicines, nutritional supplements. (NAHS)

Cleckheaton

Health Fare 27 Northgate, Cleckheaton, Yorkshire; open Monday to Saturday 9.00–17.00, closed Wednesday
A large and welcoming store where it is easy to browse and compare goods for sale: speciality groceries as well as health foods – most popular brands; free-range eggs; snacks such as wholemeal fruit scones; natural medicines, vitamin supplements. Advice freely given if you need it. (BHFTA)

Cleveleys

Cleveleys Health Centre 87/89 Beach Road, Cleveleys, Lancashire (0253 853144); 100 yards from bus station; open Monday to Saturday 10.00–17.30, closed Wednesday
This shop has recently celebrated fifty years in business. Although it stocks the usual packaged health foods, main stocks are herbal remedies and a huge range of vitamins and nutritional supplements. Full-scale botanic dispensary. Consulting herbalist available, by appointment, to advise on health and diet problems. Particularly good range of honey.

Clitheroe

C. R. Hargreaves & Sons Ltd 40/48 Parson Lane, Clitheroe, Lancashire (0200 22183/4); open Monday, Tuesday, Thursday, Friday 9.00–17.30; Wednesday, Saturday to 13.00
Family business in what has been a grocer since 1771. The health food side was incorporated in 1965. Full range of packaged health foods, diabetic and vegetarian specialities, dairy produce, free-range eggs, wholemeal bread, local compost-grown fruit, vegetables, herbs when available. Also natural remedies and cosmetics, vitamins, nutritional supplements. Deliveries within five miles.

Clynderwyn

Fedwen Stores Efailwen, Clynderwyn, Dyfed, Wales (09947 339); open Monday to Saturday 9.00–18.00
A general stores stocking some packaged health foods, fresh wholemeal bread and free-range eggs as well as ordinary groceries. The shop has a snack bar. Will deliver within radius of twenty-five miles.

Cobham

The Wholefood Centre 35 Oakdene Parade, Cobham, Surrey (266 4553); open Monday to Saturday 9.00–17.30, Wednesday to 13.30
One of the joint owners of this centre, combining shop and take-away has herself published two wholefood cookery books; the inspiration for opening this store came from a previous employer – Yehudi Menuhin. Apart from the normal comprehensive range of health foods, you will find Loseley Park yoghurts and ice-cream, books and an arts and crafts section (pottery, leather, etc). The take-away provides sweet and savoury wholefoods, salads, quiches, cakes, croquettes. Large free car park and loading bay behind the shop. When we asked if there were any plans for the future, we were told it is considered 'perfect as it is'!

Colchester

Beaumont's Health Stores 7 Pelham's Lane, Colchester, Essex (0206 46009); buses stop in High Street; open Monday to Saturday 8.00–17.30, Thursday to 14.00
In existence now for ten years, this store supplies a selection of natural vitamins and cosmetics, herbal remedies. Can answer all normal demands for health foods. (NAHS)

H. Gunton: Health Foods 81/83 Crouch Street, Colchester, Essex (0206 72200); buses to Odeon cinema; open Monday to Saturday 8.30–17.30
High-class grocer incorporating separate rather pricey health food section. Stocks packaged health foods, dairy produce, free-range eggs, natural cosmetics, vitamin and mineral supplements. Coffee-room upstairs, open only in mornings, serves decaffeinated coffee roasted on premises. Supplies for home wine-making. Deliveries within five-mile radius. (AHS)

Natrafoods 14 Short Wyre Street, Colchester, Essex (0206 63188); open Monday to Saturday 9.00–17.30
One of three shops of same name in Essex, each stocking usual range of health foods including cereals, pulses, fruit juices, dairy produce, free-range eggs. Large selection of dried fruits and nuts, herbs and spices. Honeys from all over the world. Hoping to open a restaurant on first floor in near future.

Coleraine

Health Corner 46 Railway Road, Coleraine, County Derry, Northern Ireland (0265 2697); buses to bus station, 300 yards away; open Monday to Saturday 9.00–17.30
The shop, opened in 1978, hopes to expand further and spread the wholefood gospel as far as it can. You can buy the general range of packaged wholefood groceries, vegetarian and diet foods, vitamin and mineral supplements, natural beauty products, health books, and – this being Northern Ireland – Bibles. Deliveries by rail or road.

Colne

Green's Wholefoods 25 Newmarket Street, Colne, Lancashire (0282 867971 – this is not a phone in the shop, but they have the use of it); buses to town hall; open Wednesday to Saturday 10.00–17.30
This spirited venture started life on a market barrow. The owner and staff – young and enthusiastic – try to keep the atmosphere of a meeting place where you can also buy food. They hope to extend this by combining the shop with a café and bakery, where people can gather. Wide range of goods includes numerous dried wholefoods, sold from bulk; grains, pulses, pastas, teas, nuts, dried fruits, cereals, herbs; lots of oils, juices, natural cosmetics like soap and shampoo, recycled paper goods, pottery seconds, incense, and a flourishing book and magazine section where anything you need 'within reason' will be ordered. Candles stocked, along with many items sold much more cheaply than in normal shops – there is a very low profit margin and the staff works voluntarily. They try to find time to bake bread every day and usually manage it. Customers are encouraged to serve themselves, but not because the shop is not interested.

Colwyn Bay

Colwyn Bay Health Food Centre 38 Sea View Road, Colwyn Bay, Clwyd, Wales (0492 30895); buses stop at St Paul's church; open Monday to Saturday 9.00–18.00, Wednesday to 13.00
A bias towards vegetarian products, as the proprietors are themselves vegetarian, and will freely offer advice on dietary matters. The normal range of health foods together with cosmetics, vitamins, dietary supplements, dairy products. (HSW)

Cork City

Ryan Health Foods 11 Castle Street, Cork City, Irish Republic (021 21866); open Monday to Saturday 10.00–18.00
Old-fashioned shop which has been in business for thirty years. The owner plans to retire soon; meanwhile you can still buy most packaged

brands of wholefoods, free-range eggs, full range of vitamins and remedies, home-brewing and wine-making supplies. Frequented by local personalities in tax exile from England.

Coventry

Drop in the Ocean 146 Lower Ford Street, Coventry, West Midlands (0203 22090); Buses 13, 17, 32, 38 open Monday to Saturday 9.30–17.30, Thursday to 14.00, Friday to 19.00
A cooperative, trying to secure more spacious premises nearer city centre. Good range of cereals, flour, rice, pasta, dried fruits, nuts and seeds, pulses, oils, juices. Also herbs and spices, yeast, miso, agaragar, cosmetics, cooking utensils.

Holland & Barrett 6 City Arcade, Coventry, West Midlands (0203 22752); buses to city-centre (stop nearest pedestrian shopping area); open Monday to Friday 9.00–17.30, Saturday 8.30–17.00
For further information on the Holland & Barrett chain see entry for Amersham (page 180).

Cowbridge

L'Epicure 42 High Street, Cowbridge, Glamorgan, Wales (02263 2387); buses to town hall; open Monday to Saturday 9.00–17.30, Wednesday half day
The shop, opened in 1969, is in former Wesleyan chapel complete with old beams and stone fireplace. A delicatessen stocking specialist confectionery, Indian food and spices, as well as general range of packaged health foods, wholemeal bread (fresh on certain days), vitamin and mineral supplements. Has an off-licence.

Cranleigh

Nuts High Street, Cranleigh, Surrey (048 66 3292); open Monday to Saturday 9.30–17.00, Wednesday to 13.00

This shop is run by the well-known (and well-married) actors Andrée Melly and Oscar Quitak, who also write the Wholefooders' Diary in *Here's Health*. Sells only vegetarian foods, including dried wholefoods, wholemeal flours, honey, raw sugar confectionery, bread, TVP, organically grown vegetables in season. A children's section – 'Peanuts' – and a sideline in unusual women's clothing. Opened in 1975; plans for expansion include branches abroad.

Crawley

Realfoods 3 Park Side, Crawley, Sussex (0293 32365); buses to bus station; open Monday to Saturday 9.00–17.30
For further information on the Holland & Barrett chain see entry for Amersham (page 180).

Crewe

Holland & Barrett 64a Market Street, Crewe, Cheshire (0270 214477); buses to nearby bus station; open Monday to Saturday 9.00–17.30, Wednesday to 13.00
For further information on the Holland & Barrett chain see entry for Amersham (page 180).

Criccieth

H. Pugh Jones 50 High Street, Criccieth, Gwynedd, Wales (076 671 2737); open Monday to Saturday 8.30–12.30, 13.30–17.30
Long-established family business – here since 1869 – which is a traditional family grocer's. Sells health foods as well as high-class provisions; freshly ground and decaffeinated coffee, raw sugar molasses, vegetarian and diabetic speciality foods, dried wholefoods, honey, herbs, spices, Indian and China tea, fruit juices, vitamin and mineral supplements, fresh bread. (NFDF)

Crowborough

Crowborough Health Food Stores Crowborough Hill, Crowborough, Sussex (089 26 61615); bus 90 to station; open Monday to Saturday 9.00–13.00, 14.00–17.30, Wednesday half day
Opened in 1973, the shop has built up a reputation for good service from its friendly and very knowledgeable proprietor. Stocks complete range of health and wholefoods. Very good stock of vitamins and food supplements. Herbal remedies for you and your pets. Some local deliveries in special circumstances. (NAHS)

Croydon

Croydon Health Foods 100 High Street, Croydon, Surrey (01 688 0970); open Monday to Saturday 9.00–17.30
This shop, trading since 1908, has recently been refitted by the owners. Has been changed to self-service, and name is different. They still take trouble over their stock and their customers, however; 100 per cent wholewheat bread and cakes (made with vegetable fat) delivered daily from local baker. Full range of packaged healthfoods, vegetarian specialities, herbs and spices, dairy goods from Loseley farms, vitamin and mineral supplements. (NAHS)

Holland & Barrett In Allders Croydon, 2 North End, Croydon, Surrey (01 681 2577, ext. 22); buses 109, 190, 405, 411, 414 to High Street; open Monday to Friday 9.00–17.30, Saturday to 18.00
For further information on the Holland & Barrett chain see entry for Amersham (page 180).

Holland & Barrett 1088 The Mall, Whitgift Centre, Croydon, Surrey (01 686 3000); buses 50, 68, 109, 119, 190, 411 to North End bus station or Wellesley Road; open Monday to Saturday 9.00–17.30
For further information on the Holland & Barrett chain see entry for Amersham (page 180).

Dartford

Dartford Health Food Centre 86 Dartford Road, Dartford, Kent (0322 23578); buses 96, 480 to Havelock Road; open Monday 8.30–13.00 Tuesday to Friday 8.30–13.00, 14.00–18.00, Saturday 8.00–17.00
A friendly, old-established shop which has sold health foods since 1965 as well as their huge delicatessen stocks of exotic foods. The selection of cheeses is wonderful; and the owners will be pleased to order your free-range turkey at Christmas. Goods include free-range eggs, yoghurt, fresh bread, organically grown fruit, vegetables and herbs when available, good selection of spices. Also beer- and wine-making supplies, natural remedies, vitamin supplements.

Food For Living Trading Post Indoor Market, 46 Lowfield Street, Dartford, Kent (0322 20295); buses (local services) to town centre; open Monday to Saturday 9.00–17.30, Wednesday to 14.00
Wholefoods, vegetarian supplies, vitamins and food supplements, cosmetics, herbs and spices. Also selection of health books and range of Chinese and Indian foods.

Darwen

Darwen Health Foods 10 Bridge Street, Darwen, Lancashire (0254 773311); bus 8 to Circus, Blackburn; open Monday to Saturday 9.30–17.30, closed Tuesday
Good range of dairy products: yoghurt, cottage cheese, goat's milk, buttermilk. Fresh wholemeal bread and cobs daily. Wide range of ginseng products. Also vitamin and mineral supplements, and over 120 medicinal and culinary herbs. They provided free sustenance recently, during an attempt (successful), by the local weightlifters club to gain entry to the *Guinness Book of Records*.

Daventry

Goodness Foods 1 Tavern Lane, Daventry, Northamptonshire (032 72 71746); open Monday to Saturday 9.00–17.30, Friday to 18.00
For further information on the Goodness Foods chain see entry for Braunston (page 203).

Deal

The Delicatessen and Health Food Centre 49 The Strand, Walmer, Deal, Kent (030 45 2028); open Monday to Saturday 9.00–13.00, 14.15–17.30
Wide selection of flours, grains, cereals, beverages, vitamins, herbal remedies.

Derby

Monk's Health and Speciality Foods 16 Osmaston Road, The Spot, Derby (0332 47946); buses to city centre; open Monday to Saturday 9.00–17.30, closed Wednesday
As well as the normal range of health foods, herbal remedies and vitamins, sells speciality foods, beer- and wine-making equipment, books on health.

Nature's Foods 19 Abbey Street, Derby; buses to city centre; open Monday to Saturday 10.00–18.00, closed Wednesday
A friendly shop where honey and oils are sold loose from tubs. Good range of dried wholefoods, herbs and spices, vegetarian foods. Owner mixes his own muesli and also sells muesli base. He is willing to proffer advice, which can include the esoteric: e.g. that chickens thrive on sesame seeds. Hopes soon to move shop – a fairly new venture – closer to city centre. Deliveries for orders over £20 within Derby and Nottingham.

Dereham

Guy's Pure Food Store 35 Norwich Street, Dereham, Norfolk (0352 3402); open Monday to Friday 9.00–13.00, 14.00–17.30, Wednesday half day
Good stock of wholefoods, honeys, juices, goat's milk products, fresh coffee ground on premises. Advice given on use of herbs and spices. Emphasis on maintenance of health – local doctors are regular customers. Natural food and remedies for pets as well as their owners. Also natural cosmetics and wide selection of books. Deliveries only in exceptional circumstances.

Dewsbury

W. Hemingway 24 Corporation Street, Dewsbury, Yorkshire (0924 465634); near bus station; open Monday to Saturday 8.45–17.30
A herbal dispensary since 1938. Herbal preparations and advice on use still willingly offered. Excellent range of natural medicines, herbs, powders; stock of general branded health foods, vitamin supplements. Deliveries within locality.

Doncaster

Health Product Supplies 43 Balby Road, Doncaster, Yorkshire (0302 25163); open Monday to Saturday 10.00–17.00
Extensive range of natural foods, herbal remedies, vitamins, nutritional supplements.

Holland & Barrett Unit 23, Arndale Centre, Doncaster, Yorkshire (0302 60082); buses to bus station, just behind Arndale Centre; open Monday to Saturday 9.00–17.30
For further information on the Holland & Barrett chain see entry for Amersham (page 180).

Sleath's Herbal and Health Food Centre 10 Netherhall Road, Doncaster, Yorkshire (0302 23508); bus 179 to Wheatly Hill; open Monday to Saturday 9.00–17.30
Opened twenty years ago, the shop has a friendly atmosphere with personal service. Stocks the general packaged health food ranges, dairy products, herbs and herbal preparations. Full range of natural cosmetics.

Dorchester

Cornucopia 34a High West Street, Dorchester (0305 2586); open Monday to Saturday 9.00–18.00
Stocks full range of wholefoods and large number of herbs and spices. Deliveries within twenty mile radius. Hoping to open wholefood restaurant in the area shortly.

Health Foods (Dorchester) Georgian House, Trinity Street, Dorchester (0305 3979); open Monday to Saturday 9.00–17.30, Thursday to 13.00
The owner, who took over in 1971, has moved the business to the present larger premises. His interest in old watermills is evident in the décor. Stock is wide-ranging: dried wholefoods weighed out for you, packaged health foods, good selection of honey, compost-grown local fruit and vegetables as available, fresh local wholemeal bread and yeast delivered daily, free-range eggs; the usual vitamin and mineral supplements, herbal remedies. Natural cosmetics selected with care for avoidance of animal suffering. (NAHS)

Dorking

Culpeper 37 High Street, Dorking, Surrey (0306 81698); open Monday to Friday 9.30–17.30
For further information on the Culpeper chain see entry for Bath (page 186).

Holland & Barrett 185 High Street, Dorking, Surrey (0305 884607); buses 414, 425, 439, 470 to nearby Woolworth's; open Monday to Saturday 9.00–17.30, Wednesday to 17.00
For further information on the Holland & Barrett chain see entry for Amersham (page 180).

Douglas

Isle of Man Health Food Centre 90 Bucks Road, Douglas, Isle of Man (0624 5647); buses to Rosemount; open Monday to Saturday 9.00–13.00, 14.00–17.30, closed Thursday
Moved to new premises in 1977. Stocks good range of dried wholefoods, honey (some from local hives), herb teas, fresh fruit and vegetables from local farms, free-range eggs, several brands of wholemeal flour, vitamin and mineral supplements, herb products including tobacco and herb teas, health aids like ginseng and biochemical tissue salts. Large range of natural cosmetics and hair care products; books on wholefoods and health. (HSW, NAHS)

Dover

Country Life Wholefoods 5 Church Street, Dover, Kent (0304 203604);
open Monday to Saturday 9.00–17.30, Wednesday to 13.00
This is a shop in the expanding Country Life chain of health food retail
outlets. For further information on this chain see entry on Country Life
Wholefoods in Ashford (page 181).

Nature's Way Ltd 20 Pencester Road, Dover, Kent (0304 201485);
opposite bus terminal; open Monday to Saturday 9.00–17.30
One of the chain of Nature's Way shops. For further information on
this chain see entry for Beckenham (page 187).

Dublin

Health and Herbal Centre 2 Trinity Street, Dublin, Irish Republic
(01 758883); open Monday to Friday 9.30–18.00, Saturday to 17.30
Groceries and dried goods, vegetarian foods, natural medicines,
nutritional supplements.

Ormond Health Centre 5 Parliament Street, Dublin, Irish Republic
(01 775929); buses 54a, 77, 78; open Tuesday to Saturday 10.00–17.30
Stocks all leading brands of prepackaged goods: cereals, breakfast
foods, pulses, dried fruits, herbal teas, juices, beverages. Diabetic products, diet and vitamin supplements. Also various brands of natural
cosmetics.

Dudley

Holland & Barrett 8/10 Fountain Arcade, Dudley, West Midlands (0384
55500); bus 558; open Monday to Saturday 9.00–17.30
For further information on the Holland & Barrett chain see entry for
Amersham (page 180).

Dumfries

Dumfries Health and Wholefood Store 2 Queensberry Street, Dumfries, Scotland (0387 61065); open Monday to Saturday 9.00–17.30, Thursday to 13.00
Opened nearly ten years ago, the shop is about to undergo modernization. Stocks general range of proprietary health foods; beverages such as herbal teas, decaffeinated and filter coffee; goat's milk and other dairy goods, including vegetarian cheese; wholemeal bread, flour, biscuits; vegetarian speciality foods; vitamins, minerals, dietary supplements; herbal remedies. Full range of home beer- and wine-making equipment. Deliveries any area, within reason.

Dundee

Tayside Health Food Stores 36 Albert Street, Dundee, Tayside, Scotland (0382 40515); buses 31, 32, 33, 34; open Monday to Friday 9.00–17.30, Wednesday to 13.00
Good range of wholefoods, including dried fruit, nuts and cereals sold loose from bulk. Goat's milk products. Also vitamins, natural cosmetics, books on health subjects. Deliveries within Dundee.

Dunfermline

Health Foods 10 Bruce Street, Dunfermline, Fife, Scotland; buses to Bruce Street; open Monday to Saturday 9.00–17.30
Near central shopping area. Stocks good range of wholefoods, with more fresh produce than usual, e.g. bread, free-range eggs, local dairy produce, herbs in season. Also vitamin and mineral supplements; some homoeopathic supplies. Deliveries anywhere in west and central Fife.

Durham

Black's Health Food Centre 91 Claypath, Durham City, County Durham (0385 3197); open Monday to Saturday 9.00–17.30
This is the flagship of the four Black's shops, opened in 1936. For further information on these, see entry for Bishop Auckland (page 195).

Eastbourne

Benefit Wholefoods 63 Susan's Road, Eastbourne, Sussex (0323 642981); buses to town centre; open Monday to Saturday 9.00–17.30, closed Wednesday

A friendly shop, priding itself on information service as well as provision of health foods. Well known for its two cats, Liquorice and Allsorts. Forms part of a growing craft community in the area. Large stocks of herbs and spices; many varieties of honey, flour, pasta.Own muesli and muesli base. Most pulses, grains and flakes. Wide variety of spreads, butters, sugarless jams, marmalades. Also a range of cosmetics; Cattier products; organic vegetables; organic dairy products; vegetarian cheese; juices, mineral/spring waters; soya bean products. Books, pottery, cooking-ware. Fresh wholewheat rolls prepared daily, with a variety of sweet rolls and biscuits. Plans to expand take-away service and possibly provide fresh juices during summer. All products packed in bulk if required.

Ceres (Health Foods Restaurant and Sales) Hartington Hall, Bolton Road, Eastbourne, Sussex (0323 28482); buses to town centre; open Monday to Saturday 9.30–17.30

An old established business, combining shop and self-service vegetarian restaurant. Shop has an extensive range of foodstuffs only; does *not* stock vitamins, remedies, etc. Personalities from the world of show business are often to be seen, including Walter Landau and Nyree Dawn-Porter; also many visiting ballerinas. Everything in restaurant prepared fresh on premises, with exception of bread. They have been asked on occasion whether or not the eggs are 'compost grown', and if they serve vegetarian fish. Answers not available for publication!

Country Life Wholefoods 4 South Street, Eastbourne, Sussex (0323 29643); open Monday to Saturday 9.00–17.30, Wednesday to 13.00

This is a shop in the expanding Country Life chain of health food retail outlets. For further information on this chain see entry on Country Life Wholefoods in Ashford (page 181).

Nature's Way Ltd 12 Station Parade, Terminus Road, Eastbourne, Sussex; open Monday to Saturday 9.00–17.30

One of the chain of Nature's Way shops. For further information on this chain see entry for Beckenham (page 187).

Nature's Way Ltd 196 Terminus Road, Eastbourne, Sussex (0323 26776); open Monday to Saturday 9.00–17.30
One of the chain of Nature's Way shops. For further information on this chain see entry for Beckenham (page 187).

Eastcote

Holland & Barrett Field End Road, Eastcote, Middlesex (01 868 6204); bus 282 to tube station, 5 minutes' walk from shop; open Monday to Saturday 9.00–17.30, Wednesday to 13.00
For further information on the Holland & Barrett chain see entry for Amersham (page 180).

East Grinstead

Country Life Wholefoods 18 London Road, East Grinstead, Sussex (0342 25089); open Monday to Saturday 9.00–17.30
This is a shop in the expanding Country Life chain of health food retail outlets. For further information on this chain see entry on Country Life Wholefoods in Ashford (page 181).

Nature's Way Ltd 125 London Road, East Grinstead, Sussex (0342 24541); open Monday to Saturday 9.00–17.30
One of the chain of Nature's Way shops. For further information on this chain see entry for Beckenham (page 187).

East Molesey

Corn Exchange (Wholefoods) 264 Walton Road, East Molesey, Surrey (01 941 2209); buses 131, 211, 716; open Monday, Tuesday, Thursday, Friday 10.00–17.30, Saturday 9.00–17.00
A normal stock range complemented by a selection of honeys, sugar-free jams, free-range eggs. Also recipe books and a pottery section. Fresh bread available daily.

Edgware

Holland & Barrett 14 The Promenade, Edgware, Middlesex (01 958 2657); buses 32, 107, 113, 142, 240, 288 to bus station; Edgware tube; open Monday to Saturday 9.00–17.30
For further information on the Holland & Barrett chain see entry for Amersham (page 180).

Edinburgh

Bristo Health Food Store 2/3 Bristo Place, Edinburgh, Scotland (031 225 2819); buses 2, 36, 41, 42; open Monday to Friday 8.30–17.30
Good stock of dried wholefoods including dried fruit and nuts, cereals, pulses, grains; honeys, raw sugar preserves; bran products; complete range of vitamin and mineral supplements; fruit juices; slimming aids. Also macrobiotic specialities, free-range eggs, some fresh fruit and vegetables in season. (NAHS)

Edinburgh Wholefoods 63 Cockburn Street, Edinburgh, Scotland (031 225 7598), 3 St Patrick Street, Edinburgh (031 667 0603); both open Monday to Saturday 9.30–18.00
Wholefood shops under same ownership as bulk warehouse in Coltbridge Avenue. They stock their own packaged grains, cereals, pulses, nuts and dried fruits, pasta, herbs and spices, own-brand peanut butter, own stoneground wholemeal flour; also fruit juices, dairy products, goods from their own bakery, spreads, condiments. At Cockburn Street there is a vegetarian restaurant serving soups, hot dishes, savouries, salads. Snacks to take away in both shops. Deliveries throughout Scotland.

Edinburgh Wholefoods 48/52 Coltbridge Avenue, Edinburgh, Scotland (031 346 1398/9); open Monday to Saturday 9.00–18.00
This is a cash-and-carry bulk warehouse; also main office of Edinburgh Wholefoods shops. They sell their own packaged dried goods in any quantities, including grains, cereals, nuts, seeds, dried fruits, pulses, own stoneground flour, own-brand peanut butter, honey, spreads, condiments. Ring for a full list. Deliveries throughout Scotland.

Good Food (Edinburgh) 255 Morningside Road, Edinburgh, Scotland (031 447 3020); buses 5, 11, 15, 16, 23, 41 to Morningside station; open Monday to Thursday 9.00–13.00, 14.30–17.30, Friday and Saturday 9.00–17.30
A small family business stocking health foods, honey, grains, flours, vitamins. Also yoghurt and goat's milk. Muesli mixed on premises. Deliveries locally.

Health Food Store 40 Hanover Street, Edinburgh, Scotland (031 225 4291); buses 23, 27, 41, 42 to Hanover Street; open Monday to Friday 9.30–17.30, Saturday 9.00–13.00
Packaged rather than fresh foods. Juices, some herbs and spices, vitamin and mineral supplements. (NAHS)

Henderson's 94 Hanover Street, Edinburgh, Scotland (031 225 6694 shop, 031 225 3400 restaurant); buses 23, 27 to door; open Monday to Saturday 8.30–17.30
A famous, family-run business, with shop (soon to become self-service) and buffet restaurant seating 200 where all dishes (strictly vegetarian) are prepared on premises. Full range of wholefoods, dairy goods, compost-grown fruit and vegetables, bakery goods from their own ovens, magazines on relevant topics. (HCIBT)

The Kernel 9 Deanhaugh Street, Stockbridge, Edinburgh, Scotland (031 332 8963); open Monday to Saturday 9.00–18.00
A newly opened shop. Excellent stock of dried wholefoods – grains, cereals, dried fruit and nuts; raw sugars, oils, juices, beverages, herbs and spices. Local deliveries.

Natural Food Larder 205 Bruntsfield Place, Edinburgh, Scotland (031 447 3033); buses 11, 16, 23, 45 stop outside; open Monday to Saturday 9.30–13.00, 14.15–17.30, early closing Wednesday
Opened in 1900 as a meal and cereal store, the shop recently passed to the ownership of two local ladies. A recommended shop with huge range of health foods including dried goods: pulses, nuts, cereals, grains, yeast, seeds, flours, sugars, herbs, spices sold loose from bulk. Also freshly ground coffee, vitamins, herbal remedies, natural cosmetics, lots of honeys, natural ice-cream, goat's milk, yoghurt and cheeses, natural

fruit juices, mineral waters. Wholemeal bread and other baked goods delivered fresh daily. Large choice of books on health, yoga, natural living and related topics. The owners hope soon to have a sure supply of organically grown fruit and vegetables. Seeds for sprouting are sold.

Egham

Cross and Herbert Ltd 23 The Precinct, Egham, Surrey (078 43 2464); open Monday to Saturday 9.00–17.30
A chemist Health food section includes large range of vitamins, minerals, biochemical tissue salts, herbal remedies and diet aids as well as packaged wholefoods, some organically grown. Much frequented by ladies of stage and screen, e.g. Diana Dors.

Ely

The Breakaway 31/33 Lynn Road, Ely, Cambridgeshire (0353 3175); buses to Market Street; open Monday to Saturday 7.30–18.30, Sunday 9.00–13.00
This pleasant shop has been open since 1972. Stocks the usual dried wholefoods, an extensive range of shelled nuts, beverages like herb teas, vegetarian special foods, honeys, vitamin and mineral supplements. Also herbal remedes, natural cosmetics.

Epsom

Epsom Health Foods 41 Waterloo Road, Epsom, Surrey (037 27 22076); open Monday to Friday 9.00–13.00, 14.00–17.50, Wednesday to 13.00, Saturday 9.00–17.30
A very well-stocked shop which became fully 'wholefood' in 1979 under energetic new ownership. Stock, mainly packed on premises from bulk, has been expanded to include pulses, herbs and spices, several kinds of dried fruits, honey and preserves, nuts, yeast and yeast extracts, soya protein preparations, various types of decaffeinated coffee, teas and tisanes, cereals, herbal remedies, natural cosmetics and fresh wholemeal bread.

Holland & Barrett 12 King's Shade Walk, Epsom, Surrey (037 27 23234); buses 408, 470, 478, 479; open Monday to Friday 9.00–17.30, Saturday 8.30–17.00

For further information on the Holland & Barrett chain see entry for Amersham (page 180).

Evesham

Beewell 3 Vine Street, Evesham, Worcestershire (0386 3757); open Monday to Friday 8.30–17.30

A pleasant shop with friendly service and a better than usual stock of fresh foodstuffs to supplement the groceries and dried wholefoods. Organically grown vegetables in season, locally grown fruit, free-range eggs, preserves and local honey, wholemeal bread, take-away snacks such as quiche and cheesecake. Also smoking mixtures, cider wine and vinegar made locally, herbs both loose and in the pot when in season, and 'the best range of cheeses in Worcestershire.'

Exeter

Cityditch Wholefoods Shop 14 South Street, Exeter, Devon (0392 50925); open Monday to Saturday 9.30–17.00, closed Wednesday 13.30–14.30

This shop was originally a restaurant; it reopened as a wholefood shop in the present premises in 1978. Excellent choice of dried goods, mostly organically grown, and sold from bulk: Grains, beans and peas, flours, breakfast cereals including muesli mixed in the shop, herbs and spices, dried fruit, nuts. Good range of juices, sauces, pickles, spreads. Also incense, Japanese cookware, books on health and diet. Continental and Oriental speciality foods always in stock. Deliveries within twelve-mile radius.

Grael 15 North Street, Exeter, Devon (0392 37782); open Monday to Saturday 10.00–18.00

Headquarters of the Grael Association, a community-based organization which prints, publishes and distributes books on health, nutrition, wholefoods and yoga. Teaches yoga and holistic psychology through its three outlets. Here in Exeter there is no retail shop, but a wholesale warehouse provides a wide range of wholefoods in bulk at discount prices. Also a wholefood vegetarian restaurant where all food, including vegetarian hot dishes, pizzas, quiches, salads, biscuits and American-style carrot and chocolate fudge cakes, is prepared daily on premises. The warehouse will deliver bulk supplies within the area.

Southern Health Foods 16 Waterbeer Street, Exeter, Devon (0392 77494), 16/18 Princesshay, Exeter, Devon (0392 51590); both open Monday to Saturday 9.00–17.30
One of the well-established Southern Health Foods chain. For further information on this chain see entry on Southern Health Foods in Aldershot (page 179).

Exmouth

Southern Health Foods 19 High Street, Exmouth, Devon (039 52 5741); open Monday to Saturday 9.00–17.00
One of the well-established Southern Health Foods chain. For further information on this chain see entry on Southern Health Foods in Aldershot (page 179).

Falkirk

Falkirk Health Store 5 Melville Street, Falkirk, Central, Scotland (0324 35918); open Monday to Saturday 9.00–17.30
Established in 1972, the shop moved to larger premises two years ago. Sells full range of vegetarian and wholefood groceries, dairy goods, vitamins, herbal medicines, and such items as herbal pillows. Also books on health.

Falmouth

The Granary 16 High Street, Falmouth, Cornwall (0326 311507); buses to The Moor; open Monday to Saturday 9.30–17.30
Run on cooperative lines by young people who aim to 'carry on in a spirit of community service'. They act as a collection point for charity

collections of recyclable goods. Community notice boards, and stall run by Friends of the Earth. Full range of loose dry goods; their own muesli and fruit and nut mix; free-range eggs, local cheese, herbs and spices, local fruit and vegetables organically grown where possible, coffee, teas, wholemeal bread.

Health Food Services (Cornwall) Ltd 57 Killigrew Street, Falmouth, Cornwall (0326 313040); buses to main bus station; open Monday to Saturday 9.00–17.30, Wednesday half day
Good stock of branded groceries, including wholemeal flour, grains, pulses, dried fruits and nuts, honey, raw sugar preserves, fresh eggs. Home beer- and wine-making supplies. Deliveries within most of Cornwall. Monthly postal service, (HSW)

Fareham

Holland & Barrett 92 West Street, Fareham, Hampshire (03292 280884); buses to West Street bus station; open Monday to Saturday 9.00–17.30
For further information on the Holland & Barrett chain see entry for Amersham (page 180).

Farnham

Holland & Barrett 5a West Street, Farnham, Surrey (0252 724209); buses 207, 214 stop outside shop; open Monday to Saturday 9.00–17.30
For further information on the Holland & Barrett chain see entry for Amersham (page 180).

Felixstowe

Truefoods 74 Hamilton Road, Felixstowe, Suffolk (03942 4903); open Monday to Saturday 9.00–17.30, Wednesday 13.00
Fresh daily supply of goat's milk and yoghurt, together with natural vitamins, health food items. (HSW)

Fleet

Savory and Moore Ltd 209 Fleet Road, Fleet, Hampshire (025 14 6226); open daily 9.00–17.30
Basically a chemist who stocks health foods: muesli, bran, honey, juices, pulses, salts, flour biochemicals.

Fleetwood

Health Food Store 92 Lord Street, Fleetwood, Lancashire; open Monday to Saturday 9.00–17.30
Emphasis on health rather than provisions; but you can buy the popular branded health food groceries as well as an extensive range of vitamin and herbal supplements and remedies, loose herbs, vegetarian products. The shop is associated with the naturopathic clinic at Cleveleys, Blackpool and owned by the same people who run the Cleveleys Health Food Store.

Folkestone

Country Life Wholefoods 16 Alexandra Gardens, Folkestone, Kent (0303 57679); open Monday to Saturday 9.00–1730
This is a shop in the expanding Country Life chain of health food retail outlets. For further information on this chain see entry on Country Life Wholefoods in Ashford (page 181).

Deva Wholefoods (Country Life Wholefoods) 12 Church Street, Folkestone, Kent (0303 53700); open Monday to Saturday 9.00–17.30, Wednesday to 13.00
This is a shop in the expanding Country Life chain of health food retail outlets. For further information on this chain see entry on Country Life Wholefoods in Ashford (page 181).

Forest Row

The Seasons Lewes Road, Forest Row, Sussex; open Monday, Tuesday, Thursday, Friday 9.00–13.00, 14.30–17.30, Wednesday and Saturday to 13.00

No vitamins and supplements, but a great deal of organically grown produce, including vegetables, fresh herbs and fruit in season, all compost-grown in Soil Association approved earth. They specialize in Demeter standard products wherever available and sell biodynamic grains from Busses Farm in East Grinstead; also fresh dairy produce from approved farms – organically produced cheese will be added to the range soon. Wide range of stock includes natural juices, cereals, flours, pastas, first-pressing unrefined oils, pulses, herbs and herb teas, breads from local Cyrnel bakery, and dried fruits unsulphured where possible. A great deal of trouble is taken to find sources of supply where goods are untreated, and grown entirely organically. At present the shop is trying to expand the space for 10-lb. family packs sold at a discount.

Forres

Findhorn Foundation Shop Findhorn Bay Caravan Park, Forres, Grampian, Scotland (030 92 2377); open Monday to Saturday 9.00–12.30, 14.00–17.30 (winter), 8.30–12.30, 15.00–19.30 (summer), closes Tuesday 12.30

Operated by the Findhorn Community. Tours given from the shop. Their aims lie in the area of spiritual enlightenment and cooperative living. Craft work is sold as well as groceries, loose dried wholefoods, juices, herbs and herbal products, natural remedies, vitamins.

Fort William

Glengarry Drug Store 123 High Street, Fort William, Highland, Scotland (0397 2635); open Monday to Saturday 9.00–18.00

The owner comments: 'The only business around here providing help on the road to health, the road to Skye, and the road to heaven!' He therefore offers a postal service within a fifty-mile radius. Stocks include packaged wholefoods, dried fruits, cereals, nuts, juices, herbal remedies, Scheussler salts, vitamin and food supplements.

Framlingham

Carley and Webb Ltd 29 Market Mill, Framlingham, near Woodbridge, Suffolk (0728 723503); bus 203; open Monday to Saturday 8.30–17.00, Wednesday to 13.00
A grocery store has operated here for over a century. The different sections of the shop retain an old-world charm and the service one associates with this. Primarily a high-class provisions shop, with charcuterie and delicatessen, cheeses and an attached wine bar, coffee shop, etc., the new section in Market Hill selling health foods is already a great success. All the popular brands are stocked, along with dairy goods and wholemeal bread, and the shop is to be enlarged to accommodate more varieties of wholefoods. Parking in the forecourt. Deliveries within a reasonable distance. Snacks sold in the wine bar, e.g. pâtés and soups, are not strictly wholefood, but you may find enough to satisfy you in this vein.

Frinton-on-Sea

The Bible Depot Connaught Avenue, Frinton-on-Sea, Essex (025 56 4106)
Principally a bookshop for Bibles and other religious literature, but they sell health foods 'for soul, mind, and body' because they believe in their good effects. Also herbal remedies.

Gainsborough

The Health Food Store 3 North Street, Gainsborough, Lincolnshire (0472 5201); open Monday to Saturday 9.00–17.50, closed Wednesday
This shop passed into new hands in 1978. Emphasis is on health and prevention of disease; owner observes that several local doctors are

regular customers. Large stock of herbal remedies, nutritional supplements, diet aids. They pack their own herbs, spices and dried wholefoods: a good selection. It is hoped that home wine-making kits will be sold shortly.

Garstang

Garstang Health Food Stores High Street, Garstang, near Preston, Lancashire; central bus stop; open Monday to Friday 9.00–17.30
Stocks complete range of wholefoods, vitamins, food supplements, herbal medicines, natural cosmetics. Deliveries within ten miles, and a full postal service. People travel some distance to shop here, and the staff are noted for their friendly, helpful service.

Gillingham

Country Life Wholefoods 49 High Street, Gillingham, Kent (0634 53651); open Monday to Saturday 9.00–17.30, Wednesday to 13.00
This is a shop in the expanding Country Life chain of health food retail outlets. For further information on this chain see entry on Country Life Wholefoods in Ashford (page 181).

Glasgow

Forrest & Niven Ltd 73 St Vincent Street, Glasgow, Scotland (041 221 7865); City centre bus stops; open Monday to Saturday 9.00–18.00
Reputedly Scotland's largest health food store. The shop is self-service and stocks packaged health foods, dairy produce, wholemeal bread, natural medicines, nutritional supplements, freshly baked foods, fresh fruit and vegetables when compost-grown are available. (NAHS)

The Grain Store 1238 Shettleston Road, Glasgow, Scotland (041 763 0596); buses 60, 62, 30; open Monday to Saturday 9.00–17.30
Hoping to extend the present range of grains, flours and pulses; also cosmetics and prepacked health foods.

The Health Food Centre 10 Dixon Avenue, Glasgow, Scotland (041 423 8675); buses 5, 14, 43, 44, 67; open Monday to Saturday 9.00–17.30
Packaged health foods, vegetarian cheese, some free-range eggs and wholemeal bread delivered twice weekly; natural medicines and vitamin supplements. Deliveries in south Glasgow. (NAHS)

The Health Food Store 14 Skirving Street, Glasgow, Scotland (041 632 0664); buses 23, 38, 48, 57; open Monday to Saturday 9.00–17.30, Tuesday to 13.00
Sells only completely vegetarian goods. Accent is on personal and efficient service, and the friendly owner stocks a comprehensive range of proprietary vegetarian health foods (including 'dairy' goods like cheese), dried fruits and nuts, natural juices, wholemeal cakes, biscuits and bread; also some compost-grown vegetables in season. 'Dogs only welcome if they intend to buy.' (NAHS)

Simpson's 808 Crow Road, Glasgow, Scotland (041 959 5089); open Monday to Saturday 9.00–17.30
Opened shortly after the war, the shop stocks general packaged health foods, vitamin and mineral supplements. (NAHS)

Glastonbury

Health Foods (E. E. Gane) 3 and 5 Benedict Street, Glastonbury, Somerset (0458 32127); buses to market place; open Monday to Saturday 9.00–13.00, 14.00–17.30, Wednesday to 13.00
This shop, opened three years ago, is run on the traditional lines of health food shops selling packaged goods, vitamin and mineral supplements: a good choice within this range, including raw sugars, sea salt, herbal teas, grains, and the usual cereal products. Deliveries locally.

Glossop

Health Foods 136 High Street West, Glossop, Derbyshire (045 74 1856); open Monday to Saturday 9.00–18.00, early closing Tuesday
Long-established health food store with the usual proprietary wholefoods and dietary specialities; also vitamin and mineral supplements,

herbs and spices, dairy goods. Owner always willing to order new lines
at the request of regular customers. Deliveries within locality. (NAHSW)

Gloucester

Country Market 52 Kings Walk, Kings Square, Gloucester (0452
35441); buses to central and local bus stations adjacent to Kings
Square; open Monday to Friday 9.00—17.30, Saturday 8.30–17.00
For further information on the Holland & Barrett chain see entry for
Amersham (page 180).

Health Food and Herbal Stores (Gloucester) Ltd 19 St John's Lane,
Gloucester (0452 24501); buses to Northgate Street; open Monday to
Friday 9.00–17.30, Saturday 8.30–17.30
This shop stocks the usual range of natural vitamins, wholefoods,
cereals, pulses, as well as wholemeal bread, flour and biscuits – and has
been doing so since June 1940. (NAHS)

Godalming

Godalming Health Foods 132 High Street, Godalming, Surrey (04868
7470)
A sister shop to the health food store in Chichester; both were opened
in 1979 under the present ownership. Stocks full range of packaged
health foods, vegetarian specialities, home-baking requisites including
wholemeal flours and fresh and dried yeast. Fruit and vegetable juices,
herb teas, dried fruit and nuts, honey and preserves. Vitamin and
herbal remedies, homoeopathic medicines, natural cosmetics. Also
wholefood confectionery, eggs, and bread baked fresh daily.

Gorleston

McLean & Sons (Graduates) Health Food Centre 90/91 High Street,
Gorleston, Great Yarmouth, Norfolk (0493 62162); buses 3, 8; open
Monday to Saturday 9.00–17.30, Wednesday to 13.00
Good range of dried fruit, nuts, honey, cereals, flour, wholemeal bread,

goat's milk, fruit juices, pulses, vegetable protein, biscuits, diabetic foods, confectionery, raw sugar, muesli. Also natural bran, food supplements, vitamins, remedies, tonics, etc. Deliveries locally.

Gosport

Holland & Garrett 113a Stoke Road, Gosport, Hampshire (07017 83201); open Monday to Saturday 9.15–17.30
For further information on the Holland & Barrett chain see entry for Amersham (page 180).

Grantham

Life and Health Foods Ltd Market Place, Grantham, Lincolnshire (0476 5224); buses to Watergate and High Street; open Monday to Thursday 9.00–17.30, Friday to 17.00, closed Saturday
Sells full range of branded health foods as well as untreated pulses, shelled nuts, dried fruits and other wholefood groceries. Also natural fruit juices, and food for those on diabetic or gluten-free diets. They specialize in natural remedies, herbal tonics, biochemical remedies, tissue salts, vitamin and mineral supplements: large stocks of all these. In addition to a good range of health books, a selection of Christian literature. It is hoped that a juice bar and take-away food section will open shortly. (NAHS)

Grassington

Craven Wholefoods Garrs Lane, Grassington, Yorkshire (0756 752421); near bus station; open Monday to Saturday 9.30–17.30
Opened in 1978, this pleasant shop supports market stalls in the region – Tuesday in Settle, Friday and Saturday in Stockport. Dried goods including pulses, grains, flakes, fruit, nuts and own mixed muesli packed on premises. Usual popular brands of health food groceries stocked, notably a full range of flours. Also fresh wholemeal bread, eggs, a wide range of herbs, and high-class provisions like Jackson's teas, fruit juices, good preserves.

Grayshott

Grayshott Health Foods Headley Road, Grayshott, Hindhead, Surrey (042 837 4046); open Monday to Saturday 9.00–17.30, Wednesday to 13.00
Quality goods are wholemeal bread, free-range eggs, English honeys, Loseley dairy goods (including goat's milk and ice-cream). Good range of vitamins and minerals; homeopathic and herbal remedies.

Great Yarmouth

Health Foods (Great Yarmouth) Ltd 43/44 Central Arcade, Great Yarmouth, Norfolk (0943 55316); open Monday to Saturday 9.00–17.30 (summer only), Thursday to 12.30
Packaged rather than fresh foods. Usual branded groceries, vegetarian special foods, herbal and natural remedies, dairy produce (including goat's milk), Schleusser biochemical remedies, vitamins, natural cosmetics. (NAHS)

Grimsby

Holland & Barrett 88 Victoria Street, Grimsby, Humberside (0472 43068); buses 2, 3a, 3c, 3f, 4. 8, 9, 16 to Baxtergate and Town Hall Street; open Monday to Saturday 9.00–17.30, Thursday to 13.00
For further information on the Holland & Barrett chain see entry for Amersham (page 180).

Guildford

Cranks Health Food Shop 25 Castle Street, Guildford, Surrey (0483 68258); open Monday to Saturday 9.00–17.30
A branch of the London Cranks, the shop is entirely vegetarian. Sells an excellent range of packaged wholefoods, some under their own label, including Loseley Park dairy goods; fresh compost-grown fruit, vegetables and herbs in season; natural remedies and some homoeopathic supplies; nutritional supplements; natural cosmetics. Whole grains can

now be bought in bulk at a discount. An excellent wholefood take-away service, selling the same food – freshly made on premises – as the small adjoining restaurant: wholemeal bread, cakes, pastries, biscuits; savouries like quiches, rissoles, pâtés; soups; fruit salads and crumbles. The restaurant has a licence. (HSW, NAHS)

Culpeper 10 Swan Lane, Guildford, Surrey (0483 60008); open Monday to Friday 9.30–17.30
For further information on the Culpeper chain see entry for Bath (page 186).

Food For Thought 17 North Street, Guildford, Surrey (0483 33841); open Monday to Saturday 9.00–17.30
Wide range of items, with nuts, dried fruit, pulses, etc. packed on premises to minimize retail prices. Also a take-away where you can obtain salads, sandwiches, baked goods. Deliveries not usually made.

Guildford Health Foods Bridge Street, Guildford, Surrey (0483 65498); open Monday, Tuesday, Thursday, Friday 9.00–17.30, Wednesday to 13.00, Saturday to 17.00
Run by a couple who adhere to the aims and ethics of the health food movement. Apart from a comprehensive range of some 800 items, they carry a large stock of herbal, vitamin, homoeopathic and mineral remedies upon which they are glad to advise in detail. Deliveries within five-mile radius.

Loseley Farm Shop Loseley Park, Guildford, Surrey (0483 71881); any bus to Godalming from Guildford; open Wednesday to Saturday 14.00–17.00 (28 May –27 September, including Bank Holidays)
This is attached to the Elizabethan Loseley House, which can be visited at the same time as the shop. The Park is the home of Loseley Dairy Products which supply so many health food shops and restaurants. You can buy, at their freshest, Jersey cream, untreated milk, soft cheese, live yoghurt, ice-cream, as well as fresh eggs (free-range, of course), honey, organically grown fruit and vegetables in season, wholemeal flour. Have tea in the historic barn and sample the delicious home-made cakes.

Hadleigh (Essex)

Hadleigh Health Foods 85 High Street, Hadleigh, Benfleet, Essex (0702 557691); buses 1, 2, 3, 5, 24, 400; open Monday to Saturday 9.00–17.30, Wednesday to 13.00

The director of this firm is a great believer in megavitamin therapy and consultations may be arranged for ailments like asthma, arthritis, etc. As well as muesli and granola prepared on premises, you can obtain fresh goat's milk and a wide range of vitamins, dietary supplements, herbal remedies. It is hoped to install a take-away and juice bar. Deliveries (bulk only) in area of Canvey, Benfleet, Thundersley, Hadleigh and Basildon. (NAHS)

Hadleigh (Suffolk)

Sunflower 101 High Street, Hadleigh, Ipswich, Suffolk (0473 823219); open Monday to Saturday 9.00–13.00, 14.15–17.30, closed Wednesday

Situated in 'Constable country' near the Essex/Suffolk border. The organically grown wholewheat flours are stoneground in the nearby watermill at Layham. Also stocked are home-grown garden produce, free-range eggs, local natural Jersey milk and yoghurt; their own extensive range of packaged quality dried fruit, nuts, lentils, beans, rice, together with other grains and pulses in 1-lb and 5-lb bags. Home-made cakes, slices and biscuits produced by local *cordon bleu* wholefood cooks.

Hailsham

Country Life Wholefoods 26a High Street, Hailsham, Sussex (0323 841904); open Monday to Saturday 9.00–17.30, Thursday to 13.00

This is a shop in the expanding Country Life chain of health food retail outlets. For further information on this chain see entry on Country Life Wholefoods in Ashford (page 181).

Halifax

Health Food Stores (Halifax) Ltd 3 Russell Street,Halifax, Yorkshire (0422 53991); buses to town centre; open Monday to Saturday 9.00–17.30, Thursday to 12.30
Packaged rather than fresh foods, with all the usual brands of health food groceries, juices, honey, confectionery, oils, diabetic foods, culinary and medicinal herbs, vitamin and mineral supplements herbal remedies, books on health. (NAHS)

Hamilton

Peter Ryder Limited Health Foods 27 Townhead Street, Hamilton, Strathclyde, Scotland (041 7 281415); open Monday to Saturday 9.15–12.00, 13.00–17.30
Prepacked wholefoods, grains, flour, etc. Also vitamin and mineral supplements, herbs, herbal shampoos and cosmetics.

Harlow

Holland & Barrett 12a Terminus Street, Harlow, Essex (0279 26639); all buses to bus terminus; open Monday to Saturday 9.00–17.30
For further information on the Holland & Barrett chain see entry for Amersham (page 180).

Harrogate

Holland & Barrett 1a James Street, Harrogate, Yorkshire (0423 503469); adjacent to bus terminal ; open Monday to Saturday 9.00–17.30
For further information on the Holland & Barrett chain see entry for Amersham (page 180).

Stoneground 64 Station Parade, Harrogate, Yorkshire (0423 65722); opposite bus station; open Monday to Saturday 9.30–17.30, Wednesday to 14.00

A very good shop which has only been open a year. Wholemeal baking is done on premises, and they mix their own muesli. Very full range of dried wholefood: pulses, cereals, several kinds of rice and other grains; flours; honeys; herbs and herbal remedies; dried fruit and nuts; vitamins and dietary supplements; fruit juices; goat's milk yoghurt. Owner plans to open a take-away sandwich bar soon.

Harrow

Holland & Barrett 22/24 College Road, Harrow, Middlesex (01 427 4794); buses 114, 140, 182, 183, 186, 258; open Monday to Saturday 9.00–17.30

For further information on the Holland & Barrett chain see entry for Amersham (page 180).

Hartlepool

T. D. Pattison (Chemists) Ltd 81 York Road, Hartlepool, Cleveland (0429 72690); open Monday to Friday 9.00–15.30, Wednesday to 13.00

This chemist, established in 1893, stocks a comprehensive range of herbal remedies and nutritional supplements. Also a wide range of packaged health foods. (NPA, PDA)

Haslemere

Health Care 78 Wey Hill, Haslemere, Surrey (0428 51001); open Monday, Tuesday, Thursday, Friday 9.00–17.30, Wednesday to 13.00, Saturday to 16.30

The shop has just been redesigned. Stocks a wide choice of packaged health foods, herbs and herbal medicines, Loseley dairy produce, fresh wholewheat bread, nutritional supplements, natural cosmetics. (NAH)

Hassocks

Utopia Nature's Corner 34 Keymer Road, Hassocks, Sussex (07918 5400); bus 051; open Monday to Saturday 8.30–18.00, Sunday 9.30–12.30
Well-stocked shop offering personal service and full range of health foods and natural remedies: specialities for diabetics and vegetarians, goat's milk products, natural cosmetics, large variety of herbs and spices. Wholewheat cakes and pastries made fresh on premises. A gifts section selling local crafts. This shop has only just opened and plans to expand as soon as possible.

Hastings

Nature's Way Ltd 23 Robertson Street, Hastings, Sussex (0424 431124); open Monday to Saturday 9.00–17.30
One of the chain of Nature's Way shops. For further information on this chain see entry for Beckenham (page 187).

Havant

Havant Health Food Store (Holland & Barrett) 7 North Street Arcade, Havant, Hampshire (0705 475736); buses 307, 317, 327, 331; open Monday to Saturday 9.00–17.30, Wednesday to 13.00
For further information on the Holland & Barrett chain see entry for Amersham (page 180).

Haverfordwest

Parry Chemists 25 High Street, Haverfordwest, Dyfed, Wales (0437 2043); open Monday to Saturday 9.00–17.30
Basically a chemist stocking packaged health foods, honeys, vegetarian specialities, herbal remedies, vitamin and nutritional supplements. Owner plans a gradual expansion of this side of the business.

Hay-on-Wye

Country Stores 14 Broad Street, Hay-on-Wye, Powys, Wales (0497 820773); open Monday to Saturday 10.00–17.30

Large range of wholefoods, herbs and spices, together with local cheeses and other dairy products. Also a little local baking.

Haywards Heath

Country Life Wholefoods 52 The Broadway, Haywards Heath, Sussex (0444 51529); open Monday to Saturday 9.00–17.30 Wednesday to 13.00
This is a shop in the expanding Country Life chain of health food retail outlets. For further information on this chain see entry on County Life Wholefoods in Ashford (page 181).

Heathfield

Health Food Store New Parade, 30 High Street, Heathfield, Sussex (043 52 3828); buses to High Street; open Monday to Saturday 9.15–17.15, closed Wednesday
The shop has been in the present ownership since 1970. The owners assure you of a friendly welcome and personal service. A completely vegetarian range of groceries and wholefoods, herbs, spices, herbal remedies, free-range eggs, local dairy and goat's milk products, natural cosmetics, books on health subjects. Deliveries for bulk orders only, within six-mile radius.

Hebden Bridge

Aurora Foods Ltd Wholefood Cooperative 54 Market Street, Hebden Bridge, Yorkshire; open Monday to Saturday 9.30–13.30, 14.30–17.30, closed Tuesday
One of the new cooperative shops selling wholefoods from bulk and seeing their role in a wider context; this one has the social service of a sale return and exchange shop on the premises for local crafts and quality secondhand goods. Situated in an attractive village – sometimes called 'St Ives of the North' since it is something of an artistic centre – in the middle of the South Pennines Park. Customers include visiting writers taking courses at the nearby Arvon Foundation. Stocks all the usual grains, cereals, dried fruits, shelled nuts, pulses, pastas, herbs,

spices, natural medicinal products, oils, honeys, preserves, vegetarian cheese, fresh eggs, and items like incense sticks. Muesli mixed and packed on premises, or you can buy your own muesli base. (NWFWC)

Hemel Hempstead

Heath & Heather (Holland & Barrett) 130 Marlowes, Hemel Hempstead, Hertfordshire (0442 51048); (next to Odeon cinema); buses H1, H2, H3, H12 stop opposite shop; open Monday to Saturday 9.00–17.30
For further information on the Holland & Barrett chain see entry for Amersham (page 180).

Holland & Barrett 16 Bridge Street, Hemel Hempstead, Hertfordshire (0442 64963); buses 42, 43, 307, 347, 348; open Monday to Saturday 9.00–17.30
For further information on the Holland & Barrett chain see entry for Amersham (page 180).

Henley-on-Thames

The High (Health Foods) Ltd 41 Market Place, Henley-on-Thames, Oxfordshire (04912 3764); all main bus routes; green line coaches 28, 390; open Monday to Saturday 9.00–17.30, closed Wednesday
A recently opened offshoot of The High in Reading, the shop is already looking for larger premises. Has soon become popular with local luminaries. You may bump into a well-known Agony Auntie, numerous members of the aristocracy, and a whole cast of showbusiness personalities, e.g. ex-Beatle George Harrison. Fully comprehensive range of all proprietary health foods, honeys, dried fruit, Loseley dairy products, free-range eggs. Good range of herbal products and vitamins. Also natural cosmetics including the Beauty Without Cruelty range.

Hereford

The Marches 24/30 Union Street, Hereford (0432 55712); all buses to city centre; open Monday to Saturday 8.30–17.30

A shop/restaurant/take-away, with plans to extend upper floor to incorporate a gift sales area. Stocks wholemeal products, cereals, honey, yoghurt and cheese; also vitamins, herbal remedies, natural cosmetics, health books. The take-away section provides a selection of salads, scones, savouries.

Hest Bank

Spice of Life 5 Marine Drive, Hest Bank, near Lancaster, Lancashire (0524 822655); buses to Hest Bank Lane and Marine Drive; open Monday to Friday 9.00–13.00, 14.15–17.15, Wednesday to 13.00
A pleasant shop with a good range of wholefoods, local wholemeal bread, fresh local dairy and vegetable produce from Fordhall Farms (including fresh herbs in season and potted herbs), and natural fruit juices. Also vitamin and mineral supplements, natural cosmetics, wine-making equipment, kitchenware. The shop believes in spreading the health food gospel and is planning to advertise in the near future. (NAHS)

High Wycombe

Holland & Barrett 6 The Arcade, Octagon Precinct, High Wycombe, Buckinghamshire (0494 26605); buses to High Street and nearby bus station; open Monday to Saturday 9.00–17.30
For further information on the Holland & Barrett chain see entry for Amersham (page 180).

Hinckley

Health Food Store 7 Edwards Buildings, Hinckley, Leicestershire (0455 923 5627); open Monday to Saturday 9.00–18.00, Thursday to 13.00
Packaged health foods rather than fresh produce; vitamins and minerals natural medicines; home beer- and wine-making equipment.

Hitchin

Health Fare (Holland & Barrett) 94 Hermitage Road, Hitchin, Hertfordshire (0462 51643); bus 2 to stop outside shop or 4 to Bancroft two minutes' walk away; open Monday to Saturday 9.00–17.00
For further information on the Holland & Barrett chain see entry for Amersham (page 180).

Hoddesdon

Burchnall's Health Foods 33 Rye Road, Hoddesdon, Hertfordshire; buses 316, 327 to The Old Highway; open Monday to Saturday 8.30–13.00, 14.00–18.00
A small shop in which friendly advice is given. Stocks good range of dried packaged groceries, macrobiotic special foods, goat's milk and yoghurt, biochemical, herbal and homoeopathic remedies. Also natural cosmetics, books on health-related subjects. (NFDA)

Holt

Larner's of Holt 10 Market Place, Holt, Norfolk (026 371 2244); open Monday to Friday 8.30–17.00, Thursday to 13.00
Principally a delicatessen and buttery specializing in own-baked ham and lunchtime foods. Huge retail outlet for Prewett's fruit juices, backed by comprehensive range of health foods.

Holywood

Jack McClelland Naturally 27 Church Road, Holywood, County Down, Northern Ireland (023 17 5655); open Monday to Saturday 9.00–17.30, closed Wednesday
The stock here is vegetarian and preferably vegan: wholefoods both loose and packaged (with muesli mixed in the shop); dairy goods including vegetarian cheese; free-range eggs; fresh compost-grown fruit and vegetables when available; bread from main shop in Belfast; natural medicines; nutritional supplements. (HSW, NAHS)

Horsham

M. J. Dench 46 Queen Street, Horsham, Sussex (0403 3403); open Monday to Saturday 8.45–13.00, 14.15–17.30, Monday and Thursday half days
A good greengrocer who sells his own organically grown vegetables and salad goods. Look out for his labels – some goods are imported. (SA)

Holland & Barrett 23 Freshwater Parade, Horsham, Sussex (0403 61739); bus 298 stops 25 yards from shop; open Monday to Saturday 9.00–17.30
For further information on the Holland & Barrett chain see entry for Amersham (page 180).

Nature's Way Ltd 34 Carfax, Horsham, Sussex; open Monday to Saturday 9.00–17.30
For further information on this chain see entry for Beckenham (page 187).

Hounslow

Hounslow Health Foods 103 Hanworth Road, Hounslow, Middlesex (01 572 2375); buses 110, 111, tube Hounslow East or Central; open Monday to Saturday 9.00–17.30
A shop incorporating snacks and sandwiches prepared on premises. Good range of dairy produce including goat's milk and cheese, free-range eggs. Grains, pulses, fruit juices, herbal remedies and teas also stocked. Books and magazines. Deliveries locally.

Hove

Healthy and Wise 108 Western Road, Hove, Sussex (0273 731324); buses to Brunswick Road; open Monday to Saturday 9.00–18.00
Has changed hands since the last edition, but carries much the same stock – branded wholefood groceries, vitamin and mineral supplements, dairy goods, an extensive range of honeys, books on health and diet. Deliveries throughout Brighton and Hove.

Holland & Barrett 31 George Street, Hove, Sussex (0703 731418); buses 1, 5, 5b, 6 19, 37, 49 to George Street; open Monday to Saturday 9.00–17.30
For further information on the Holland & Barrett chain see entry for Amersham (page 180).

Nature's Way Ltd 122 Church Road, Hove, Sussex (0273 722641); open Monday to Saturday 9.00–17.30
One of the chain of Nature's Way shops. For further information on this chain see entry for Beckenham (page 187).

Huddersfield

Coletta's Fine Food Health Store 123/125 Market Hall, Huddersfield, Yorkshire (0484 30211); open Monday and Tuesday 9.00–17.30, Wednesday 9.00–13.00, Thursday to Saturday 9.00–18.00
General selection of packaged health foods, wheat germs, raw sugars, fruit snacks, nuts, bran, muesli, honeys, herbs, teas and coffees, spices, compost-grown flours, pulses, cereals, vitamins, herbal medicines. (NAHS)

The Health Food Stores 20 Byram Arcade, off Westgate, Huddersfield, Yorkshire (0484 22991); near bus station; open Monday, Tuesday, Thursday, Friday 9.00–15.30, Saturday 9.00–17.00, closed Wednesday
This shop sells branded health foods, proprietary herbal products, vitamin and mineral supplements, honey, wholemeal bread, free-range eggs, and dairy goods including vegetarian cheeses. Deliveries in Huddersfield and surrounding district.

Lifespan 72a Westbourne Road, Marsh, Huddersfield, Yorkshire (0484 43050); buses 70, 71, 73 to Marsh Shops; open Monday, Tuesday, Thursday, Friday 9.30–18.00, Wednesday 9.00–12.30, Saturday 9.30–17.00
A pleasant shop, with good stocks of wholefoods sold loose, including their own muesli mix. Also juices, oils, raw cane sugars, cereal beverages, free-range eggs, wholemeal bread, lots of herbs and spices, natural toiletries. The shop has recently changed hands, and the new owners

welcome suggestions from their customers to increase the range of stock.

Peaceworks 58 Wakefield Road, Aspley, Huddersfield, W. Yorkshire (0484 23915); on most local bus routes; open Monday to Friday 10.00–18.00, Saturday 10.00–20.00
A cooperatively run venture. Selling wholefoods is only one of its many activities. As the name implies, the shop is a centre for 'alternative' activities – these include the local radical magazine, local Safe Energy group, and Men Against Sexism group. Stocks a large range of dried goods sold from bulk, e.g. grains and grain products, cereals including their own muesli mixes, pulses, teas, nuts, dried fruits, pasta, and over 100 herbs and spices. Also fruit juices, oils, honey, yeast, malt, spreads, fresh wholemeal bread and cakes. Lots of other things: natural soaps and shampoos, recycled paper goods, activist badges; a good choice of books on wholefoods, gardening, alternative lifestyles and religions; magazines of radical and ecological interest. (FNWC)

Hull

Holland & Barrett 62 Bond Street, Hull, Humberside (0482 27912); buses 37, 48, 57, 58; open Monday to Saturday 9.00–17.30
For further information on the Holland & Barrett chain see entry for Amersham (page 180).

Nature's Larder 10/14 Paragon Square, Hull, Humberside (0482 27853); buses to city centre; open Monday 8.30–17.30, Tuesday to Saturday 8.30–19.30
Hoping to extend their premises in the near future. A comprehensively stocked shop together with restaurant/café seating 140 persons; everything freshly prepared on premises.

T. H. Newton and Son 1028/1034 Anlaby High Road, Hull, Humberside (0482 561605); bus 63c to Anlaby Park; open Monday to Saturday 8.45–18.00, Thursday to 13.00, Saturday to 17.30
One of the old-style healthfood shops selling proprietary dried goods, vegetarian specialities, honeys, vitamin and nutritional supplements.

What Comes Naturally 513 Anlaby Road, Hull, Humberside (0482 564676); open Monday to Friday 9.00–17.30, Thursday to 18.00, Friday to 19.30, Saturday 8.00–17.30, Sunday 12.00–16.00

This shop specializes in homoeopathy, vitamin treatments, biochemical and herbal remedies. They are also skin and hair-care specialists using natural methods. Full range of wholefoods, natural juices, vegan foods, wholemeal bread and pizza fresh daily. Plans for wholefood snack bar. Books also sold. Deliveries within ten-mile radius, or goods sent anywhere by post.

J. Whiteman Home Brews and Health Foods 51 Princes Avenue, Hull, Humberside (0482 42092); buses 14, 20; open Monday 9.00–13.00, Tuesday to Saturday 9.30–17.00

Full range of proprietary health foods, including vitamins, herbs and spices; books; all types of home wine- and beer-making equipment. (HFA)

Hungerford

E. P. Spackman, 25 High Street, Hungerford, Berkshire (048 86 2593); open Monday to Friday 8.30–17.00, Saturday 8.30–13.00, closed Thursday

Established twenty years, this family business sells everything from caviar to soap powder and dog-food. Health food section with branded groceries, dairy produce, free-range eggs, wholemeal bread, vitamin and mineral supplements. Delicious fish and meat pâtés made on premises. Deliveries within ten-mile radius.

Hunstanton

B. & N. Alcock 32 High Street, Hunstanton, Norfolk (048 53 33524); buses to bus station; open Monday to Saturday 9.00–13.00, 14.15–16.00, Monday and Thursday to 13.00

Items like oats, bran, muesli, rice, barley, dried fruits sold from bulk. Also herbs and spices (sold by the ounce), fresh fruits and vegetables, vegetarian foods, prepacked food supplements, etc. Health books and magazines; natural cosmetics. You can always rely on a personal service, and items you may find unobtainable elsewhere will be posted.

Ilford

W. G. Findlay 131 Ilford Lane, Ilford, Essex (01 478 1337); tube to Barking; open Monday to Saturday 9.00–19.00
Sells all main brands of popular healthfoods; spring waters; Vingard wine-making supplies.

Food For Thought 4 Cameron Road, Seven Kings, Ilford, Essex (01 597 4388); buses 86, 193, 721; open Monday to Saturday 10.00–17.30, closed Thursday
Opened in 1977, one of the new-style wholefood shops selling from bulk rather than packaged food. Macrobiotic special foodstuffs; good range of dried wholefoods: like grains, cereals, herbs, spices; free-range eggs; dairy produce such as goat's milk and yoghurt, vegetarian cheeses. The shop mixes its own muesli. Deliveries within fifteen-mile radius.

Holland & Barrett 3 Centre Way, Ilford, Essex (01 553 3677); buses 25, 86, 129, 145, 148, 193, 199; open Monday to Saturday 9.00–17.30
For further information on the Holland & Barrett chain see entry for Amersham (page 180).

Ilkeston

W. R. Evans (Chemist) Ltd 69 Bath Street, Ilkeston, Derbyshire (0602 324748); open Monday to Saturday 9.00–17.30
A chemist selling some health foods, including wholemeal flour, grain coffees, mineral waters and juices, herbal medicines, vitamin supplements. (NPA)

Food For Thought 142 Station Road, Ilkeston, Derbyshire (0602 302485); open Monday to Saturday 9.00–17.30, closed Wednesday
Full, wide range of usual groceries including dry goods sold in bulk. Muesli, vegetable protein mixes and bread prepared on premises. The proprietors are interested in dietary reform: apart from literature on the subject, advice given on use of herbs in healing. Diet sheets have been prepared for some customers with successful results. Herbal mixtures also prepared on premises, (NAHS)

Inverness

Gordon's Health Food Shop 20 Baron Taylor Street, Inverness, Highland, Scotland (0463 33104);
This shop was originally a meal store; now stocks a good range of wholefoods including four kinds of freshly milled oatmeal, local honey, free-range eggs, locally baked bread, local dairy produce, fresh fruit and vegetables in season, vitamin and mineral supplements, herbal remedies.

Ipswich

Holland & Barrett 24 Westgate Street, Ipswich, Suffolk (0473 212271); buses to town hall 200 yards away; open Monday to Saturday 9.00–17.30, Wednesday 9.30–17.30
For further information on the Holland & Barrett chain see entry for Amersham (page 180).

Southern Health Foods 14 Dial Lane, Ipswich, Suffolk (0473 53165); open Monday to Saturday 9.00–17.30
One of the well-established Southern Health Foods chain. For further information on this chain see entry on Southern Health Foods in Aldershot (page 179).

Jarrow

Health Foods 1 Arndale Arcade, Jarrow, Tyne and Wear (0632 892308); open Monday to Saturday 9.00–17.00
Groceries and dried goods, full range of vitamin aids, diabetic products, gluten-free products. Wholemeal bread and cakes. Dairy products, including fresh goat's milk. (NAHS)

Keighley

'Be Healthy' Health Store New Market Hall, Low Street, Keighley, Yorkshire (0535 604125); buses to bus station; open Monday to Saturday 8.45–17.30, closed Thursday
This shop was bought by the present owner in 1970. General range of dried wholefoods, vitamins and minerals, with emphasis on herbal

remedies, herb teas, local honey. Big expansion into bulk sale of grains and pulses is planned. For some reason the shop is popular with local wrestlers. Deliveries within five-mile radius. (HSW)

Jewitt's Health Stores 52 Church Street, Keighley, Yorkshire (0535 604125); adjacent to bus station; open Monday to Saturday 9.00–17.30, early closing Tuesday
A herbalist since 1865; the present owner has been in the shop almost twenty-five years. He is still enthusiastic enough to be planning a full extension of the health services offered, which include the supply of National Health surgical appliances as well as herbal remedies – you are welcome to take the leaflets and try the samples available. Popular health food brands, grains, pulses, fresh wholemeal bread baked daily, vitamins, minerals, and a large section selling home-brew supplies. Local deliveries by arrangement. (HSW)

Kendal

The Lakes Health Food Store 105 Highgate, Kendal, Cumbria (0539 23439); buses to town centre; open Monday to Saturday 9.00–17.30
Opened in 1969, the shop caters for home-brewing specialists but also stocks packaged health foods, vitamin and mineral dietary supplements, dried fruit and nuts, etc. Deliveries in South Lakeland district.

Kenilworth

Country Market (Holland & Barrett) 44 The Square, Kenilworth, Warwickshire (0926 57764); buses 517, 537 to Randall Road; open Monday to Friday 9.00–17.30, Saturday 8.30–17.30
For further information on the Holland & Barrett chain see entry for Amersham (page 180).

Kettering

Holland & Barrett 54 Gold Street, Kettering, Northamptonshire (0536 2079); open Monday to Saturday 9.00–17.30
For further information on the Holland & Barrett chain see entry for Amersham (page 180).

Kidderminster

Holland & Barrett 8 King Charles Square, The Swan Centre, Kidderminster, Worcestershire (0562 4527); open Monday to Saturday 9.00–14.30
For further information on the Holland & Barrett chain see entry for Amersham (page 180).

Kings Lynn

Holland & Barrett 13 St Dominic Square, Kings Lynn, Norfolk (0553 5188); buses 336 (Wisbech), 337 (Peterborough), 401 (Hunstanton); open Monday to Saturday 8.45–17.15
For further information on the Holland & Barrett chain see entry for Amersham (page 180).

Kingston-upon-Thames

Fire and Rose Wholefoods 132 London Road, Kingston-upon-Thames, Surrey (01 549 6072); buses 85, 131, 213, 213a, 285 stop outside shop; open Monday to Saturday 10.00–18.00, Wednesday to 14.00
A new but very progressive partnership with many activities and a promise of more for the future. Apart from a very comprehensive range of all health foods from seeds to seaweed, beans to buckwheat, this partnership operates a stall in Kingston Monday market. Wholefood cookery demonstrations are organized; fresh food and wholewheat bread available three days a week. Possibility of starting a local wholefood magazine with hints, news, reviews, recipes, etc: support of course is needed from local people. A section is devoted to books, pamphlets,

badges, T-shirts, kitchenware, henna and other interesting things. Another possible project is expansion into a wholefood restaurant. (CHF, FOE, Veg. Soc.)

Holland & Barrett 4 Fife Road, Kingston-upon-Thames, Surrey (01 546 2394); buses 65, 71, 211, 281, 418 to Clarence/Eden Street; open Monday to Saturday 9.00–17.30, Friday to 18.00
For further information on the Holland & Barrett chain see entry for Amersham (page 180).

Lampeter

Mulberry Bush Wholefood Stores 2 Bridge Street, Lampeter, Dyfed, Wales (057 046 380); open Monday to Saturday 9.30–15.30, Wednesday and Saturday to 13.30.
This shop has been open since 1974 and has steadily expanded to provide excellent service. Goods sold in bulk to order; cakes, pies and wholemeal bread baked daily for the shop from locally ground flour; organically grown vegetables on sale in season. General stock includes grains, pulses, honeys, loose herbs, oils, confectionery, teas, olives, dried fruit and nuts, spices, biscuits; also tissue salts, vitamins, some natural medicines and cosmetics, incense, books and magazines on wholefoods and health.

Lancaster

Community Supplies The Warehouse, 78a Penny Street, Lancaster (0524 63021); open Monday to Saturday 10.00–18.00, closed Wednesday
A registered cooperative which is part of a community centre. They aim soon to open a café on one floor of the building which will also serve as a performance space. Meanwhile the shop sells from bulk in any quantity. Huge stock of dried goods including twenty kinds of bean, fifty of grain, fifteen of dried fruit, twenty of shelled nuts and seeds, fifteen spreads and sauces. Cold-pressed oils; beverages such as herb teas; special foods like palm sugar and dried mushrooms. Also dairy produce from a local farm, eighty herbs and spices, natural toiletries, kitchen hardware, recycled paper products. Deliveries, for a small charge, within fifteen-mile radius.

Largs

Largs Health Food Store 29 Nelson Street, Largs, Strathclyde, Scotland (0475 686167); open Monday to Saturday 9.00–17.30
Good range of bulk wholefoods, juices, yoghurt. Also vitamins, proteins, ginseng, etc. With increase of local interest on the Clyde coast, they hope to open a separate wholefood section should premises become available.

Leamington Spa

Holland & Barrett 55 Warwick Street, Leamington Spa, Warwickshire (0926 21775); buses 518, 544, 546, 547, 565, 566, 567; open Monday to Saturday 9.00–17.30
For further information on the Holland & Barrett chain see entry for Amersham (page 180).

Leatherhead

Leatherhead Wholefood Shop 34 High Street, Leatherhead, Surrey (037 23 73128); open Monday to Saturday 9.00–17.30, Wednesday to 13.00
Small bright shop with good stock of dried wholefood groceries; some fresh herbs, fruit and vegetables as available; dairy goods; free-range eggs; home-baked wholemeal cakes, scones, savouries; vitamins, minerals, herbal remedies, natural cosmetics; books on health subjects. (HSW, NAHS)

Leeds

Curtis Health Food Shop 4 Fish Street, Leeds, Yorkshire (0532 457948); buses 10, 19, 31 to Vicar Lane; open Monday to Saturday 9.00–17.30
A family-run business established in 1932. Sells packaged wholefood groceries, fresh bread, fresh farm eggs, yeast, raw sugars, herbs and spices (including good range of medicinal herbs sold by weight), vitamins, body-building and slimmers' supplies, natural remedies, ginseng, natural toiletries, books (especially on herbalism). Local deliveries

every Tuesday; postal service all over Britain. Bulk prices quoted on request. (NAHS)

The Health Stores 20 East Gate, Leeds, Yorkshire (0532 451095); most buses; open Monday to Saturday 9.00–18.00
A well-stocked shop. As well as the usual groceries and dry goods, a certain amount of dairy produce; bread; free-range eggs; natural remedies; nutritional supplements; a few fresh vegetables, locally grown. Deliveries in Leeds area. (NAHS)

Holland & Barrett 22 The Bond Street Centre, Leeds, Yorkshire; open Monday to Saturday 9.00–17.30
For further information on the Holland & Barrett chain see entry for Amersham (page 180).

Leeds Wholefoods Ltd 182 Woodhouse Lane, Leeds, Yorkshire (0532 33213); buses 1, 4, 30, 93, 96; open Monday to Friday 8.30–17.30, Saturday 10.00–14.00
This shop, opened in 1973, specializes in real wholefoods rather than packaged panaceas: no tablets, potions or other preparations sold. Extended range of dried goods. Useful take-away selling soups, curries, sandwiches and other snacks freshly made on premises. A café may soon be opened upstairs.

Realfoods (Holland & Barrett) 11 Arndale Centre, Crossgates, Leeds, Yorkshire (0532 601409); buses 24, 35, 38, 40 to Arndale Centre; open Monday to Saturday 9.00–17.30
For further information on the Holland & Barrett chain see entry for Amersham (page 180).

Leicester

Downey's Health and Wholefood Centre 143 Evington Road, Leicester (0533 736108); open Monday to Friday 9.00–17.30
Stocks the usual range of health food products.

Leicester Health Food Centre 2 and 4 Charles Street, Leicester; 3 Odeon Arcade, Market Place, Leicester (0533 21939); buses to town centre; both open Tuesday to Saturday 9.00–17.30
In operation since 1935. Stocks include proprietary health foods, wholewheat breads, flour, dried fruit, organically grown products. A range of over sixty different types of honey. (NAHS)

Leigh

Healthylife Centre 51 King Street, Leigh, Lancashire (0942 606355); open Monday to Saturday 9.00–17.30
Open since 1969, the shop stocks packaged rather than fresh foods; honey, flours, cereal products, grains and pulses, oils, herbal teas, slimming aids, diet foods for diabetics and vegetarians, vitamins and natural remedies, body-building products. The business manufactures the Nature Boy range. (NAHS)

Leigh Health Food Store 4a Lord Street, Leigh, Lancashire (0942 671068); bus 599 stops opposite; open Monday to Saturday 9.00–17.30
This self-service store came into new ownership in 1977. Stocks a full range of dried wholefood goods: muesli mixed on premises, organically grown flours and breakfast cereals, beans, pulses, bran, wheat germ, dried fruits, herbs. Also pure fruit juices, mineral waters, herb teas, decaffeinated coffees, raw sugar preserves, diabetic foods, vitamin and mineral supplements, dairy goods including goat's milk. Diet sheets prepared by the owner and supplied to customers; deliveries in Leigh, Atherton and Hindley Green.

Leominster

Nitty Gritty Grain Store 9 School Lane, Leominster, Herefordshire (056 887 670); open Monday to Saturday 10.00–17.00, Thursday to 13.00
As well as your purchases you can also get information on local events and functions. Wholefoods sold loose. Wide range of flours, tea, coffee. Fresh vegetables available in season together with local honey, homemade jam, wholemeal bread fresh daily. Plans for a snack bar sometime in the future.

Lerwick

Health Food Store 12 Commercial Road, Lerwick, Shetland (0959 2924); open Monday to Saturday 9.00–13.00, 14.15–17.30, Wednesday to 13.00

Self-service shop near busy harbour. Stocks groceries, dried goods, bread from local bakery, free-range eggs when available, fresh fruit, fresh locally grown vegetables, natural medicines, nutritional supplements. Appointments can be made through the shop for regular visits from an acupuncturist and advocate of megavitamin therapy.

Lewes

Lansdown House Health Foods 10 Lansdown Place, Lewes, Sussex (079 16 4681); open Monday to Friday 9.00–17.30, Saturday to 13.00

This shop, pretty and informal, looks more like a kitchen than a store. They sell all kinds of organically grown flour, dried fruits, raw sugars; dairy products from local farm, including yoghurt and goat's milk; herbs and herbal remedies, with some fresh herbs in season; wholemeal bread, pastries, savouries; compost-grown fruit and vegetables in season from local suppliers; vitamins, mineral supplements, natural remedies, natural cosmetics.

Lichfield

The Granary 17 Market Street, Lichfield, Staffordshire (054 32 23404); buses, including country bus 112, to Birmingham Road; open Monday to Saturday 9.00–17.30, Wednesday to 13.00

An attractive shop, housed in an old Georgian pharmacy with bow windows. Now wholly a health food shop. Stocks usual choice of packaged health foods, dietary and vegetarian specialities, vitamins, mineral supplements, herbs, spices. The shop is being extended.

Lincoln

Green's Health Foods Ltd 4 Exchange Arcade, Lincoln (0522 24874); buses to city centre; open Monday to Friday 8.30–17.30, Wednesday half day

Open on the same site since 1910, this shop is now owned by a couple who see their role as encouraging the ecological and medicinal benefits of wholefoods. Frequent additions to stock include wholemeal cereals and flours (compost-grown where possible), yoghurt, cheese, pulses, shelled nuts, preserves, dried fruit, vitamins and other nutritional supplements. Also natural cosmetics; herbal, biochemical and some anthroposophical medicines; slimming, diabetic and body-building products; home-brewing equipment; incense. Deliveries within Lincoln and arrangements for sending goods by bus or post to outlying districts. (BHMA, HSW, NAHS)

Pulse 25 Corporation Street, Lincoln (0522 28666); buses to St Mary's Street depot; open Monday to Saturday 9.00–17.30, Wednesday half day

The shop, originally opened by the Buddha Maitreyas Sangha, has recently been taken over by people who intend to carry on the policy of selling wholefoods and Zen macrobiotic foods as economically as possible. Stocks all the usual flours, cereals and grains (including several kinds of brown rice), organically grown where possible; most kinds of pulses and seeds; cooking oils; yeast; herbs and spices; nuts and dried fruits; natural teas; a good range of honey; wholemeal bread; dairy produce; yoghurt. Also specializes in Japanese foods like teas, misos, soya and other sauces, sea vegetables, umeboshi plums. A selection of magazines and books on health and Buddhist subjects.

Liskeard

Ough & Sons 10 Market Street, Liskeard, Cornwall (0579 43253); buses to Bay Tree Hill; open Monday to Saturday 8.30–17.30, closed Wednesday

A truly traditional shop which, opened in 1847, still dispenses with such modern notions as a refrigerator. Provisions are stored in the cool of the shop's cellar. As well as general groceries like green and smoked bacon, freshly ground coffee, they sell a very wide range of whole and health foods both packaged and from bulk: grains, pulses, cereals, herbs, spices, dairy goods (including cheese and goat's milk yoghurt delivered from a local farm), and a wide selection of teas. Deliveries within a ten-mile radius.

Liverpool

Atherton's Herbal Stores 15 St Oswald's Street, Old Swan, Liverpool, Merseyside (051 228 9785); open Monday to Saturday 9.00–13.00, 14.15–17.30, Wednesday 9.30–13.00
Established in 1933, specializes in herbs and herbal remedies, as the name implies. Advice available on diet and nutrition. Also medicinal supplements, natural remedies, packaged health food groceries. (BHMA).

The Happy Nut House 27 Warbreck Moor, Liverpool, Merseyside (051 523 3111); open Monday to Saturday 9.00–17.30, Wednesday to 13.00
This is a branch of Hillstart Ltd's chain of health food shops in the Midlands and the North of England. Further information on this chain see entry on The Happy Nut House in Birkenhead (page 192).

The Happy Nut House 48/50 Whitechapel, Liverpool, Merseyside (051 708 7418); open Monday to Saturday 9.00–17.30
This is a branch of Hillstart Ltd's chain of health food shops in the Midlands and the North of England. For further information on this chain see entry on The Happy Nut House in Birkenhead (page 192).

Holland & Barrett In Owen & Owen, Clayton Square, Liverpool, Merseyside (051 709 6060 ext. 91); open Monday to Saturday 9.00–17.30
For further information on the Holland & Barrett chain see entry for Amersham (page 180).

Horsfield's Health Foods 81/85 Dale Street, Liverpool, Merseyside (051 236 2338); open Monday to Saturday 9.00–17.30
Claims to be largest health food store in Liverpool. Sells cereals from its own watermill in North Wales, as well as dairy goods, fresh wholemeal bread, packaged healthfood groceries, herbal remedies and cosmetics,

vitamin supplements. A counter for take-away snacks, with freshly prepared wholemeal sandwiches, soup, squeezed fruit juices.

Reform Health Foods (Holland & Barrett) 17 Whitechapel, Liverpool, Merseyside (051 236 8911); buses to city centre (nearest stop Church Street or Lord Street); open Monday to Friday 9.00–17.30, Saturday 8.30–17.00
For further information on the Holland & Barrett chain see entry for Amersham (page 180).

Llandeilo

The Regency Health Shop 109 Rhosmaen Street, Llandeilo, Dyfed, Wales (05582 3556); buses to Llandeilo church; open Monday to Friday 9.00–17.30, Thursday half day
Principally a delicatessen. Owner has expanded the health food side of the business as a service to the community. Accent is on health. Large range of vitamins and minerals, as well as packaged wholefoods, some vegetarian specialities. Intends to expand health aspect by opening chemist's next door. Deliveries locally and to Byfleet.

Llandrindod Wells

Van's Good Food Shop Clovelly, High Street, Llandrindod Wells, Powys, Wales (0597 3320); all buses and trains; open Monday to Saturday 9.30–13.30, 14.00–17.30, Wednesday 14.00
A restaurant and shop combined with wholesale deliveries to northeast, Midlands, south-west Wales and West Midlands from Bristol and Birmingham to Chester. The restaurant, open 9.30–16.30, offers a three-course meal daily, with main course either vegetarian or meat; all food prepared daily on premises. Usual range of wholefoods, herbs, cheeses pulses, grains, etc.; also daily supplies of fresh vegetables, own muesli and soup mixes.

Llandudno

Lovell's Health Store 8 Gloddaeth Street, Llandudno, Wales (0492 76546); open Monday to Saturday 9.00–17.30, Wednesday to 13.00 (winter)
Stocks most brand-name prepacked goods. Will try to order anything asked for, if possible.

Llanelli

Jelf's Health and Herbal Store 17 Central Precinct, Llanelli, Dyfed, Wales (055 42 3071); open Monday to Saturday 9.00–17.30
Opened early this century by Professor Price, this is one of the oldest known herbalists in South Wales. Very wide range of dried as well as culinary herbs and spices. Full range of all well-known brands of health foods. Under the present ownership, there is now also a wholefood and wine-making shop in Bridge Street.

London E2

Friends Foods 51 Roman Road, Bethnal Green, London E2 (01 981 1255); buses 8, 106 to Roman Road; tube to Bethnal Green; open Monday to Friday 10.00–18.00, Saturday 9.00–18.00
The main shop of this enterprise run on cooperative lines is in the old Bethnal Green fire-station. They sell a very large variety of goods, now including speciality items like flaked almonds, mixed fruit peel, miso, muscovado sugar. Large range of teas and grain coffees. Free-range eggs and goat's milk products. Opened in 1978, has expanded along the street to include a bulk sales wholefood shop, with a huge range of dried goods (take your own containers if possible; a deposit is charged on jars), and a wholefood café open all day. The entire business offers good service and very low prices. They are just starting deliveries, so please ring to inquire. New price and stock lists provided monthly.

E7

Health-wise 310a Romford Road, Forest Gate, London E7 (01 534 8534); buses 25, 86, 162 outside door; tube Forest Gate and Upton Park; open Tuesday to Saturday 11.00–17.00
Apart from the normal range of wholefoods, e.g. grains, bread, eggs, honey, this firm has earned a reputation for its specialist knowledge of nutrition/vitamin therapy: stocks, herbal remedies, vitamins, minerals, natural cosmetic products, vegetarian and specialist dietary items.

Phillips Stores 162 Boleyn Road, Forest Gate, London E7 (01 472 3190); buses 40, 51, 58 to Green Street; tube to Upton Park; open Monday to Friday 8.30–19.00
This long-established shop is well stocked with all the normal health food lines; dairy produce; fresh wholemeal bread; free-range eggs; fruit, vegetables and herbs from a farm in Essex. Good choice of honeys, vitamins, dietary supplements, fruit juices, natural cosmetics. Although self-service, the staff are always willing to help or advise. Deliveries within ten-mile radius.

E8

Towards Jupiter 191 Mare Street, Hackney, London E8 (01 985 5394); buses 253, 106, 55; nearest tube, Bethnal Green; open Monday to Saturday 9.30–17.30
The food in this shop is grown according to the principles laid down in 1925 by Rudolf Steiner: the Demeter system of agriculture incorporates the forces of the moon and planets into crops by paying attention to sowing and harvest times. All food here is at least organically grown. They also sell Weleda and Wala natural cosmetics, wooden crafts and toys, Rudolf Steiner books.

E10

Harvan's Health Food Store 565 Lea Bridge Road, Leyton, London E10 (01 539 3245); buses to Bakers Arms corner; tube to Leyton/Walthamstow; open Monday to Saturday 9.00–18.00

'If we haven't got it, we'll get it' is the policy of this enthusiastically run store. They claim to have the largest range of foods, dairy produce, vitamins, etc. in the area. Sarsparilla wine made on premises. Plans to expand in the near future. (NAHS)

E12

J. Baker 690 Romford Road, Manor Park, London E12 (01 514 2800); buses 25, 86; tube to Manor Park; open Monday to Saturday 9.00–17.30, Thursday to 13.00
A well-stocked shop with full range of packaged health foods, honey, oils, dairy products (including goat's milk and yoghurt), wholewheat bread delivered daily, herbal remedies, vitamin supplements. Deliveries 'anywhere, within reason'.

E17

Tate's Health Food Stores 613 Forest Road, London E17 (01 527 1356); buses 21, 34, 55, 69, 123, 275, 276, 278 all within 50 yards; tube to Leyton; open Monday to Saturday 9.00–17.30, Thursday to 13.00
Normal range of branded whole and health foods, vegetarian specialities, vitamins, natural cosmetics, fruit juices, herbal tobacco and cigarettes.

E18

Woodford Health Foods 83 High Road, South Woodford, London E18 (01 989 5134); bus 20; open Monday to Saturday 8.30–17.30
The shop has a handy car park at the rear. Sells full range of packaged wholefoods, free-range eggs, natural ice-cream and other dairy products, vitamin and mineral supplements, natural cosmetics. A good take-away service selling salads, soups, sandwiches, nut cutlets, wholefood cakes, all prepared fresh on premises. Deliveries within five-mile radius.

EC1

The Health and Beauty Bar 56/58 Leather Lane, London EC1 (01 242 9685); all buses through Holborn; tube to Chancery Lane or Farringdon Street; open Monday to Thursday 8.30–17.30, Friday 9.00–18.00
Quite useful if you have to shop in the City because you work there, but concentrates more on health aids than good food. Stocks packaged groceries, large range of vitamin and mineral supplements, diet aids. Good choice of natural toiletries and cosmetics.

Sunwheel Nature Foods 196 Old Street, London EC1 (01 250 1708); bus or tube to Old Street; open Monday to Saturday 10.00–19.00
Situated in a beautiful converted industrial building which somehow has the feeling of a mock-gothic church because of expanses of dark wood, arched doorways and stained-glass front windows, the shop sells produce to the highest standards at prices unfortunately a little above average. Some of the staff are of Eastern origin and this is reflected in the stock (and the previous name, East-West Foods). Only genuine wholefoods sold, mostly packaged on premises, and emphasis is on organically grown staple foods: nothing contains preservatives, chemicals, or sugar. Macrobiotic special foods; organically grown fruit and vegetables; fresh tofu and bean sprouts; wholegrain breads, pastries and sandwiches; oils and spreads; books on health food and cooking; some very attractive cookware. Extension of stock is planned; it would be a pity if the shop's spaciousness were lost. The connection of the shop with the East-West Institute next door ensures a provision of advice and recipe leaflets, and access to classes on cooking, yoga, nutrition, etc. A restaurant is run by the shop's owners at 3 Chalk Farm Road NW1; the East-West Institute next door to the shop also has two health food restaurants. Free off-street parking and loading.

EC3

Merebonn Ltd 1 Philpot Lane, London EC3 (01 626 0900); buses 8a, 35 40, 47, 48; tube to Bank or Momument; open Monday to Friday 9.00–17.30
Packaged wholefoods, thirty brands of honey, and nutritional supplements in this little shop. No perishable goods.

EC4

City Wholefoods 73 Queen Victoria Street, London EC4 (01 248 3170); buses 6, 9, 11, 15, 23, 76, 141 stop outside; tube to Mansion House, St Paul's, Cannon Street; open Monday to Saturday 8.30–17.30
Hoping to expand to new premises in the near future. A good takeaway service of wholewheat rolls, sandwiches, cakes. In addition: the normal stocks of herbs, remedies, vitamins, naturally grown vegetables.

N1

Barnsbury Health Foods 285 Caledonian Road, London N1 (01 607 7344); buses 14, 45, 168a, 221, 259, 263; King's Cross station or Caledonian Road station; open Monday to Saturday 10.00–18.00
Possibly the newest shop in the book. In the new mould of cheap bulk wholefoods sold loose. Stock is wide: grains, pulses, fruits, nuts, flours, herbs, herbal remedies, juices, preserves, beverages, dairy goods (including vegetarian cheese), cooking utensils, books on health. Also a take-away counter serving freshly prepared savouries, cakes and snacks baked on premises. Deliveries within three-mile radius. (HSW)

N6

Earth Exchange 213 Archway Road, Highgate, London N6 (01 340 6407); between Archway and Highgate tube stations; buses 43, 104, 134, 263 to Shaftesbury pub; open (shop) every day 12.00–19.00 except Wednesday and Thursday; (café) Monday and Tuesday 12.00–19.00, Friday to Sunday 12.00–22.00
This collective, comprising several enterprises on the same site, is run by young people heavily into creative situations and fulfilling possibilities. Many plans for expansion which will 'unfold when the time is right'. For the present, their wholefood shop is well stocked with dried goods sold in bulk: medicinal and culinary herbs, pulses, grains, cereals, flours, teas, grain coffees, nuts, dried fruits; honeys and other spreads; sea vegetables; free-range eggs; wholemeal bread baked locally; plant milk; fresh fruit and vegetables; oils, seeds, salts. Vegan special foods are stocked. Also natural cosmetics; incense; some wines; a large, com-

prehensive book and magazine section on subjects related to health, wholefood cooking, ecology and alternative politics. A craftshop with local and Third World crafts, and a workshop for weaving and macrame. Massage service available if you book in advance. Meetings on various related topics are arranged. Or you can relax in the café, which is exclusively vegetarian and serves vegan dishes. All food prepared freshly on premises from organically grown produce: main dishes, soups, snacks, desserts (including wholemeal cakes). For summer there is a garden with some seating.

N7

Holland & Barrett 452 Holloway Road, London N7 (01 607 3933); buses 43, 104, 172, 271; tube to Holloway Road; open Monday to Saturday 9.00–17.30
For further information on the Holland & Barrett chain see entry for Amersham (page 180).

N8

The Haelan Centre 39 Park Road, Crouch End, London N8 (01 340 4258); buses 41, W2 W7 to Crouch End Broadway (clock tower); tube to Finsbury Park
Natural medicines sold here and advice freely given by staff who retain some of the better qualities of the 1960s lifestyle and attitudes. Wholefoods and packaged health foods, macrobiotic supplies, free-range eggs, cow and goat dairy produce, some fresh compost-grown vegetables, herbs, herbal remedies, natural cosmetics and toiletries. Also aids to a better life in the form of incense, prayer candles. Books for physical and spiritual improvement. Local deliveries by arrangement.

N9

Holland & Barrett 7 North Mall, Edmonton, London N9 (01 803 4315); buses 149, 259, 279, 283 to Edmonton Green shopping precinct; tube to Seven Sisters; open Monday to Saturday 9.00–17.30, Thursday to 13.00

For further information on the Holland & Barrett chain see entry for
Amersham (page 180).

N10

Holland & Barrett 121 Muswell Hill Road, London N10 (01 444 8126);
buses 102, 144 stop outside shop; tube to East Finchley; open Monday
to Saturday 9.00–17.30
For further information on the Holland & Barrett chain see entry for
Amersham (page 180).

N11

Warehouse Community Foods Ltd Unit 1, Brunswick Industrial Park,
Brunswick Park Road, London N11 (01 368 9215); tube to Arnos
Grove; open Tuesday to Friday 11.00–12.30, 14.00–17.30, Saturday
12.00–16.00
Completely comprehensive coverage as expected from a wholesale
organization dealing with shops, cooperatives and bulk-buying groups.
Cash-and-carry basis only, with a minimum order value of £30 at the
time of writing.

N13

Holland & Barrett 332 Green Lanes, Palmers Green, London N13 (01
886 6769); buses 29, 123, W4; tube to Wood Green; open Monday to
Saturday 9.00–17.30
For further information on the Holland & Barrett chain see entry for
Amersham (page 180).

N14

Health and Beauty Food Store 42 High Street, Southgate, London N14
(01 886 1990); buses 298, 298a to Medway Corner; tube to Southgate;

open Monday to Wednesday 9.00–17.30, Thursday to 13.00, Friday to 18.00, Saturday to 17.00

There has been a health food store here for thirty years; the present management took over two years ago. The shop is still run with a true dedication to health: a Mecca for health addicts and recommended to novices who will be warmly welcomed and advised. A herbal-healing centre, and an enormous range of goods which you may not find elsewhere. Wholemeal bread fresh daily (with some speciality breads), rolls and scones at weekends; Loseley dairy goods, including goat's milk, yoghurt, cheese; free-range eggs; compost-grown fruit and vegetables in season; macrobiotic foods; herbs for culinary as well as medicinal use. Huge choice of nutritional supplements, natural remedies, natural cosmetics.

N21

The Lima Shop 65 Station Road, Winchmore Hill, London N21 (01 360 7143); train to Hertford from King's Cross; open Monday, Tuesday 9.30–16.30, Thursday, Friday 9.30–17.00, Saturday 9.30–14.30, closed Wednesday

An energetically run shop, opened in the early 1960s by a lady who believes in her produce: she emphasizes that the eggs are genuinely free-range. Stocks vegetarian cheese, goat's milk, yoghurt, gluten-free foods and other diet specialities, dried fruit, nuts, herbal and biochemical remedies, natural cosmetics, and the normal packaged groceries. Also books on health and wholefoods, and a growing gifts and crafts section with items like hop pillows, potpourri, Liberty print aprons.

NW1

Chalk Farm Nutrition Centre 40/42 Chalk Farm Road, London NW1 (01 485 0116); bus 24; tube to Chalk Farm or Camden Town; train to Primrose Hill (Broad Street line); open every day 9.00–21.00

Opened in 1977, this store helps to fill the gap left by the community stores after dispersal of the Prince of Wales Crescent squat. Health food section stocks cereals, pulses, grains, wholefoods for babies, macrobiotic special foods, raw sugar preserves, honeys, dried fruit, shelled nuts, herbs both dried and growing, soya-derivative foods, natural cosmetics and toiletries, vitamins, herbal and homoeopathic remedies, medicinal plant preparations, and a selection of books on health and health food topics.

Sesame 128 Regents Park Road, London NW1 (01 586 3779); buses 31, 74, 74b; tube to Chalk Farm; open Monday to Saturday 9.00–19.00
An informal and enterprising shop with tables in front where the large take-away range of soups, salads, quiches, sandwiches, cakes, etc. may be eaten. They are trying to expand the take-away choice to include a larger variety of beverages, such as herb teas, as well as more food. Convenient for the popular picnicking areas of Regents Park, Primrose Hill, and the Zoo. Very full range of wholefoods, especially dried goods flours, grains, pulses, nuts. Also macrobiotic speciality foods, dairy produce (including ice-cream), herbs, teas, coffee, vitamins, natural cosmetics, organically grown vegetables, books. (NAHS)

NW2

Holland & Barrett 86 Camden High Street, London NW2 (01 388 0808) buses 29, 253, tube to Camden Town; open Monday to Saturday 9.00–17.30
For further information on the Holland & Barrett chain see entry for Amersham (page 180).

Holland & Barrett 63 Cricklewood Broadway, London, NW2 (01 450 8359); buses 16, 32, C11 stop 100 yards from shop; tube to Kilburn; open Monday to Saturday 9.00–17.30, Thursday to 13.00
For further information on the Holland & Barrett chain see entry for Amersham (page 180).

NW3

Culpeper 9 Flask Walk, Hampstead, London NW3 (01 794 7263); open Monday to Saturday 10.00–18.00

For further information on the Culpeper chain see entry for Bath (page 186).

Holland & Barrett 14 Northways Parade, Swiss Cottage, London NW3 (01 722 5920); buses 268, C11 stop just outside; tube to Swiss Cottage; open Monday to Saturday 9.00–17.30
For further information on the Holland & Barrett chain see entry for Amersham (page 180).

Lodders at the Coffee and Tea Warehouse 2 Flask Walk, Hampstead London NW3 (01 435 0959); bus 268, or tube to Hampstead; open Monday to Saturday 9.30–17.30, Thursday to 13.00
A traditional shop with village charm, filled with the aroma of freshly ground coffee. Tea and coffee are the speciality, but they also stock a huge variety of other top-quality goods: over 100 kinds of herbs, several sorts of bread, dairy produce, grains, beans, flours (all packed on premises), home-made preserves, compost-grown fruit and vegetables in season, crispbreads, free-range eggs, and items like natural cosmetics – a particularly good selection. Vitamins, honey and yoghurt of rare kinds. Also coffee-making equipment. Salad-filled pitta bread, vegetables quiches, cakes freshly made to take away. The new owners are continuing the tradition of personal service and high quality, and the shop continues to be thronged with the flower of the British literary and theatrical worlds.

Pippin 83/84 Hampstead High Street, London NW3 (01 435 6434); open Monday to Saturday 9.30–18.30, Sunday 10.30–17.30
This shop offers the usual range of health food supplies but also has its own bakery; various types of bread and cakes available fresh daily, including salt-free loaves, sugarless and eggless cakes. A vegetarian restaurant adjoins the shop and is open seven days a week, 10.00–12.00. Willing to cater for parties.

NW4

Holland & Barrett Unit W16, Brent Cross Shopping Centre, London NW4 (01 202 8669); tube to Brent Cross; open Monday to Friday 10.00–20.00, Saturday 9.00–18.00

For further information on the Holland & Barrett chain see entry for Amersham (page 180).

NW7

Healthways 36a Broadways, Mill Hill, London NW7 (01 959 0771); buses 62, 221, 240 or tube to Mill Hill Broadway; open Monday to Saturday 9.15–17.30, Thursday to 13.00
Vitamin and mineral supplements, herbal remedies and cosmetics. Also a variety of yoghurt, ice-cream, goat's milk, free-range eggs, wholemeal bread. (NAHS).

NW8

Holland & Barrett 55 St John's Wood High Street, London NW8 (01 586 5494); buses 2, 2b, 13, 26, 113 to Wellington Road; tube to St John's Wood; open Monday to Saturday 9.00–17.30
For further information on the Holland & Barrett chain see entry for Amersham (page 180).

NW10

Whole Earth 88 Cobbold Road, London NW10 (01 459 7024); buses 52, 206, 266, 297 to Willesden Lane, Willesden Garage; tube to Dollis Hill; open Monday to Friday 9.00–17.30, Saturday 10.00–17.00
Of interest particularly to retailers, cooperatives, or bulk-buying associations as they sell only in bulk, 'mini-bulk', or cases of prepacks. Extensive range of cereals, pulses, pasta, flakes, flours, honeys, juices, spreads, nuts, dried fruits. Other items include Japanese foods, cosmetics, condiments, seeds, literature, seaweed, household equipment, packaging materials.

NW11

Holland & Barrett 81 Golders Green, London NW11 (01 455 5811); buses 13, 26, 83, 183 240; tube Golders Green; open Monday to Saturday 9.00–17.30
For further information on the Holland & Barrett chain see entry for Amersham (page 180).

Holland & Barrett 17 Temple Fortune Parade, London NW11 (01 458 6087); buses 2, 26, 102, 260; tube Golders Green; open Monday to Saturday 9.00–17.30, Tuesday 9.30–17.30
For further information on the Holland & Barrett chain see entry for Amersham (page 180).

SE17

G. Baldwin & Co 171 Walworth Road, London SE17 (01 703 5550); buses 12, 45, 68, 171; tube to Elephant and Castle; open Tuesday to Saturday 9.00–17.30
Opened more than six years ago, this shop has built up a comprehensive stock of health foods. Sells goat's milk but no other farm produce or fresh vegetables. (NAHF)

SE18

The Pharmacy 6 Woolwich New Road, Woolwich, London SE18 (01 854 3684); train to Woolwich Arsenal; open Monday to Saturday 9.00–17.30
A general chemist stocking in the interest of health a few packaged health foods such as groceries; some natural remedies and nutritional supplements.

SE23

Provender 103 Dartmouth Road, Forest Hill, London SE23 (01 699 4046); buses 12, 22, 176, 185; open Monday, Tuesday, Thursday, Friday 10.00–17.30, Saturday 9.30–17.30

This is a welcome and relatively new shop selling food from bulk. Very good range of dried goods: muesli, flours, pulses, nuts, cereals, grains, herbs, spices. Also free-range eggs, all kinds of honey, yeast, raw sugar, live yoghurt, fresh wholemeal bread. Good selection of take-away food prepared on premises daily: salads, soups, savouries, and snacks like hummus.

SE25

Healthmart 27 Portland Road, South Norwood, London SE25 (01 654 5362); buses 12, 12a, 196, 196a, 197, 197a; train to Norwood Junction; open Monday to Friday 9.00–19.30
Large stocks of wide range of wholefoods and health foods. Nutritional and therapeutic preparations. A careful selection of herbal products for cooking and medicinal use. Planning to open a wholefood restaurant shortly to extend to the international market the wholesale deliveries already covering all of London.

SE26

Holland & Barrett 33 Winslade Way, Catford, London SE6 (01 690 3903); buses 1, 36, 36b, 47, 54, 75, 108b, 124, 160, 180, 185 to Catford High Street (200 yards away); open Monday to Saturday 9.00–17.30
For further information on the Holland & Barrett chain see entry for Amersham (page 180).

Lewisham World Shop 1 and 201 Sydenham Road, Sydenham, London SE26 (01 778 6665); train to Sydenham; open (1) Tuesday, Thursday, Friday 10.00–13.00, 14.00–18.00, Saturday 10.00–17.00; (201) irregular hours – sometimes Monday 10.00–17.00
Less a food shop than a volunteer educational project which hopes, by drawing in people to buy goods, to promote an awareness of ecological and development issues. Attempts being made to find larger premises where all activities can be united; even so, volunteers still needed to help staff existing enterprises. Sell a fair range of dried wholefoods (but not bread), as well as recycled paper products, books, badges, magazines, posters and campaigning T-shirts.

SE27

Nature's Larder 340 Norwood Road, West Norwood, London SE27 (01 670 0288); buses 2b, 68, 172, 191; open Monday to Friday 9.15–17.20, Wednesday to 13.00
The general range of packaged health foods, good selection of honeys, Loseley dairy produce, Marriage's flours, locally baked wholemeal bread, vitamins. Also herbal remedies and cosmetics; books on health subjects. (HSW, NAHS)

SW1

Harrods (Health Food Counter, Ground Floor) Brompton Road, London SW1 (01 730 1234); open Monday to Saturday 9.00–17.00, Wednesday to 19.00, Saturday to 18.00
This small but extremely handy counter can be found at the end of the fruit and vegetable department. Stocks a very wide range of the usual health food shop items with, it seems, an unusual choice of imported remedies, cures and other medicaments for exhausted Harrods shoppers. Wholemeal bread at nearby bakery counter (they make their own).

Holland & Barrett 10 Warwick Way, Pimlico, London SW1 (01 834 4796); buses 2, 36; tube to Victoria and Pimlico; open Monday to Saturday 9.00–17.30
For further information on the Holland & Barrett chain see entry for Amersham (page 180).

SW4

Linicers 90 Clapham High Street, London SW4 (01 720 4567); bus 19 or tube to Clapham Common; open Monday to Saturday 9.00–19.00
Useful late opening hours. Sells branded health foods, natural medicines and vitamins, general and continental provisions.

SW6

Windmill Wholefoods 486 Fulham Road, London SW6 (01 381 1281);

buses 11, 14, 28; tube to Fulham Broadway (2 minutes); open Monday to Saturday 9.30–18.00, Friday to 19.00

Situated right on the Broadway, this shop is well worth the bus fare from Putney or the top end of Fulham Road; not only bulk prices are cheap. Goods rather haphazardly packed on shelves, in 1 lb and 3 lb bags; sacks out at the back for bigger quantities. Ask if you can't find what you want – it may well be there somewhere! Brown rice cheaper by 17p per lb than in the nearest Holland and Barrett branch, and the other grains, pulses, raw sugars, etc. are also a bargain. Packs own unsweetened muesli and muesli base, the former full of ingredients like chopped figs and dates. Over 200 herbs and spices for medicinal and culinary use; homoeopathic remedies; usual vitamin and mineral supplements. Also natural soaps and cosmetics, wholemeal cakes to take away, fresh fruit and vegetables in season, dairy goods (including goat's yoghurt), macrobiotic supplies. Books sold – but our researcher did not know this until she inquired for the purpose of this entry; similarly, the existence of a wholefood vegetarian restaurant had escaped her notice although she uses the shop frequently! The owners, who opened the shop last year, should stop hiding their light under a bushel. The bulk business is to be extended. Deliveries for large orders all over London.

SW7

Holland & Barrett 12 Gloucester Road, London SW7 (01 584 0372); buses 49, 52 to Palace Gate; tube to Gloucester Road; open Monday Wednesday, Saturday 9.00–18.00, Thursday to 13.00, Friday to 18.30
For further information on the Holland & Barrett chain see entry for Amersham (page 180).

SW10

Holland & Barrett 220 Fulham Road, London SW10 (01 351 3904); bus 14 to St Stephen's hospital; tube to South Kensington; open Monday to Saturday 9.00–17.30
For further information on the Holland & Barrett chain see entry for Amersham (page 180).

L'Herbier de Provence 341 Fulham Road, London SW10 (01 352 0012); bus 14 to ABC cinema; open Monday to Friday 10.00–13.00, 14.00–18.00, Saturday 10.00–18.00

A welcome addition to the Fulham Road, the shop exudes exotic scents for several yards along the pavement, enticing customers. Passers-by stop to look through the plate-glass window, surveying the sacks of herbs displayed all over the floor: more than 200 varieties for medicinal and culinary use, from the shop's supplier in Provence. Not cheap, but delicious; very useful for gifts, since packs with instructions are attractively wrapped and displayed. A mixed packet useful for pot roasts is thyme – rather than rosemary – based. Ranged on shelves all round the shop are jars of olive oil, honey, comfits and jams, syrups, essential oils; gifts like lavender bags; a good range of natural cosmetics including Marseilles soap.

SW11

All Manna Natural Foods 256 Battersea Park Road, London SW11 (01 223 9211); open Monday to Saturday 10.00–18.00

Recently opened by the hardworking pair who run the All Manna shop in Richmond. Their policy is to sell 'the best-quality goods at competitive prices'. Goods are bought wholesale for packing on the spot. Very good range of dried health and wholefoods. Wholemeal bread and confectionery baked daily for both shops. Further expansion planned for next year. Deliveries in locality for orders over £10. The staff care about their work and are always eager to be helpful.

SW15

Florian of Putney 75 Upper Richmond Road, London SW15 (01 874 0398); bus 37; tube to East Putney; open daily 9.00–18.00

A health shop specializing in shampoos and hair lotions made on the premises; also vitamins, herbs soaps, beauty preparations.

Holland & Barrett 26 High Street, Putney, London SW15 (01 789 0504); buses 14, 30, 74, 93; tube to Putney Bridge; open Monday to Saturday 9.00–17.30

For further information on the Holland & Barrett chain see entry for Amersham (page 180).

SW16

Holland & Barrett 110 High Road, Streatham, London SW16 (01 769 1418); buses 57, 95, 109, 118, 133, 159 stop outside; tube to Streatham Hill; open Monday to Saturday 9.00–17.30
For further information on the Holland & Barrett chain see entry for Amersham (page 180).

Nature's Store 334–336 Streatham High Road, London SW16 (01 677 4429); buses 50, 109, 113, 159 or tube to Tooting Bec; open Monday to Friday 9.30–18.00, Saturday 9.30–17.30
Sells the usual range of branded health foods, vitamins, minerals, natural remedies, toiletries.

SW17

James John 211 Upper Tooting Road, London SW17; buses to Tooting; tube to Tooting Broadway; open Monday to Saturday 9.00–18.00
Full range of health foods including wholemeal bread, goat's milk, yoghurt, free-range eggs. Also herbal remedies and New Era bio-chemicals. (HSW, NAHS)

SW18

Holland & Barrett 5 The Arndale Centre, Wandsworth, London SW18 (01 874 3598); open Monday to Saturday 9.00–17.30
For further information on the Holland & Barrett chain see entry for Amersham (page 180).

W1

Cranks Farm Bar In Heal's, 196 Tottenham Court Road, London W1

(01 637 2230); tube to Goodge Street; open Monday to Saturday 10.00–17.00

This tiny shop, adjacent to the main restaurant counter, enables customers to buy the food served in all Cranks restaurants, to take away. In addition: a range of honeys, preserves, vegetable and fruit juices, dairy produce, yoghurt, Loseley ice-cream. On premises they prepare bread, cakes, savouries, biscuits, scones, soups, salads. (NAHS)

Cranks Health Food Shop 8 Marshall Street, London W1 (01 437 2915); buses 6, 12, 15, 88, 153 to Regent Street; tube to Oxford Circus or Piccadilly Circus; open Monday to Friday 9.00–18.00 Saturday 9.00–16.30

This is the original Cranks store, opened in 1961. It adjoins the famous vegetarian restaurant, a haven from the burger bars of Regent Street and Oxford Street. Entirely vegetarian, sells packaged wholefood groceries; bread, cakes, biscuits made on premises; honey and preserves; all dairy products; fresh organically grown fruit and vegetables (at a price); natural cosmetics, natural remedies; vitamins; drinks, e.g. fruit juice, wine, cider. Also books on related subjects. A take-away service sells freshly made wholemeal bread, savouries, quiches, pastries, soup, salads, drinks: all these and other hot dishes are served all day until 20.30 in the adjoining licensed restaurant. (HSW, NAHS)

Cranks Health Foods The Market Covent Garden London W1; tube to Covent Garden; open Monday to Saturday 10.00–20.30

The most recently opened branch of the marvellous Cranks chain of shops and restaurants is in three sections. The first is an extensive counter of food to take away, offering their usual large range of health foods made on premises. Downstairs, a grain shop sells all the cereals, beans, grains, pulses, nuts and fruit you would expect to find in a first-class shop. Also a small juice bar, to seat about thirty people; this offers a similar range of fresh fruit and vegetables, drinks, salads. (NAHS)

Cranks Wholegrain Shop 37 Marshall Street, London W1 (01 439 1809); open Monday to Friday 9.30–18.00, Saturday 9.30–16.30

Opposite the main Cranks vegetarian restaurant and grocery shop, this branch caters for bulk buyers of wholefood grains, cereals, pulses, nuts, dried fruits, herbs, spices. Also bread-making kits, grain mills, recipe books, and 'kitchenwear' with 'I like Cranks' motifs. Prices lower than in the main retail shop. (HSW, NAHS)

Culpeper 21 Bruton Street, Berkeley Square, London W1 (01 629 4559); bus 25; tube to Green Park or Bond Street; open Monday to Saturday 9.30–17.30

This is the original shop (opened in 1927) of the famous herbalists, who have now opened several more branches, mainly in the provinces. This shop, unlike its new sisters, does not sell live herbs; but the range of the dried variety is immense. All the spices you can think of and many of which you have never heard. Condiments, sauces, mustards, natural curry powders, vinegars, flavourings like natural vanilla, some natural foods (including raw sugars), honey, organically grown wholewheat flour, beverages such as tea and apple juices, and preserves – the marmalade is memorable. Advice given on use of goods, including medicinal use of herbs. Herbal remedies and tonics, herbal beer extracts, potpourri and pomanders, herbal cushions and sachets, seeds. The cosmetic range is huge: hair and bath preparations, toilet waters and colognes, essential oils, real sponges. Also delights like real liquorice, crystallized seeds and leaves from France. Dried flowers and flower cards; a good range of books on herbs. No deliveries, but goods valued at over £3 will be sent by post anywhere.

Holland & Barrett 78 Baker Street, London W1 (01 935 3544); buses 1, 2, 30, 74, 113, 159 to Baker Street post office; tube to Baker Street; open Monday to Friday 9.00–17.30, Saturday 9.00–17.00

For further information on the Holland & Barrett chain see entry for Amersham (page 180).

Holland & Barrett 19 Goodge Street, London W1 (01 580 2886); tube to Goodge Street; open Monday to Friday 9.00–17.30, Saturday 9.00–13.00

For further information on the Holland & Barrett chain see entry for Amersham page 180).

Holland & Barrett In Bournes, 116 Oxford Street, London W1 (01 636 1515); buses 25, 73 to Oxford Circus; tube to Oxford Circus and Tottenham Court Road; open Monday to Saturday 9.30–18.00, Thursday 10.00–20.00

For further information on the Holland & Barrett chain see entry for Amersham (page 180).

Wholefood 112 Baker Street, London W1 (01 935 3924); buses 18, 27, 30, 176 to Baker Street station, or any bus to Baker Street; open Monday 8.45–18.00, Tuesday to Friday 8.45–18.30, Saturday 8.45–13.00
Quite simply, one of the best wholefood shops in existence. All produce is organically produced and unsprayed; perishables are absolutely fresh. If a fruit or vegetable is not in season, it is not in stock. They sell salads, fruit and vegetables when available, bread, cakes and biscuit-type snacks from Mayall's organically grown unsprayed wheat and rye, dried fruits, ice-cream made from real cream, free-range eggs, dairy produce, macrobiotic foods, a good selection of packaged wholefoods, natural remedies, nutritional supplements. Attached bookshop has a huge range of relevant reading matter, from gardening to vitamins. Also natural cosmetics. For butcher's, see below.

Wholefood Butchers 24 Paddington Street, London W1 (01 486 1390); buses 18, 27, 30, 176 to Baker Street station, or any bus to Baker Street post office; open
This may be the only butcher's in London – or in England – where all meat is guaranteed to be from animals reared naturally, at a normal pace and without artificial stimulants or chemical injections. Farmers providing the animals are inspected by the shop. Range includes poultry and an excellent delicatessen counter selling pâtés and brawn, cooked beef, and cold roasted poultry prepared in the shop to the highest standards.

W2

Holland & Barrett 62 Edgware Road, London W2 (01 723 2339); buses 6, 7, 8, 16, 16a, 36, 36b to Marble Arch; tube to Edgware Road; open Monday to Saturday 9.00–17.30
For further information on the Holland & Barrett chain see entry for Amersham (page 180).

W3

Acton Health Foods Ltd 8 Market Place, Acton, London W3 (01 993

3848); bus 207 to High Street; tube to Acton Town/Acton Central;
open Monday to Saturday 9.00–18.00
Maintains a large stock of prepackaged goods and is prepared to give
attractive discounts on bulk purchases of any product in the shop.
Herbal remedies, cosmetics, microbiotic products, vegetarian foods and
confectionery; magazines and books. Deliveries in Acton area.

Old Oak Health Food Shop 195 Old Oak Road, London W3 (01 743
2348); buses 7, 12, 15 to Mecca bingo hall, East Acton, tube to East
Acton; open Monday to Saturday 9.00–18.30
An attractive, independent store, with a knowledgeable and friendly
staff. Well known for wholewheat bread, cakes and pastries, baked in
own accredited kitchen. Stocks natural yoghurt, cheeses, honeys, jams,
juices, mineral waters. A section devoted to health books and magazines; also culinary and medicinal herbs, spices, vitamins, tonics,
slimming aids. Deliveries by arrangement, and discount given for
bulk buying.

W4

Holland & Barrett 416 High Road, Chiswick, London W4 (01 994 1683);
buses to Turnham Green church; tube to Chiswick Park; open Monday
to Saturday 9.00–17.30
For further information on the Holland & Barrett chain see entry for
Amersham (page 180).

W5

Cornucopia 64 St Mary's Road, Ealing, London W5 (01 579 9431); open
Monday to Saturday 9.00–17.50
A self-service store stocking a wide range of groceries. Grains, beans,
pulses, cereals sold loose from bins. Various dried fruits and nuts;
wholewheat flours and speciality flours. Imported honeys, local conserves; fruit juices, vegetable juices, Loseley ice-cream and yoghurt. Also
English and continental cheeses; Japanese foods, including miso, tamari,
sea vegetables. In season, organically grown fruit and vegetables. Free-
range eggs (FREGG approved). Willing to sell in bulk. Free price list on

request. Plans in progress to open a snack bar/vegetarian restaurant and to increase the take-away food section, at present offering freshly made corn-bean pie, honey cake, carob and walnut brownies, cheese scones.

Holland & Barrett 6 Ealing Broadway, Ealing, London W5 (01 579 3017); buses 65, 83, 112, 273, 274, E1, E2 stop outside shop; tube to Ealing Broadway; open Monday to Saturday 09.00–17.30
For further information on the Holland & Barrett chain see entry for Amersham (page 180).

Holland & Barrett 3 High Street, The Green, Ealing, London W5 (01 567 3310); buses 65, 274, E1, E2 to Uxbridge Road, 2 minutes' walk from shop; tube to Ealing Broadway; open Monday to Saturday 9.00–17.30
For further information on the Holland & Barrett chain see entry for Amersham (page 180).

W6

Holland & Barrett 232 King Street, Hammersmith, London W6 (01 748 1975); tube to Ravenscourt Park; open Monday to Saturday 9.00–17.30
For further information on the Holland & Barrett chain see entry for Amersham (page 180).

Holland & Barrett King's Mall Hammersmith London W6; open Monday to Saturday 9.00–17.30
For further information on the Holland & Barrett chain see entry for Amersham (page 180).

W8

Holland & Barrett 139 Kensington Church Street, London W8 (01 727 9011); buses 27, 28, 31, 52; tube to Notting Hill Gate; open Monday to Saturday 9.00–17.30
For further information on the Holland & Barrett chain see entry for Amersham (page 180).

Holland & Barrett 260 Kensington High Street, London W8 (01 602 3627); buses 27, 28, 49 to Commonwealth Institute; tube to High Street Kensington; open Monday to Saturday 9.00–17.30
For further information on the Holland & Barrett chain see entry for Amersham (page 180).

W11

Ceres Grain Shop 269 Portobello Road, London W11 (01 229 5571); buses 15, 28, 52; tube to Ladbroke Grove or Westbourne Park; open Monday to Saturday 10.00–18.00, Friday to 18.30
Launched on the wave of the 1960s swing to an alternative lifestyle Ceres now has a rather tattered feel. The shelves are not so well stocked and it is annoying not to be able to find spices, for example, in an area, where they are much in demand but not easy to come across. Nevertheless, still sells a wide range of wholefoods packed in the shop, including organically grown grains, shelled nuts, dried fruits; also some organic vegetables, natural cosmetic products such as soap and shampoo, teas, some dairy products. Most items sold in medium weights – 5-lb or 10-lb packs. Fruit juices to take away are served from a refrigerated dispenser, and wholemeal rolls, sandwiches, pizzas. In the bakery shop, there is a big selection of the Ceres wholegrain bread and cakes (but our researcher reports that these have recently become rather heavy and indigestible); they are made without dairy products, sugars, or preservatives. Friends of the owners have reported that the shop may soon be sold, so check if it is open if you are making a special trip. (NAMB)

W12

Holland & Barrett 112 Shepherds Bush Centre, Shepherds Bush, London W12 (01 743 1045); buses 12, 49, 88, 295 to Shepherds Bush Green; tube to Shepherds Bush; open Monday to Saturday 9.00–1730.
For further information on the Holland & Barrett chain see entry for Amersham (page 180).

W13

West Ealing Healthfoods, Herbalist and Beauty Parlour 3 Leeland Road, London W13 (01 567 4638); buses to West Ealing Broadway; tube to Northfields and Ealing Broadway; open Monday to Saturday 9.00–17.30, Wednesday to 17.30, Friday to 18.00
A very comprehensive range of health foods, vitamin supplements, herbal remedies, hand-made creams and cosmetics. Knowledgeable staff answer queries and give advice on diet and beauty preparations. Very large range of loose herbs. Honey fresh from the hive; eggs fresh from the free ranges.

W14

Westken Wholefoods 6 Charleville Road, West Kensington, London W14 (01 385 0956); bus 28 to North End Road; tube to West Kensington; open Monday to Saturday 9.00–18.30
Large general range of health goods: varieties of juices, honeys, grains, pulses, nuts, seeds, granolas, dried fruits. Minerals, herbal remedies and cosmetics, macrobiotic products, vegetarian foods, wholemeal bread and confectionery. Also pottery. Famous regular customers are Virginia Wade and Ralph Michael. Attractive discounts on bulk orders and cases of any product. Deliveries in the immediate area. (NAHS)

WC1

Holland & Barrett Unit 36, Brunswick Shopping Centre, Bloomsbury, London WC1 (01 278 4640); buses 68, 77 to Marshmouth Street, 18, 30, 73, 177 to Euston Road; tube to Russell Square; open Monday to Friday 9.00–17.30, Saturday 9.00–16.00
For further information on the Holland & Barrett chain see entry for Amersham (page 180).

WC2

Golden Orient 17 Earlham Street, London WC2 (01 836 5545); tube to

Leicester Square/Covent Garden; open Monday to Friday 10.00–18.30, Saturday 10.00–17.30
This newly established shop has already become renowned for its own mixed curry powder and spices. The owner firmly believes in selling goods at realistic, not exorbitant, prices. Stocks a comprehensive range of wholefoods, including wholewheat bread, rye, a variety of herbal teas, chutneys, preserves. Also Beauty Without Cruelty cosmetics and vitamins. Honeys of various kinds supplied in large or small quantities. Hopes to expand the existing range of goods and introduce many more of own items: in particular, pickles, fresh herbs and savouries.

Neal's Yard Wholefood Warehouse 2 Neal's Yard, Covent Garden, London WC2 (shop 01 240 1154; bakery 01 836 5199; dairy 01 379 7646); buses to Charing Cross Road and Shaftesbury Avenue, tube to Covent Garden; open Monday to Friday 10.00–18.30, Saturday 10.00–17.30
The warehouse shop specializes in mid-weight quantities – dry goods sold in 5-lb, 10-lb and 20-lb measures are cheap. Larger and smaller measures work out slightly more expensive than in some other shops. Oils, tahini, tamari, peanut butter and miso are sold cheaply, in half gallon measures. The flourmill on the premises grinds a good variety of flours (telephone number: 01 836 1082). The bakery, in the same premises, is completely wholefood, and there is a tearoom where you can have a snack away from the bustle of the streets. The dairy, also in the yard, produces (daily) soft cheese, yoghurt, ice-cream. As you approach the yard from the Monmouth Street side, you will find the associate coffee shop (01 836 5272) which sells, economically, a good range of coffees roasted on premises. All these are fairly recent developments, but they function with great economy and efficiency. Service is helpful and welcoming.

Scott's Health Foods 27 Chancery Lane, London WC2 (01 405 5696); buses or tube to Chancery Lane; open Monday to Friday 9.00–17.30
A boon to those who work in the Holborn or Fleet Street area, this recently opened shop sells muesli and other breakfast cereals, yoghurt, cheese, a variety of packaged whole and health foods, vitamins, nutritional supplements, preserves, dairy products. Also beverages, e.g tea, fruit juices. A wholefood snack bar caters for the take-away trade.

Longton

Your Health 65 Market Street, Longton, Stoke-on-Trent, Staffordshire (0782 313120); bus to Longton bus depot; open Monday to Friday 9.00–17.45
Steam and massage parlour, sun-ray and facial facilities are provided above the shop selling the normal range of health and vegetarian foods and juices. Bread, scones, yoghurt and ice-cream freshly made on premises. Departments for books and magazines; wine- and beer-making; marital aids. Also a hobbies corner. (HSW)

Loughborough

Combine Wholefood Collective 7 Leicester Road, Loughborough, Leicestershire (0509 213865); open Monday to Saturday 9.00–17.30, Wednesday to 13.00
A collective opened in 1978, the shop encourages people to treat it as a meeting place for discussion and debate, and supports the setting-up of 'alternative' ways of life. Books and magazines sold. A very good supply of dried wholefoods, including carob and coconut, cold-pressed oils, muesli, wholemeal bread, herbs and spices. With very few exceptions goods are sold loose from bulk. Your own containers are appreciated. The collective is trying to ensure a more regular supply of compost-grown vegetables.

Taylor's Health Foods 31 Biggin Street, Loughborough, Leicestershire (0509 214923); adjacent to main bus station; open Monday to Saturday 9.00–17.30, Wednesday to 13.00, Saturday to 17.00
This shop was formerly in Derby Square. In the new premises the owners hope to incorporate a wholefood snack bar and a centre for 'alternative medicine'. All the full-time staff have the NAHS certificate in retailing. Stocks dried wholefoods, both branded and loose; over 100 herbs for culinary and medicinal use; some fresh organically grown vegetables; fresh dairy produce; a comprehensive range of herbal remedies and food supplements. Will obtain goods or bulk supplies for customers where these are not part of the normal stock. Bread, fresh daily, from three local wholefood bakeries. (HSW, NAHS)

Loughton

Holland & Barrett 285 High Road, Loughton, Essex (01 508 4009);
buses 20a, 167 stop outside shop; tube to Loughton; open Monday to
Saturday 9.00–17.30
For further information on the Holland & Barrett chain see entry for
Amersham (page 180).

Lowestoft

Oregano 88 High Street, Lowestoft, Suffolk (0502 82907); buses 601,
602, 603, 680 to market place
This attractive and relatively new shop is very welcome in Lowestoft
with its poor communications. An abundance of dried wholefoods:
grains, flakes, pulses, flours, breakfast cereals, nuts, dried fruits, herbs,
spices, herbal teas, decaffeinated coffees, rices, vegetarian special foods,
confectionery; also oils, spreads, vitamins, minerals, natural cosmetics,
herbal tobacco products, pet food, homoeopathic remedies. The owners
are trying to find a regular and reliable supply of dairy goods, especially
goat's milk products. They hope to begin baking their own wholefood
cakes and pastries soon. Deliveries within Lowestoft.

Ludlow

Ludlow Health Food Shop 3 Tower Street, Ludlow, Shropshire (0584
2221); buses to Corve Street; open Monday to Saturday 9.00–17.30
Packaged and canned health food groceries, dairy goods, honey and
preserves, bread, vitamins; also natural remedies and cosmetics.

Luton

Health and Heather (Holland & Barrett) 53 Wellington Street, Luton,
Bedfordshire 0582 27324); buses 20, 21, 22 to Stuart Street; open Mon-
day to Saturday 9.00–17.30
For further information on the Holland & Barrett chain see entry for
Amersham (page 180).

Lydney

Mieux Vivre Health Foods Newerne Post Office, Lydney, Gloucestershire (0594 42816); open Monday to Saturday 9.00–13.00, 14.00–17.30, Wednesday to 13.00
A wide range of groceries and dry goods; fresh vegetables and fruit; wholewheat bread; beers and ciders; free-range eggs. Also minerals and vitamins; herbal remedies for pets; beauty preparations/toiletries. Plans for selling fresh wholewheat flour ground on premises and yoghurt ices in summer. (NAHS)

Lymington

Gerard House 31 St Thomas Street, Lymington, Hampshire (0202 36466)
This shop, with its sister store in Bournemouth, has one of the best collections of books on health, wholefoods and ecology-related subjects on sale in the South. Good range of wholefoods, vegetarian and vegan foods, herbs and herbal preparations, vitamins, minerals, natural cosmetics. (BHMA, NAHS)

Lytham St Anne's

Health Food Store 12 Park Street, Lytham St Anne's, Lancashire (0253 724912); buses 11, 11a, 13a to the Square; open Monday to Saturday 9.00–17.30, Wednesday to 13.00 ,Saturday to 17.00
Packaged rather than fresh foods, including loose cereals and pulses, raw sugar preserves, herbs, spices, oils, juices, mineral waters; also biochemical remedies, vitamin supplements, natural cosmetics, books on health. Deliveries within five miles. (HSW, HNAS)

Macclesfield

The Granary 1 Brook Street, Macclesfield, Cheshire (0625 23825); open Monday to Saturday 9.00–17.00
Opened in 1976, this is an offshoot of The Granary in Chester. Goods,

sold from bulk in any quantity, include the usual wide range of grains and flakes, pulses, cereals, flours. Dairy produce – yoghurts, organically produced butter, cottage and vegetarian cheese, goat's milk; a good stock of herbs and spices, with over 100 medicinal herbs, herb teas, herbal oils; vitamin and mineral supplements; an attractive selection of general goods like bean sprouts, rock and sea salts, honey, raw sugar preserves, spreads, oils, vinegars, pickles. Over 500 titles in the book section. Although the shop specializes in bulk sales, it does not undertake deliveries.

Maidenhead

Country Market 122 High Street, Maidenhead, Berkshire (0628 32725); buses to nearby shopping precinct; open Monday to Saturday 9.00–17.30, Tuesday 9.30–17.30
For further information on the Holland & Barrett chain see entry for Amersham (page 180).

Reading Wholefoods Maidenhead Market, Berkshire; open Fridays only 10.00–17.30
Market stall supplied by Reading Wholefoods, who visit the surrounding markets with their own packed goods from bulk. An excellent, economical range of dried wholefoods – grains, cereals, pulses, nuts, seeds, dried fruits, sugars, pasta, yeast, herbs, spices; also oils, juices, honey, preserves, fresh bread, some food supplements. Cookware, natural cosmetics, books on related subjects.

Maidstone

Country Life Wholefoods 25 Pudding Lane, Maidstone, Kent (0622 56139); open Monday to Saturday 9.00–17.30
This is a shop in the expanding Country Life chain of health food retail outlets. For further information on this chain see entry on Country Life Wholefoods in Ashford (page 181).

Nature's Way Ltd Unit 247, Stoneborough Centre, Maidstone, Kent (0622 63224); open Monday to Saturday 9.00–17.30
For further information on this chain see entry for Beckenham (page 187).

Malvern

The Whole Earth 65 Church Street, Malvern, Worcestershire (068 45 4952); open Monday to Friday 9.00–17.30, closed Wednesday
Formerly belonged to Southern Health Foods; the full range of wholefoods is now stocked. A bright modern shop opposite the church, it continues to sell fresh local produce where possible. Also a full range of herbal remedies.

Manchester

Country Life Wholefoods 121 Oldham Street, Manchester (061 832 8791); open Monday to Saturday 9.00–17.30
This is a branch of Hillstart Ltd's chain of health food shops in the Midlands and the North of England. For further information on this chain see entry on The Happy Nut House in Birkenhead (page 192).

Country Life Wholefoods 9 South King Street, Manchester (061 834 5923); open Monday to Saturday 9.00–17.30
This is a branch of Hillstart Ltd's chain of health food shops in the Midlands and the North of England. For further information on this chain see entry on The Happy Nut House in Birkenhead (page 192).

Eccles Health Food Centre 190 Church Street, Eccles, Greater Manchester (061 789 1098); buses 10, 20, 67, 166 to The Crown; open Monday to Friday 9.30–17.00; closed Wednesday
The normal range of packaged health foods and nutritional supplements, including tissue salts, herbal remedies, diabetic diet products.

Fruitfresh Healthfood Store 14 Albert Road, Levenshulme, Greater Manchester (061 224 8247); buses 169, 170, 171; open Monday to Friday 9.00–18.00, Wednesday to 13.00, Saturday 9.00–17.30

Sells strictly vegetarian goods. Most produce is prepacked and includes wholefood groceries, fruit, fruit juices, mineral waters, vitamins. (RFTA)

Healthcare Foods 1041 Oldham Road, Newton Heath, Greater Manchester (061 682 7288); buses 71, 77, 82, 168, 181, 182, 183; open Monday to Saturday 9.30–17.30
A pleasant shop which still provides personal service. Sells wholefood cereals, grains, pulses, fruit and nuts, a good selection of herbs, free-range eggs, vegetarian cheese, fresh wholemeal bread, preserves, ginseng, slimming aids. Also natural cosmetics, herbal medicines, vitamin supplements. (NAHS)

Health Food Shop 25 Cheetham Parade, Manchester (061 740 5766); numerous buses to Crescent Road; open Monday and Tuesday 9.00–17.45, Wednesday 9.00–13.00, Thursday and Friday 9.00–18.00, Saturday 9.00–17.30, closed for lunch 13.00–13.30 every day except Saturday
Relatively inexpensive, this shop gives discounts on some lines, including Granose ranges. Sells dried goods in bulk: dried fruit, nuts, cereals. Also stocks vegetarian special foods, four kinds of vegetarian cheese, wholemeal bread (fresh daily), loose herbs and spices, pulses, preserves and honey, juices, vitamins, herbal remedies, bio-chemicals, tissue salts, books on health subjects. Deliveries arranged for a fee, subject to value, for a minimum amount of goods. (HSW, NAHS)

Holland & Barrett Unit 122, Arndale Centre, Manchester (061 834 5975); open Monday to Saturday 9.00–17.30
For further information on the Holland & Barrett chain see entry for Amersham (page 180).

M. Lees Health Food Centre 190 Church Street, Eccles, Greater Manchester (061 789 1098); buses 10, 67, 166 to Church Street or The Crown; open Monday, Tuesday, Thursday, Friday 9.30–17.00, Saturday 9.30–16.30
Long-established, old-fashioned shop selling most popular brands of health foods, fresh wholewheat bread, eggs (free-range when obtainable), a good choice of honeys. Full range of herbal and other remedies, vitamins, natural cosmetics. Deliveries within locality.

Manchester Delicatessen 58 Wilmslow Road, Rusholme, Greater Manchester (061 224 3062); open Monday to Saturday 9.30–18.00
Primarily a delicatessen and continental grocer's, this has a very good stock of health foods, especially since the new owner took over in 1978. Stocks wholemeal bread, twenty varieties of continental bread baked freshly every day, free-range eggs, English and continental cheeses, vegetarian special foods, vitamins, herbal remedies, and the usual dried wholefoods.

Nature's Grace 37 Barlow Moor Road, Didsbury, Greater Manchester (061 434 6784); buses 42, 45, 46, 606, 370 to Cavalcade; open Monday to Friday 9.00–18.00
This new store run by young people is much frequented by local students. Well stocked with dried wholefood goods, baking requisites, honeys, preserves, chutneys, meatless savouries, ground and shelled nuts, a good range of cheeses, natural cosmetics, vitamins, herbs, spices, nutritional supplements. Also a good range of yoghurts and other dairy produce. Wholefood bread and take-away snacks fresh daily.

Nut'n'Meg 444 Palatine Road, Northenden, Greater Manchester (061 998 4589); buses 44, 45, 370; open Monday to Saturday 9.30–18.30
Comprises a restaurant serving a choice of six dishes and a 'special each day; a take-away section offering wholemeal cakes and sandwiches, pizzas, flans, soup, baked potatoes; and a well-stocked health food store. The shop itself offers a complete range of dried fruits, nuts, cereals, grains, pulses, vegetables, juices, fresh wholemeal bread; vitamins, minerals and a comprehensive selection of supplements; natural cosmetics, herbs, spices, herb teas, breakfast foods. Situated close to the Post House in Manchester, has received visits from show business personalities: James Last, The Drifters and Lindisfarne. (HCIMA, NAHS)

On The Eighth Day Cooperative Ltd 111 Oxford Road, All Saints, Manchester (061 273 4878); buses, all numbers in the 40s, to BBC, Oxford Road; open Monday to Saturday 10.00–17.30, restaurant 12.00–15.00
This cooperative has now been open some years and continues to thrive in spite of having had to close the herb shop because of fire regulations. The huge stock of medicinal herbs is scattered around the main shop

now; you should ask the knowledgeable staff for anything you might need. A good range of culinary herbs and spices; grains, flours, flakes; beans and pulses; nuts, seeds, dried fruits; honey and other spreads – bring your own jar for honey and peanut butter from bulk; cold-pressed oils; dairy produce including vegetarian cheese, goat's milk products; free-range eggs; seasonings, e.g. miso, seaweed. Some natural cosmetics. A lunchtime café serves vegetarian and some vegan meals, prepared fresh on premises each day; dishes include soups, salads, pâtés, flans, cakes, biscuits and a changing 'dish of the day'. The collective hopes to run the general shop, herb shop, and café as separate entities in the same building before too long. This will give more seating in the crowded café. (FNWC).

Mansfield

Storey's Health Foods 7/8 Handley Arcade, Mansfield, Nottinghamshire (0623 20154); near Four Seasons bus station; open Monday to Saturday 9.00–17.00, Wednesday to 13.00
A recently sold shop, under new ownership. Soon to be expanded to include new stock. What you can buy now is a good range of medicinal and culinary herbs, spices, packaged health food groceries, fresh wholemeal bread, preserves, oils, coffee, tea, body-building products, vitamin and mineral supplements, natural cosmetics and toiletries. (NAHS)

Margate

Southern Health Foods 128 High Street, Margate, Kent (0843 291316); open Monday to Saturday 9.00–17.30
One of the well-established Southern Health Foods chain. For further information on this chain see entry on Southern Health Foods in Aldershot (page 179).

Marlborough

Doves Farm Ham, Marlborough, Wiltshire (04884 374); open: by appointment only

A working farm which has been in the same family for over 160 years. They specialize in growing and milling a complete range of wholemeal and gluten-free flours, wheatgerm, bran. The grain grown in several varieties, is stoneground at the farm. Will deliver in bulk anywhere south of Birmingham. This attractive new enterprise – opened in 1978 – has plans to expand, but the details are so far secret.

Marple

Health Foods 72a Stockport Road, Marple, Cheshire (061 427 5662); open Monday to Saturday 9.00–18.00
Long-established health food store with the usual packaged wholefoods and dietary specialities; vitamin and mineral supplements; herbs and spices. The owner tries to give friendly and personal service, and is always willing to stock new lines at customers' suggestions. Deliveries anywhere in locality, within reason. (NAHSW)

Matlock

Health Food and Beauty Centre 50 Dale Road, Matlock, Derbyshire (0629 4898); open Monday to Saturday 9.00–17.30, closed Thursday
This business is run by a couple one of whom is a fully trained beauty therapist. The salon offers sauna, Slendertone, massage and sun-ray treatments, and uses a wide range of natural beauty products which are also sold in the shop. Health foods include all the leading brands. Good choice of herbs and spices; vitamin and mineral supplements; natural medicines.

Melton Mowbray

Gourmet's Pantry 19 Leicester Street, Melton Mowbray, Leicestershire (0664 2415); adjacent to bus station
Opened in 1972, the shop stocks the full range of proprietary health foods as well as grains, shelled nuts, dried fruits and spices from bulk. Herbal remedies and nutritional supplements. Wholemeal cakes baked on premises. Plans to extend the business as soon as finance allows.

Middlesbrough

Black's Health Food Centre 295 Linthorpe Road, Middlesborough, Cleveland (0642 247975); open Monday to Saturday 9.00–17.30
One of four shops owned by the athlete G. B. Black. For further information on these, see entry for Bishop Auckland (page 195).

Impulse Wholefoods 47 Roman Road, Linthorpe, Middlesbrough, Cleveland (0642 826561); open Monday to Saturday 9.30–13.15, 14.15–17.30
A wide range of wholefoods, herbs, spices, free-range eggs, vegetarian cheese, vegetables, wholemeal bread daily. The owners hope to move to larger premises where they can realize plans for a restaurant.

Middleton

Holland & Barrett Unit F10, Arndale Centre, Middleton, Greater Manchester (061 643 6941); buses 17, 59, 122, 138, 163, 415; open Monday to Saturday 9.00–17.30
For further information on the Holland & Barrett chain see entry for Amersham (page 180).

Milton Keynes

Holland & Barrett 26 Midsummer Arcade, Milton Keynes, Buckinghamshire (0908 604281); open Monday to Saturday 9.00–17.30
For further information on the Holland & Barrett chain see entry for Amersham (page 180).

Minehead

Minehead Health Stores St Andrew's Lane, Friday Street, Minehead, Somerset (0643 4179); open Monday to Saturday 9.00–13.00, 14.00–17.00
Opened six years ago, the shop sells the normal range of packaged healthfoods, vitamins, herbal remedies, natural cosmetics.

Morecambe

Morecambe Health Food Centre 9 Pedder Street, Morecambe, Lancashire (0524 413733); buses to Central Promenade and Ribble bus station; open Monday to Saturday 9.30–12.45, 13.45–17.30, Wednesday to 12.45

Originally opened in the 1920s as a herbalist. Now a well-stocked healthfood store, dealing mainly in prepacked goods, including many well-known brand names. Vitamin and mineral supplements, herbal and biochemical remedies, vegetarian products, natural cosmetics. (NAHS)

Natural Health Store 23a Alexandra Road, Morecambe, Lancashire (0524 419558); buses to Alexandra Road and Battery Promenade; open Monday to Friday 9.00–12.30, 13.30–17.00, Saturday 9.00–12.30, 13.30–16.00

There has been a herbalist and health shop here since 1936; the present owner continues the tradition of friendly personal service. A wide choice of herbs for medical and culinary use; vitamin and mineral supplements; most popular brands of packaged health foods. Deliveries within Morecambe, Heysham and the immediate area. (NAHS)

Nailsworth

Nailsworth Health Store 3 Market Street, Nailsworth, Stroud, Gloucestershire (045 383 2919); open Monday to Friday 9.15–17.30

Sells a good variety of proprietary health foods and an excellent range of herbal remedies. The owner, a member of the Vegetarian Society and of the Cotswold Health Centre, will give herbal consultations by appointment. Stocks natural cosmetics; Dene's pet products. A small snack bar serves varied refreshments prepared on premises.

Narberth

Daisan Health Foods High Street Narberth Dyfed Wales (0834 860185); open Monday to Saturday 9.00–17.45

Groceries, dried goods, natural cosmetics, nutritional supplements. 'Alternative medicine' clinic, with free advice on food and health matters. (NAHS)

Neath

Neath Health Foods 61/63 Market Hall, Neath, Glamorgan, Wales (0639 2665); open Monday to Saturday 9.00–17.00, Thursday to 13.00
Small market stall with an enthusiastic proprietor who stocks wholefoods, vegetarian, gluten- and salt-free special diet foods; honey, fruit juices, vitamins, minerals; free-range eggs; yoghurt; wholemeal bread. Also herbal and homoeopathic remedies.

Newark

Wright's Herbal and Health Food Stores 42 Middlegate, Newark, Nottinghamshire (0636 702875); open Monday to Friday 9.00–17.30, Thursday to 13.00
A new branch of the main shop at Worksop, with similar emphasis on selling herbs for both medicinal and culinary use. Also stocks the normal range of packaged health foods.

New Ash Green

New Ash Green Health Food Centre 5 Upper Street North, New Ash Green, near Dartford, Kent (0474 874512); buses 489, 490; open Tuesday to Saturday 9.00–13.00, 14.00–17.30
Opened in 1979, the shop sells from bulk a wide range of grains, pulses, cereals, and other dried wholefoods; herbs and spices; free-range eggs; honey and preserves; wholemeal bread. Also natural cosmetics, local pottery, home-brew supplies.

Newbury

Varley Health Foods Ltd 6 Eight Bells Arcade, Bartholemew Street, Newbury, Berkshire (0635 44324); open Monday to Saturday 9.00–17.00, closed Wednesday
Sells the general range of wholefood groceries, vegetarian specialities, vitamins, herbal remedies, cosmetics. Plans soon to open a snack and coffee bar.

Newcastle-under-Lyme

Kermase Wholefoods 64 Liverpool Road, Newcastle-under-Lyme, Staffordshire (0782 613255); near bus station; open Monday to Saturday 9.00–17.30, Thursday to 14.00
This collective is known affectionately in the town as 'Kermit Whalefoods'. They aim to sell as cheaply as possible by dispensing from bulk, like most collectives. Oils, honeys and shampoos are sold loose from the drum, so take your own containers. Dried wholefoods include grains, cereals, pastas, dried fruits, pulses, nuts, grain coffees, flours, over 100 herbs and spices, and teas. Also peanut butter and other spreads, natural soaps, local cheese, fresh fruit and vegetables, fruit juices, and muesli packed on the premises from their own mix. Bran loaves baked daily. Bean sprouts and take-away salads on Saturdays only; other snacks available everyday. The collective hopes soon to extend this service by opening a vegetarian café in town. Local craft pottery, books and magazines also on sale. (ICOM, FNWC)

Newcastle-upon-Tyne

Mandala Wholefoods 43 Manor House Road, Jesmond, Newcastle-upon-Tyne, Tyne and Wear; buses 33, 38; open Monday 11.00–18.00, Tuesday to Saturday 10.00–18.00
This shop is run by a collective in somewhat cramped conditions; they have been refused planning permission for further extension, and are managing to operate without a telephone. Nevertheless, a good basic stock is kept of goods from bulk including grains, flour, cereals and pulses at low prices, dried fruit, shelled nuts. Honey, fruit juice, herbs and spices, ginseng and seaweed, incense. Natural cosmetics, e.g. soap. (ICOM, NWC)

Milburn Health Foods Newgate Shopping Centre, Newcastle-upon-Tyne, Tyne and Wear (0632 29091), 13 Mistletoe Road, Jesmond, Newcastle-upon-Tyne (0632 813533); both open Monday to Saturday 8.30–17.30
Both shops carry a very extensive range of wholefoods, vitamins, food supplements, natural cosmetics, herbs, jams. Deliveries twice weekly within twenty-mile radius. (HSW, NAHS)

New Mills

Natural Foods 37 Union Road, New Mills, near Stockport, Greater Manchester; open Monday to Friday 9.00–17.30, Wednesday to 12.30, Saturday 9.00–17.00
A fairly new shop selling the popular brands of wholefood groceries, fresh bread, honey and preserves, vegetarian savouries, juices, dairy products, herbs and remedies, vitamins, home beer- and wine-making supplies.

New Milton

Scoltock's Natural Food and Beauty Shop 35 Station Road, New Milton, Hampshire (0425 619090); open Monday to Saturday 9.00–17.00
One of three shops in Hampshire bearing this name, each with own specialization. Here at New Milton, as at Ringwood, grains and pulses are combined with a range of traditional health foods. Prepares own-brand muesli on premises, and makes and markets a range of herbal facial creams. This is the only one of the three outlets with a wholemeal take-away facility. Deliveries within five -mile radius. (NAHS)

Newport (Isle of Wight)

The Newport Health Food Centre St James's Square, Newport, Isle of Wight (0983 522121); open Monday to Saturday 8.30–17.30, Thursday to 13.00
Sells a full range of branded dried wholefoods and the usual vitamins, nutritional supplements, natural cosmetics. Also locally produced bread, honey, flour, yoghurt. Deliveries throughout the island. (BHMA, NAHS)

Newport (Wales)

Southern Health Foods 15 John Frost Square, Newport, Gwent, Wales (0633 66215); open Monday to Saturday 9.00–17.30
One of the well-established Southern Health Foods chain. For further

information on this chain see entry on Southern Health Foods in Aldershot (page 179).

Newquay

David C. Sharpe (Chemists) Ltd 2 Cliff Road, Newquay, Cornwall (06373 2957); open Monday to Saturday 9.00–18.30.
Hoping to extend the space available for health foods: vitamin and mineral supplements, flours, cereals, pasta, soya proteins, fruit juices, mineral waters. Deliveries within Newquay and district.

Newton Abbot

Southern Health Foods 17 Union Street, Newton Abbot, Devon (0626 3141); open Monday to Saturday 9.00–17.30
One of the well-established Southern Health Foods chain. For further information on this chain see entry on Southern Health Foods in Aldershot (page 179).

Newtown

Sparrow Wholefood 14 The Bank, Gas Street, Newtown, Powys, Wales (0686 27002); open Monday to Saturday 9.00–17.30, Thursday half day
This enterprising new shop, in the second oldest building in the town, takes its name from sparrows calling in from their nests below the eaves. The atmosphere is described as 'Dickensian', and there is an open fire. New premises being sought so that a snack bar can be opened and stock widened to include books and kitchenware. A comprehensive range of wholefoods and cheeses, Greek and Indian foodstuffs. Quiches, cheesecake, biscuits, stuffed parathas and similar snacks as available. Also medicinal herbs and some natural cosmetics. A herbal surgery once a month. Deliveries within twelve-mile radius for bulk orders. (WWFF)

Newtownards

Scrabo Health Products Ltd 19 Regent Street, Newtownards, County Down, Northern Ireland (0247 817775); next to bus station; open Monday to Saturday 9.15–17.15, closed Thursday

The usual range of health foods together with natural vitamins, herbal products, biochemicals, German homoeopathic remedies.

Northampton

Goodness Foods 59 and 61 Gold Street, Northampton (0604 39243); 4 minutes from bus station; open Monday to Saturday 9.00–17.30, Friday to 18.00 (see entry for Braunston, page 203).
The same concern runs similar shops elsewhere in Northamptonshire (see entry for Braunston, page 203). This shop has a good take-away section, separate from the check-out, where food is prepared daily for this and other shops' take-away counters; it includes filled rolls, pitta, hot soup and pies (precooked and heated in a microwave oven), rice dishes, Indian-style foods, delicious cakes and sweets.

Holland & Barrett 4 Peacock Avenue, Northampton (0604 36149); open Monday to Saturday 9.00–17.30
For further information on the Holland & Barrett chain see entry for Amersham (page 180).

Northwich

Northwich Health Food Centre 117 Witton Street, Northwich, Cheshire (0606 3061); open Monday to Saturday 9.00–17.30, Wednesday to 13.00
A well-established shop with a wide range of dried wholefoods which can be sold in bulk; dried herbs and spices; honey; vitamin supplements; natural cosmetics; herbal remedies. Dairy goods include frozen yoghurt which you eat like ice-cream on a cone. Also a good book section.

Norwich

Culpeper 14 Bridewell Alley, Norwich, Norfolk (0603 618911); open Monday to Saturday 9.30–17.30
For further information on the Culpeper chain see entry for Bath (page 186).

Lloyds Drug Store 12 Earlham House Shopping Centre, Earlham Road, Norwich, Norfolk; buses 510, 511, 512; open Monday 14.00–17.30, Tuesday, Wednesday, Thursday, Saturday 9.00–17.30, Friday 9.00–19.30
At present has Health Craft and New Era stock together with a small amount of American Nutrition. However, the health food side of the business is due for expansion in the near future.

Natural Food Store 4 Exchange Street, Norwich, Norfolk (0603 613228); buses to city centre; open Monday to Saturday 8.30–17.30
This family-run business tries to maintain an informal and courteous service in the old-fashioned manner. The staff are knowledgeable in their field and able to offer advice on health and diet problems. Items will be stocked if requested by regular customers. Free-range eggs, including duck and geese eggs; a variety of fresh shelled nuts and dried fruits; all the general health foods; wholemeal bread baked locally to the shop's own recipe and delivered daily. Also a very large range of vitamin and mineral supplements; natural remedies. Local deliveries anywhere, within reason.

Rainbow Wholefoods 16 Dove Street, Norwich, Norfolk (0603 25560); open Monday to Saturday 9.30–17.30
A non-profit making concern, cooperatively run. Stocks all the general wholefoods and prepares its own muesli on premises. Also handmade objects. Deliveries, free of charge, within Norfolk and Suffolk.

Taylor's Wholefoods 5 Orford Hill, Norwich, Norfolk (0603 21831); open Monday to Saturday 8.30–17.30
Taylor's was the first established herbalist in the city and has kept its identity during several moves. Some of the medicinal goods are own brand of nerve pills and tonic. There are the usual vitamin and mineral supplements; a good range of wholefoods – muesli, grains, pulses, nuts,

etc. – sold loose from boxes; flour; popular branded wholefood groceries; yoghurt and other dairy produce; fresh wholemeal bread; herbs and spices; books on health and health food cooking. A first-floor restaurant is planned for later this year. (NAHS)

Nottingham

March Wholefoods 11 Brightmoor Street, Nottingham; open Monday to Friday 10.00–13.00, 14.00–17.00, Thursday to 13.00
This is the wholesale section of the Ouroboros collective, providing wholefoods in bulk at very good prices; see the shop list for goods offered. Deliveries (wholesale only) are free within Nottingham. Price list sent on request. (COM, FNWC)

The Old Herb Shop 62 Peveril Street, Nottingham (0602 785811); buses 21, 23, 62, 44, 71, 85 to Bentink Road; open Monday to Saturday 9.00–17.30, closed Thursday
Founded in 1862 as a herbalist. Stocks health foods; vitamin supplements; herbal, homoeopathic and biochemical remedies. Consultations by appointment. (HSW, NAHS)

Ouroboros Wholefood Collective Ltd 37a Mansfield Road Nottingham (0602 49016); buses to city centre; open Monday to Saturday 9.00–17.30, Thursday to 14.00
One of the new collectives which seeks to provide best-quality wholefoods as cheaply as possible by selling from bulk. Customers are encouraged to bring in their own bags and jars, whether for their own or others' use. An excellent range of dried wholefoods – whole and flaked grains, flours, beans and peas, nuts, pasta, muesli; cold-pressed and refined oils; spreads – miso, molasses, tahini, honey; herbs, spices, aromatic oils; dairy produce, including goat's milk. Information on how to use the foodstuffs. Deliveries on weekly basis, in radius of thirty miles, charged according to distance. (FNWC, ICOM)

Nuneaton

Health Food Store 6 Coventry Street, Nuneaton, Warwickshire (0682 66200); open Monday to Saturday 9.00–18.00
Sells complete range of packaged health foods and herbal remedies. Also vitamin supplements; home wine- and beer-making supplies. The owner would like to expand the herbal medicine service. Some deliveries, in cases of difficulty.

Oldham

The Alternative 103 Union Street, Oldham, Greater Manchester; buses 82, 98 (from Manchester); open Monday to Saturday 8.30–17.30; (restaurant) Monday, Tuesday 9.00–15.00, Wednesday to Saturday 9.00–15.00, 17.30–22.30
Maintains that it sells healthy food 'rather than pills or potions'. Stocks organically grown vegetables. Sandwiches and pies made on premises. The restaurant is vegetarian; food can be eaten there or taken away. Policy is to sell good food at the lowest possible price in order to introduce people to organic and vegetarian eating.

Health Food Centre 16 Union Street, Oldham, Greater Manchester (061 624 4838); buses to Union Street or George Street; open Monday to Friday 9.00–17.30, Saturday 9.00–17.00, closed Tuesday
New owners took over in 1979 and seem anxious to help customers and respond to their needs. So far the shop is still in the 'vitamins and remedies' tradition rather than a place to buy fresh food; but stock includes fruit juices, wholewheat bread, goat's milk and yoghurt, cheeses. Also the usual packaged health foods and natural cosmetics. (NAHS)

Ormskirk

Wheatsheaf Health Foods 23 Burscough Street, Ormskirk, Lancashire; open Monday to Friday 9.15–17.30, closed Wednesday
Packaged health foods, vitamins, herbal remedies, natural cosmetics, books and magazines on health subjects, supplies for home wine- and beer-making.

Orpington

Food Reform 111 High Street, Orpington, Kent (0689 66 37172); open
Monday to Saturday 9.00–13.15, 14.15–17.30, early closing Thursday
A well-stocked shop open now for ten years. Sells free-range eggs;
wholewheat bread, biscuits, cakes; dairy goods, including yoghurt and
Jersey cream (but not milk); molasses and raw cane sugar; fruit juices;
some fresh natural fruits in season; herbs and spices; honey and conserves; the usual range of wholefood groceries. Also natural cosmetics,
Denes pet products, vitamins and minerals, books and magazines on
health topics. Brian Jacks the Olympic judoist is a regular customer.
One client lives according to the shop's diet plan for the week and still
loses weight although she indulges at weekends!

Ossett

Funny Foods (Wholefoods and Healthfoods) 42 Station Road, Ossett,
near Wakefield, Yorkshire (0924 273255); buses to Prospect Road and
Town centre; open Tuesday to Thursday 9.15–13.00, 14.00–17.30,
Friday 9.15–13.00, 14.00–18.00, Saturday 9.15–14.00
An enormous selection of products sold loose, including flour, unrefined sugar, peas, beans, lentils, brown rice, dried fruit. Fresh goat's
milk, yoghurt, cheeses, cream, with a good supply of free-range eggs
from the local farm. Plans to produce home-made wholefood bakery
items and to sell locally organically grown fruit and vegetables. Deliveries of large orders within five-mile radius, but a visit to the shop could
find you being served by Mrs Stan Barstow (*A Kind of Loving*) who
helps out in emergencies.

Oswestry

Dutton's Health Foods 2 Bailey Street, Oswestry, Shropshire (0691
3120); open Monday to Saturday 9.00–17.30, Thursday to 13.00
One of three shops owned by the Dutton family who have many years
experience in the business; the younger generation has emphasized the
health food side of the business. Sells dried wholefoods, muesli mixed
on premises, decaffeinated wholeberry coffee ground to your choice,

fresh wholewheat bread, dried fruit and nuts sold loose, herbs, vegetarian and slimmers' diet foods, free-range eggs, yoghurt and goat's milk, honey, juices, preserves, vitamin and mineral supplements, natural remedies, books on health. (NAHS)

Honeysuckle 53 Church Street, Oswestry, Shropshire (0691 3125); open Monday to Saturday 9.00–15.00, Thursday to 13.00
This cooperatively run shop was opened in 1978. The workers get a nominal wage which keeps prices low; any profits are ploughed back for expansion which should soon lead to the opening of a wholefood café. The enormous range of dried wholefoods includes three kinds of muesli mixed in the shop, many kinds of dried fruit – all free from mineral oils, lots of beans and pulses, nine kinds of flour, flakes and grains, raw sugars, shelled nuts. Honeys, a huge selection of herbs and spices, cold-pressed oils. Toilet rolls from recycled paper, henna, and complete choice of Beauty Without Cruelty natural cosmetics. Full range of home brewing and wine-making requisites. Books and magazines on health, wholefoods and the environment. Deliveries in the area for a small charge and you are invited to phone in your order for delivery or collection. (ICOM)

Otley

The Health Stores 28/30 Gay Lane, Otley, Yorkshire (0943 462328); open Monday to Saturday 9.00–17.30
Open since 1945, is about to undergo extensive renovation. Stocks most popular packaged health foods, herbs, spices, fresh eggs, wholemeal bread, vitamin and mineral supplements. Deliveries within twenty miles.

Oxford

The Food Centre 47 Walton Street, Oxford (0865 50772); open Monday and Friday 10.00–19.00, Tuesday to Thursday 10.00–18.00, Saturday 9.00–18.00
A fairly new shop which tries to stock a comprehensive, interesting range in the limited space available. Deliveries being discussed.

Holland & Barrett 3 King Edward Street, Oxford (0865 43407); open Monday to Saturday 9.00–17.30
For further information on the Holland & Barrett chain see entry for Amersham (page 180).

Uhuru Foods 48 Cowley Road, Oxford (0865 48249); open Monday to Saturday 9.30–18.00, Wednesday 10.00–14.00
This collective was originally founded in 1973 to promote crafts from the Third World – the name is from the Swahili, meaning freedom – in such a way as to educate the public about exploitation in underdeveloped countries. They then decided to lure customers by selling food and opening a café (closed temporarily). The atmosphere is such that you can sit and browse over the publications, including their newsheet *Indigestion* which discusses problems in relation to big business, food production, the source of food and related exploitation, as well as giving recipes and gardening hints. Goods are sold in a variety of sizes, packed in biodegradable bags (which can be bought by other shops). The excellent range of dried wholefoods includes two kinds of muesli – the deluxe variety is delicious – all sold at very competitive prices. Any profits go towards local community projects or the kind of research mentioned above. Craft goods are no longer sold, but there is cookware and a good selection of books and periodicals on relevant topics. It would take too long to list the numerous products on sale, but they include bread and goat's milk. Volunteers are always welcome to help out; the structure ensures that they will be accorded some say in the changing nature of the organization, which is much involved in community work of all kinds.

Wholefoods (Health Foods) 6 Suffolk House, Banbury Road, Summertown, Oxford (0865 52523); buses 421, 521 (50 yards away); open Monday to Saturday 8.30–17.30

Claims to have the largest stock in Oxford. Carries all the main brands of health foods, as well as free-range eggs, fresh dairy produce, wholemeal bread. Some organically grown vegetables, fruit and herbs in season. Also the usual vitamin and mineral supplements. (NAHS)

Paignton

Preston Health Stores 300 Torquay Road, Preston, Paignton, Devon (0803 522233); bus 120 to Preston shelter; open Monday to Friday 9.00–17.30, Wednesday to 13.00
A well-run, well-stocked shop with a huge range of herbs, goat's milk and cheese, compost-grown fruit and vegetables when available, and the usual packed wholefoods. Stocks Natrodale products and Beauty Without Cruelty cosmetics. Full range of health foods for your pets. Deliveries within Torbay area.

Southern Health Foods 52 Torquay Road, Paignton, Devon (0803 559193); open Monday to Saturday 9.00–17.30
One of the well-established Southern Health Foods chain. For further information on this chain see entry on Southern Health Foods in Aldershot (page 179).

Paisley

Hill's Health Foods 9 New Street, Paisley, Strathclyde, Scotland (041 889 4669); open Monday to Friday 9.00–17.30, Saturday 9.00–18.00
Founded in 1914, this shop was taken over by the present owners in 1975. The stock has been expanded and now includes a very good choice of dried wholefoods and health foods with the emphasis on quality. Large range of vegetarian foods and huge variety of vitamin and mineral nutritional supplements. Daily deliveries of 100 per cent wholemeal bread and rolls, plain and fruit scones, free-range eggs, fresh goat's milk, The owners are enthusiastic, and willing to offer advice should you need it.

Parkstone

Vitality Fare 11 Bournemouth Road, Parkstone, Poole, Dorset (0202 741732); buses 101, 102, 103, 104, 166, 167 stop outside; open Monday to Saturday 9.00–17.30, Wednesday to 13.00
This is in effect a new shop, the present owner having acquired the good-will in April 1979. Top-quality health foods, vitamins, herbal remedies,

cosmetics, nuts, dried and fresh fruit, wholemeal bread, ice-cream, herbs and spices. Also health magazines and herbal cigarettes. The present range of home-made cakes is, hopefully, soon to be extended.

Penarth

Dairy and Delicatessen 101 Glebe Street, Penarth, Glamorgan, Wales (0222 709189); buses to Windsor Road; open Monday to Saturday 9.00–13.00, 14.15–17.30
A general delicatessen, taken over by the present owners in 1971. They are interested in promoting wholefoods and continue to extend this side of the business: books on related subjects should soon be in stock. Vegetarian and branded health foods, herbs and spices sold loose, English and foreign cheeses, cream and milk fresh daily, vitamin and mineral supplements.

Penzance

The Granary Traditional Foods 39a Causeway Head, Penzance, Cornwall (0736 61869); buses to Greenmarket; open Monday to Saturday 9.00–17.00, closed Wednesday
Wholefoods packed from bulk include cereals, flakes and grains, pulses, flours, pasta, raw sugars, dried fruit, herbs and spices, honeys. Also teas and coffee, fresh yeast, dairy goods, vegetables – both organically grown and exotic – and bee-keeping equipment. Expansion of the shop is planned and will include a snack and juice bar. Deliveries within Penwith area.

Richards Health Foods Bread Street, Penzance, Cornwall (0736 2828); buses to Greenmarket; open Monday to Friday 9.00–15.30, Wednesday to 13.00
A centrally situated shop. Reputedly one of the best-stocked in the West Country. Mr Richards and his staff give a friendly, helpful welcome to residents (including members of the local artistic community) and visitors alike. Stocks the usual dried goods, loose herbs and herbal remedies, honeys, goat's milk yoghurt, wholemeal bread, spring waters, natural cosmetics, books on health. Hoping soon to expand to include a juice bar and sell kitchenware. (NAHS)

Perth

Edinburgh Wholefoods 35 South Street, Perth, Tayside, Scotland (0738 25755); open Monday to Saturday 9.30–18.00
There are two other shops and a warehouse under the same ownership, in Edinburgh. Sells its own packaged dried goods including grains, cereals, nuts and seeds, dried fruits, pulses, pasta, herbs and spices. Also dairy products, fruit juices, bread and savouries from its own bakeries, own stoneground wholemeal flour, honey, preserves, spreads, condiments. Deliveries throughout Scotland.

Highlands Health Store 7 St John Street, Perth, Tayside, Scotland (0738 28102); buses to High Street; open Monday to Saturday 9.30–17.30, Wednesday to 13.00
Good range of groceries and dried goods, some dairy products, wholemeal bread, twenty varieties of honey, vitamins, supplements, natural cosmetics, health books. (NAHS)

Petersfield

Spice of Life 15 High Street, Petersfield, Hampshire (0730 3925); open Monday to Saturday 9.30–13.00, Monday, Tuesday, Wednesday, Friday 14.15–17.00
This very attractive shop, decorated with natural woods and fibres, was converted from the owner's own Georgian house and retains the original frontage. Stocks a good range of packaged wholefoods, free-range eggs, live yoghurt, wholemeal bread, natural remedies. Also speciality foods which are not specifically health foods but cannot be obtained in other local shops.

Pinner

Crunch Wholefoods 61 Bridge Street, Pinner, Middlesex (01 429 1336); Buses 96b, 189. 209; tube to Pinner; open Monday to Saturday 9.30–17.00, Wednesday to 13.00
Although it sells only the usual range of proprietary brands of health foods, vitamins, herbal remedies, etc., the shop is a welcome oasis in the supermarket suburbs of west London.

Health Foods Pinner Green, Middlesex (01 866 8212); bus 183 to The Starling pub; tube to Pinner; open Monday to Saturday 9.00–17.00 Wednesday to 13.00

The present owners took over the shop in 1971. They have extended the stock to include free-range eggs, a good variety of flours, fresh yeast, stoneground wholemeal bread, pulses, honeys, herbs, spices, vegetarian special foods. A good range of vitamin and mineral supplements, herbal remedies, coffee substitutes, dairy produce (including yoghurt), and Denes natural pet products. Deliveries within locality. (NAHS)

Plymouth

Beggars' Banquet 58a Regent Street, Plymouth, Devon (0752 28449); buses to city centre

The shop is small and has a pungent smell from the wide range of herbs and spices available – mainly for culinary use or for making herb teas. Stocks a wide range of dried wholefoods like grains, pulses, cereals, flours; and a variety of Japanese foods including shoyu, umeboshi plums, miso, seaweed. A good choice of shelled nuts and dried fruits, various cooking oils, honey, coffee substitutes, take-away snacks. Also incense, natural toiletries, cook books.

Rickard-Lanes 47 Mayflower Street, Plymouth, Devon (0752 65175); open Monday to Saturday 9.00–17.30, Wednesday to 13.00

In existence for over 100 years, keeps health and vegetarian foods, juices, yoghurt, vitamins, herbal remedies, etc. (BHMA)

Southern Health Foods 40 Eastlake Walk, Drake Circus, Plymouth, Devon (0752 21681), 9 Frankfort Gate, Plymouth (0752 61822); both open Monday to Saturday 9.00–17.30

For further information on this chain see entry on Southern Health Foods in Aldershot (page 179).

Pontypridd

Health and Herbs 4a Mill Street, Pontypridd, Glamorgan, Wales (0443 405091); open Monday to Saturday 9.00–17.30, early closing Thursday

This shop was opened in 1972 by the broadcaster and author V.Lloyd-Jones. As one would expect, the emphasis is on herbs and herbal medicines. Private consultations are given. A comprehensive range of vitamin and mineral supplements, and the usual varieties of packaged health foods. Ultra-violet and infra-red treatments. A sauna and solarium is in preparation.

Natural Health Foods 6 Church Street, Pontypridd, Glamorgan, Wales (0443 402429); open Monday to Friday 9.00–17.00
Stocks the full range of wholefoods, vitamins, herbal remedies, vegetarian specialities and sells dried fruit, nuts, sugar and other goods loose from bulk.

Poole

Earthfoods 113 Commercial Road, Poole, Dorset (0202 733393); buses 101, 102, 103; open Monday to Saturday 9.00–17.30, Wednesday to 13.00
An offshot, opened in 1979, of the main shops in Bournemouth (see page 200). A very good selection of wholefoods sold loose from bulk.

Holland & Barrett 14 Kingsland Crescent, Poole, Dorset (02013 71449); open Monday to Saturday 9.00–17.30
For further information on the Holland & Barrett chain see entry for Amersham (page 180).

Porth

Porth Health Foods 43a Hannah Street, Porth, Rhondda, Glamorgan, Wales (044 361 2638); open Monday to Friday 9.30–17.00
Opened in autumn 1979, this is an offshot of Natural Health Foods at Pontypridd (see above). Stocks the full range of health foods, vitamins, herbal remedies, vegetarian specialities. Dried fruit, nuts and sugar sold loose.

Portmadoc

Natural Foods Park Dairy, 37 High Street, Portmadoc, Gwynedd, Wales (0766 2295); open every day 9.00–18.00, except Wednesday and Sunday (to 13.00); July and Aguust, every day 9.00–21.00
Open now for a year. The owner continues to demonstrate the enthusiasm displayed by his phenomenal opening hours. A refit is to be followed by further expansion. The attractively displayed stock includes fresh fruit and vegetables, dairy products, a wide range of dried wholefoods – flours, pulses, grains – vegan foods, vitamin and mineral supplements, herbal cigarettes and tobacco, natural cosmetics. Good choice of healthfood publications. Wholefood sandwiches in summer. In all, a boon to resident and visitor alike. Deliveries within Portmadoc.

Portsmouth

Holland & Barrett 126 London Road, North End, Portsmouth, Hampshire (0705 61517); open Monday to Saturday 9.00–17.30
For further information on the Holland and Barrett chain see entry for Amersham (page 180).

Prestatyn

Health Food Store 25 Meliden Road, Prestatyn, Clwyd, Wales; open Monday to Saturday 9.00–13.00, 14.00–17.30
Has been trading for twenty years. Stocks the general range of branded wholefoods, free-range eggs, natural cosmetics, herbal remedies, nutritional supplements. Wholemeal bread delivered daily. Deliveries within Prestatyn. (HFA)

Preston

Herbal and Health Food Stores (Preston) Ltd 26 Guildhall Street, Preston, Lancashire (0772 57617); open Monday to Saturday 9.00–17.30, closed Thursday
A very friendly, cheerful atmosphere. The stores continually update

ideas on all matters concerning health foods and herbal products. A comprehensive range of health foods, vitamin supplements, herbs and herbal medicines. Also beer- and wine-making ingredients and equipment. (BHMA, NAHS)

Princes Risborough

Bonkers The Garden Centre, Church Street, Princes Risborough, Aylesbury Buckinghamshire (084 44 5106); bus to Aylesbury/High Wycombe stops 200 yards away; open Monday to Saturday 9.00–17.30, Wednesday to 13.00
Health food shop and delicatessen with a large selection of cheeses, cooked meats, seafood. Also vitamins, herbal remedies, natural cosmetics. The staff are able to give expert advice in all departments. No delivery service as such, but will send items by post. (HTA, IGC)

Purley

Holland & Barrett 2 Brighton Road, Purley, Surrey (01 668 4046); buses 109, 166, 190, 197, 234, 234a, 403, 409, 411, 414; open Monday to Saturday 9.00–17.30
For further information on the Holland & Barrett chain see entry for Amersham (page 180).

Purley Wholefoods 48 High Street, Purley, Surrey (01 668 1293; buses 109, 115, 166, 190, 197, 234, 405, 409, 411 to Purley Cross; open Tuesday to Saturday 9.00–17.00
A well-established business with a comprehensive range of health foods. Free-range eggs, sea salts, teas, herbal cosmetics. Also seasonal fresh fruits. Deliveries within five-mile radius.

Ramsey

Herbex Ltd 7 Market Hill, Ramsey, Isle of Man (0624 812694); 5 minutes' walk from central bus station; open Monday to Saturday 9.00–17.30, Wednesday to 13.00

The speciality is herbs and herbal products, as the name implies. An extensive choice of herbal remedies and toiletries as well as loose herbs. Also health food groceries, dairy goods, frozen fruit and vegetables, home beer- and wine-making kits. The helpful owner will try to deliver anywhere in the island.

Ramsgate

New Seasons 2a Addington Street, Ramsgate, Kent (0843 57685); open Monday to Saturday 9.00–13.00, 14.00–17.30
Large range of prepacked goods, together with grains, beans, etc. sold from bulk. Also herbal remedies, brand-name cosmetics. Plans for a restaurant/take-away. Deliveries in Thanet area.

Rayleigh

Natrafoods 11 High Street, Rayleigh, Essex (0268 770166); open Monday to Saturday 9.00–17.30
An offshoot of Natrafoods in Southend-on-Sea (see page 349), stocking branded health foods, vegetarian and slimming diet foods, honey and preserves, dairy goods, beverages, fruit juices, herbs and herbal remedies, books on health subjects. Wholemeal snacks, fresh daily, include bread pudding, oatcakes, carob cake .(NAHS)

Reading

The High 8 High Street, Reading, Berkshire (0734 54621); buses 17, 18, 29, 40, 43, 43a, 44, 44a, 45 to Jackson's Corner; open Monday to Saturday 9.00–17.30, Wednesday to 13.00;(buttery) Monday to Friday 9.00–14.30
This well-run and popular shop now has an offshoot in Henley. Sells both proprietary health foods – an extensive range – and loose dried goods: grains, cereals, nuts, dried fruits. Wholemeal flour, milled on the premises, is sold in any quantity. Loseley products, free-range eggs, sugars, honeys, most leading brands of herbal and nutritional supplements. Also a comprehensive selection of natural cosmetics. Herbs and

spices can be bought loose. Hopes to resume delivery service soon. The very good buttery attached to the shop is strictly vegetarian, wholefood and non-smoking – all the better since it is a basement; serves vegetable soup, quiches, savoury flans, salads, cheesecake, flapjacks and sandwiches prepared daily on premises. (HSW, NAHS)

Jelly's Stores 69 Whitley Street, Reading, Berkshire (0734 81097); buses 15, 23, 26; open Monday to Saturday 8.30–20.00
Very well-run. Useful early and late opening hours. A normal grocers' with strong emphasis on wholefoods, including wholemeal bread and Loseley dairy goods; also a wide range of Indian and continental foods – pickles, spices, cheeses, salamis, sweets. A section selling books on food; and a full off-licence with a variety of continental wines, canned beers.

Reading Wholefoods 7 London Road, Reading, Berkshire (0734 55175), Traders Shopping Hall, Station Road, Reading; both open Monday to Saturday 10.00–17.30
The main shop at London Road also supplies the market unit and stalls at markets in the surrounding towns. Excellent stock of dried wholefoods sold loose in any quantity: cereals, grains, pulses, nuts, seeds, dried fruit, sugars, pasta, yeast, herbs, spices. Also oils, honey, juices, sauces, preserves; a few food supplements. Fresh bread delivered daily, and a lady at the Market Hall prepares wholemeal cakes, savouries and rolls every day. Cookware; natural cosmetics; books on related subjects. Deliveries free around Reading, for orders over £6.

Southern Health Foods 58 The Butts Centre, Reading, Berkshire (0734 52787); open Monday to Saturday 9.00–17.30
One of the well-established Southern Health Foods chain. For further information on this chain see entry on Southern Health Foods in Aldershot (page 179).

Reddish

Health Food Store 562 Gorton Road, Reddish, near Stockport, Greater Manchester (061 223 6719); buses to Bull's Head Hotel; open Monday to Saturday 9.00–18.00, Wednesday to 13.00
Full range of packaged health foods, dietary supplements, vitamins,

minerals, herbs, natural cosmetics, some local dairy produce. Also books and magazines on health subjects. Deliveries locally. (NAH)

Redditch

Prana Wholefoods (Holland & Barrett) 13 Royal Square, Redditch, Worcestershire (0527 64384); open Monday to Saturday 9.00–17.30, Wednesday to 13.00
For further information on the Holland & Barrett chain see entry for Amersham (page 180).

Reigate

Holland & Barrett 68 High Street, Reigate, Surrey (07372 48260); buses 410, 411, 414, 430; open Monday to Saturday 9.00–17.30
For further information on the Holland & Barrett chain see entry for Amersham (page 180).

Retford

Good Health 43 Grove Street, Retford, Nottinghamshire (0777 6384) open Tuesday to Saturday 9.30–17.30, early closing Wednesday
Partly delicatessen, partly health foods. Stocks the usual range of proprietary wholefoods; wholemeal flours and cereals; wholemeal bread baked locally and delivered daily; vitamins and herbal remedies; Indian foods and spices; home-brew requisites; a range of delicatessen-type foods. Hopes to have refrigeration facilities shortly, to include dairy produce and ice-cream.

Richmond

All Manna Natural Foods 179 Sheen Road, Richmond, Surrey (01 948 3633); open Monday to Saturday 10.00–18.00, restaurant (summer only) 11.00–20.00
Opened in 1976, the shop now has a sister in Battersea and plans further

expansion next year. Buying wholesale and packing on the premises keeps prices as low as possible for the excellent range of dried wholefoods, dairy goods, and wholemeal bread and confectionery baked daily at the Battersea shop. An outdoor restaurant serves wholefood dishes and snacks in summer only. Deliveries within locality for orders of £10 and over.

Holland & Barrett 46 George Street, Richmond, Surrey (01 940 3103); buses 27, 33, 37, 65, 90; tube to Richmond; open Monday to Saturday 9.00–17.30
For further information on the Holland & Barrett chain see entry for Amersham (page 180).

Ringwood

Savoury House 34 Christchurch Road, Ringwood, Hampshire (04254 3134); buses to town centre; open Monday to Friday 9.00–17.00, Thursday to 14.00
Large delicatessen and healthfood store with a modern snack bar selling salads, pizzas, quiches, etc. made on premises. Take-away service. Huge variety of goods including free-range eggs, local compost-grown fruit and vegetables, local herbs, natural remedies, nutritional supplements. Goods can be bought in bulk at a discount. Described as 'just like a mini-Fortnum's' by a local personality. Makes a point of good service. (Proprietor FHCIMA and FRSH)

Scoltocks Natural Food and Beauty Shop 22 Lynes Lane, Ringwood, Hampshire (04254 3787); open Monday to Saturday 9.00–17.00
One of three shops bearing this name in Hampshire, each with own specialization. Here at Ringwood grains and pulses are combined with a range of traditional health foods. Own brand muesli prepared on premises, and own herbal facial creams. Featured in *The Good Food Shop Guide*. Deliveries within five-mile radius. (NAHS)

Rochdale

Rochdale Health Food Centre 113 Yorkshire Street, Rochdale, Greater Manchester (0706 48141); 5 minutes' walk from main bus station; open Monday to Saturday 9.00–17.30
With a slogan 'Come and see, our advice is free', has recently been modernized for the second time, which can't be bad. Selects and buys dried fruit, nuts, pulses and rice in bulk for repackaging on the premises. Stocks vitamins, herbal remedies, protein supplements. Good range of flours, dried fruits, preserves, raw sugar, pasta – and smoking mixtures. (C of T, NAHS)

Romford

Holland & Barrett 21/23 Laurie Walk, Romford, Essex (0808 21176); buses 66, 103, 174, 175, 247, 247a, 252, 294 stop at nearby library; open Monday to Saturday 9.00–17.30
For further information on the Holland & Barrett chain see entry for Amersham (page 180).

Rotherham

Storey's Health Foods 26 Wellgate, Rotherham, Yorkshire (0709 77885); open Monday to Saturday 9.00–17.00
Popular health food groceries, dried wholefoods, herbs, herbal remedies, vitamin and mineral supplements, yeast, natural cosmetics. Bread delivered fresh daily. (NAHS)

Wicker Herbal Store Ltd 9 Eastwood Lane, Rotherham, Yorkshire (0709 72219)
Main shops of this group are in Sheffield. The company has its own factory packing a huge range of herbs and herb products, dried wholefoods like pulses, mueslis, bran preparations and other cereal products, dried fruits, nuts, vitamin capsules. Although not geared to bulk sales, vitamins can be obtained by mail order. (BHFSA, BHMA)

Rustington

Rustington Health Foods 121 The Street, Rustington, Sussex (090 62 4101); buses 212, 232 to Ash Lane and Rustington church; open Monday to Saturday 9.00–13.00, 14.15–17.30
Normal range of wholefoods, vitamin and mineral dietary supplements, honey, vegetarian foods. Quiches, pizzas, vegetable pasties and date slices prepared fresh on premises, with take-away snacks and salads during summer. (BHMA)

Ryde

Health Food Stores 103 High Street, Ryde, Isle of Wight (0983 62055); buses 3, 7 to West Street, open Monday to Saturday 9.00–13.00, 14.15–17.30
Very long-established store, carrying the usual range of wholesome groceries and herbal remedies. Always a wide variety of cheeses. Deliveries in Ryde only. (NAHS)

Rye

Miller's (Rye Health Foods) 92 High Street, Rye, Sussex (079 73 2118); nearest bus stop is at depot in railway station approach; open Monday to Friday 9.00–13.15, 14.15–17.30; Saturday 9.00–13.15, 14.15–17.00, closed Tuesday
A store within a store. Rye Healthfoods was originally a shop in its own right, but is now incorporated in Miller's – essentially a delicatessen. So you now have speciality foods from around the world on one hand and a health food department on the other. Very wide range of branded and prepacked products; flours, cereals, pulses, oils, vinegar. Also tissue salts, vitamins, herbal remedies, tisanes, natural cosmetics. (C of T)

Saffron Walden

Nuts in May 32 King Street, Saffron Walden, Essex (0799 23573); buses to High Street; open Monday to Friday 9.00–17.30, Thursday to 13.00, Saturday 8.30–17.00
Small but very well-stocked shop (described by the owner as a strait-jacket). Wide range of loose dried wholefoods: grains, cereals, rice, pulses, nuts, dried fruits, several kinds of flour, herbs, spices. Also tahini, miso, fresh yeast, free-range eggs. Jersey and local dairy goods – cheese, yoghurts, buttermilk, goat's milk, cream. Wholewheat bread baked daily in the town and delivered fresh. Muesli mixed on the premises. Good stock of Asian foods like basmatti, fresh chillies, chapatty flour, and poppadums. Usual wholefood packaged goods, e.g. teas, honey. Natural cosmetics; vitamin and mineral supplements. (NAHS)

St Albans

Health Food Shop (Holland & Barrett) 13 French Row, St Albans, Hertfordshire (0727 66562); buses 221 from Luton, 330 from Hemel Hempstead, 343 from Markyate, and many others stop in busy French Row; open Monday 9.30–17.30, Tuesday to Saturday 9.00–17.30
For further information on the Holland & Barrett chain see entry for Amersham (page 180).

Tudor Stores 2 Castle Road, St Albans, Hertfordshire (0727 58435); buses 320, 330, 341, 342, 343, 724 to Fleetville; open Monday to Friday 7.30–17.15, Saturday 9.00–13.00
Opened in 1977, and run by jolly staff who enjoy the educational aspect of their work – answering questions on ,e.g. how long soya beans have to be cooked before they turn into mincemeat, and whether sea salt is fishy. The shop mixes its own muesli with chopped figs, dates and apricots. Large range of dried goods sold from bulk includes pulses, grains, cereals, several kinds of flour, shelled nuts, dried fruits, herbs, spices, herb teas, salts, oils, juices, vitamin and mineral supplements. Additional storage space will provide facilities for bulk supplies on demand. Hopes to arrange delivery service in the future.

St Andrews

Lemnos Wholefoods 183 South Street, St Andrews, Fife, Scotland; buses to bus station; open Monday to Saturday 9.15–13.00, 14.15–17.30, Thursday to 13.00
Friendly atmosphere. Advice freely proffered by the proprietors on dietary and other matters. Stocks many types of beans, lentils, wholewheat flours, flakes, dried fruit, oils, coffee, seaweed, carob powder, syrup, chocolate, culinary herbs and spices. A section devoted to natural cosmetics, vitamins, literature. During the summer they sell home-grown vegetables. Deliveries in and around St Andrews.

St Anne's

Health Food Store 39 St David's Road South, St Anne's, Lancashire (0253 724912), The Shopping Arcade, Garden Street, St Anne's (0253 24912); both open Monday to Friday 9.00–17.30, Wednesday to 13.00, Saturday 9.00–17.00
Packaged rather than fresh foods: loose cereals and pulses, raw sugar preserves, herbs, spices, oils, juices, mineral waters, biochemical remedies, vitamin supplements, natural cosmetics, books on health. Deliveries within five miles. (HSW, NAHS)

St Austell

Goodness Gracious Globe Yard, St Austell, Cornwall (0726 63555); open Monday to Saturday 9.30–13.30, 14.00–17.00
Fairly new shop, run by people who care about the food they sell and are happy to give advice. Good stock of pulses, grains, wholemeal flours, mueslis, dried fruit, nuts, nut butters, honeys, oils, herbs, spices, wholemeal bread. Also teas, decaffeinated drinks, fruit juices, henna, incense, locally made pottery. The expanding book section has a wide choice of books on cookery, alternative technology, yoga, eastern religions; some novels. All food in the take-away section is freshly baked from wholefood products.

Health Foods (St Austell) New Town Centre, St Austell, Cornwall (0726 4685); open Monday to Saturday 9.00–17.15, Thursday to 13.00
This bright, modern shop has been in existence for more than ten years. A good range of whole and vegetarian foods. Also herbs, spices, herbal remedies, tissue salts. Well known for their range of honeys: more than fifty flavours from all over the world. (BHMA)

St Columb

Goodness Gracious 54 Fore Street, St Columb, Cornwall (0637 880602); open Monday to Saturday 9.30–13.30, 14.00–17.00
A new branch of the enterprising shop at St Austell (see page 339). Good range of all the usual wholefoods. Items like henna and incense. A variety of books. Locally made pottery. Staff are knowledgeable and willing to give advice. All food in the take-away section is freshly baked from wholefood ingredients.

St Helens

Country Life Wholefoods 19 Barrow Street, St Helens, Merseyside (0744 20821); open Monday to Saturday 9.00–17.30, Thursday to 13.00
For further information on this chain see entry on The Happy Nut House in Birkenhead (page 192).

Only Natural 40 Baldwin Street St Helens Merseyside; buses to town centre; open Monday to Saturday 9.00–17.30 Thursday to 13.00
New branch of the famous Platt's Herbal Store in Wigan (see page 374), opened in 1979 by the grandchild of Harold Platt. Emphasis here is more on wholefoods. Very good range of grains, flours, muesli and other cereals, honey, fruit juices, etc. The owners welcome suggestions from local residents to expand the stock. Dairy produce; wholemeal bread; home beer- and wine-making equipment; vitamins; minerals; special diet foods; a good selection of books on herbs and health; the famous Platt's selection of loose herbs, mainly for medicinal use. Herbal remedies include Platt's ointment. Goods sent anywhere by post. (BHMS)

St Helier (Jersey)

Leetian Health Foods House of Dupré, 15 Halkett Street, St Helier, Jersey, Channel Isles (0534 71588); open Monday to Saturday 9.00–17.30

A welcome addition to the scene in St Helier; has been open since 1976. Most goods packed from bulk on the premises; the owners claim to have the most comprehensive range in the UK! Goods include grains, pulses, nuts, seeds, bread, oils, dairy goods, free-range eggs, coffee, teas, herbs, spices, goat's milk, vitamin and mineral supplements, herbal and biochemical remedies, natural cosmetics, books and magazines, hand-made pottery. Home-made cakes and biscuits on sale, and a juice and snack bar will be opening soon. The shop is popular with heavyweight celebrities – Harry Walker and Pat Phoenix have been seen there. Deliveries throughout the island. (NAHS)

Omega Health Stores 64 Colomberie, St Helier, Jersey, Channel Isles (0534 24086); open Monday to Friday 9.00–17.00)

Stocks the usual brands of proprietary health foods, including Zimbabwe goat's milk products, Allinson's 100 per cent wholewheat bread, herbs, spices, vitamins, ginseng, herbal remedies, gluten-free products, vegetarian foods and cosmetics. Diabetic foods sold only at the new branch, Isabella House. (See below.)

Omega Health Stores Isabella House, St Aubin's Road, St Helier, Jersey, Channel Isles (0534 71295)

New branch of the Omega Stores in the Colomberie, selling a similar range of dried wholefood groceries, dairy goods (including goat's milk produce), vegetarian foods, beauty preparations, free-range eggs. Wholewheat bread, from Allinson's flour, fresh daily from local bakery. (NAHS)

St Leonards

Healthcare 18 Marine Court, St Leonards, Sussex (0424 430712); buses stop opposite shop; open Monday to Friday 9.00–17.30, Wednesday to 13.00, Saturday 9.00–17.00

Health foods, herbal remedies, vitamin supplements. Also goat's milk, yoghurt, fats, cheeses. Dried fruit and nuts sold loose. Large selection of health books.

St Peter Port (Guernsey)

Hansa Wholefood 16 Fountain Street, St Peter, Guernsey, Channel Isles (0481 23412); buses to town centre; open Monday to Saturday 9.00–17.00, early closing Thursday
This ambitious enterprise has been trading for only a year. They sell their own wholefood bakery range (baked on premises), wholefood dried goods – both in bulk and prepacked dairy products, macrobiotic foods, organically grown vegetables, natural cosmetics, literature on health foods. Wine bar adjacent sells wholemeal snacks and vegetarian dishes. Planning soon to open a health-care centre devoted to natural medicines and health through diet.

Sale

Modern Health Foods (Sale) 182 Northendon Road, Sale Moor, Sale, Greater Manchester (061 962 6551); buses 40, 41, 99 stop opposite; open Monday and Tuesday 10.00–17.30, Thursday and Friday 9.30–18.00, Saturday 9.30–17.30, closed Wednesday
Very wide range of wholefoods: many customers have said that it is the best-stocked shop for miles around. Stocks include vitamins and cosmetics but accent here is on vegetarian items. The owner says she isn't getting any younger and might pass over the business to a keen and enthusiastic buyer before long.

Salisbury

Culpeper 33 High Street, Salisbury, Wiltshire (0722 26159); open Monday to Friday 9.30–17.30
For further information on the Culpeper chain see entry for Bath (page 186).

Southern Health Foods 2 Old George Mall, Salisbury, Wiltshire (0722 29422); open Monday to Saturday 9.00–17.30
One of the well-established Southern Health Foods chain. For further information on this chain see entry on Southern Health Foods in Aldershot (page 179).

Scarborough

The Health Food Stores 8 Valley Bridge Parade, Scarborough, Yorkshire; buses to bus station; open Monday to Saturday 9.00–17.15, Monday to 12.00
Comprehensive range of health foods, together with medicinal herbs, herbal remedies, toiletries, vitamins. Deliveries in Scarborough area. (NAHS)
Victoria Health Food Centre 91 Victoria Road, Scarborough, Yorkshire (0723 69895); buses 104, 105, 106, 107, 114 to railway station; open Monday to Saturday 9.00–13.00, 14.00–17.30
The proprietor states: 'I supply a missionary in Peking and an unknown in California and any thinking person in between, including Thomas Laughton, brother of actor Charles.' Stocks most leading brands of health foods as well as wholewheat bread, vegetarian foods, herbal remedies. Nuts, dried fruits, herbs, muesli, bran and sugars sold loose. Also Denes natural dog foods. Deliveries in Scarborough, Filey, Burniston, Cloughton.

Scunthorpe

Health Food Store 11 Oswald Road, Scunthorpe, Humberside (0724 3983); buses to Mary Street; open Monday to Friday 9.00–17.30, Saturday 9.00–16.00
Established by one of the partners over twenty-five years ago. Comprehensive stock of nuts, fruit, honey, wholemeal flour (and products), fruit juices, vitamins, minerals, tissue salts. Wholemeal scones, pastries and fruit loaves baked on premises. (HSW, NAHS)

Seaford

The Health Store 14 Place Lane, Seaford, Sussex (0323 893473); buses 111, 112 to Southdown; open Monday to Saturday 9.00–13.00, 14.00–17.15, Wednesday to 13.00
Comprehensive range of groceries, dried goods, vitamin and herbal preparations. (NAHS)

Sevenoaks

Health Foods 116 High Street, Sevenoaks, Kent (0732 54705); buses to bus station; open Monday to Saturday 8.30–17.30, Wednesday to 13.00
A recently opened store. Wide range of wholefoods, grains in bulk pulses, nuts. Also cosmetics, herbal remedies, vitamin and mineral waters, and a range of English wines. Wholemeal cakes, flans and nut rissoles made on premises.

Shanklin

The Health Food Centre 25 High Street, Shanklin, Isle of Wight; open Monday to Saturday 8.30–17.30
An offshoot of the popular Newport Health Food Centre (see page 316). Full range of branded dried wholefoods; the usual vitamins and nutritional supplements; natural cosmetics. Also locally produced wholemeal bread; honey, flour, yoghurt. Deliveries throughout the island. (BHMA, NAHS)

Sheffield

Down to Earth Wholefood Collective 406 Sharrow Vale Road, Sheffield, Yorkshire (0742 685220); buses 8, 9, 81, 82, 83, 84, 88; open Tuesday to Saturday 9.30–13.00, 14.45–17.30
This is a collective (see page 14 for details), with the emphasis very much on ideals, not profits. The atmosphere is strongly political – you can buy alternative society publications. After staff receive the same subsistence wage, all profits go to local community projects. Recycled

materials used in packaging. Oils, shampoo, soya sauce, apple juice sold only from bulk, so please take your own containers. Reductions made only on normal prices for a whole sack of produce. The collective tries to avoid selling produce from countries where there is an especially repressive régime or where labour conditions are very poor. All produce organically grown and produced, as far as possible. Huge stock of grains, pulses, flours, pastas, nuts, seeds, cereals, dried fruit, spreads, drinks, seaweeds. Enormous range of loose herbs and spices; speciality foods like bean sprouts and vegetarian cheese; free-range eggs; compost-grown fruit and vegetables; natural cosmetics. Also toilet rolls from recycled paper, posters, kitchenware, wholefood cook books. Orders may be phoned in advance. Old age pensioners offered 10 per cent discount on Wednesdays and Thursdays. Advice freely given on cooking unfamiliar items. (FNWC)

Sunshine Foods Orchard Street and Chapel Walk, Sheffield, Yorkshire (0742 21869); open Monday to Saturday 9.00–17.30
Natural foods, supplements and cosmetics; free-range eggs; vegetarian foods. Fresh daily supplies of wholemeal bread and pastries.

Wicker Herbal Stores Ltd 174 Norfolk Street, Sheffield, Yorkshire (0742 21608), Stall 80, Sheaf Market, Sheffield; both open Monday to Saturday, closed Thursday
This company owns three shops in Sheffield and one in Rotherham. They have their own factory which enables them to package a wide range of dried goods, keeping quality high and prices low. Mueslis, specialized bran preparations, a full range of pulses, vitamins, dried fruits, nuts, vitamin capsules. The packs are not geared to bulk sales, but there is a mail order service for vitamins. Full range of health foods. A separate branch next door but one to the Norfolk Street shop sells a very good range of herbs and herbal products packed in the factory. (BHFSA, BHMA)

Shipston on Stour

Health and Speciality Foods 36 Sheep Street, Shipston on Stour, Warwickshire (0608 61851); open Monday to Friday 9.00–17.30
Good wide range of delicatessen-type foods and wholefoods such as

grains and pulses, stoneground flours, wholemeal and granary breads, cereals. Also coffee, speciality teas, vitamins, herbal remedies, natural beauty preparations, home beer- and wine-making equipment. Deliveries to surrounding villages.

Shoreham-by-Sea

Shoreham Health Food Ltd 5 Buckingham Road, Shoreham-by-Sea, Sussex (079 17 3705); buses 2, 21 stop outside shop; open Monday to Friday 9.00–13.00, 14.15–17.30, Saturday 9.00–13.00, 14.15–17.00, closed Wednesday
Groceries, dried goods, bread, free-range eggs, natural medicines, nutritional supplements and remedies.

Shrewsbury

Crabapple Natural Foods 16 St Mary's Street, Shrewsbury, Shropshire (0743 64559); open Monday to Saturday 9.30–17.45, Thursday to 13.00°
Supplies of dried goods in bulk and small sizes. Bread, cheese, eggs, vegetables, yoghurt, over 100 herbs and spices. Interested in and sells books and leaflets concerning self-sufficiency, herbs health, alternative technology, civil rights. Recycled paper, rennet for cheese-making, pottery, vegetable knives and other items.

Healthiway Stores Riverside, Smithfield Road, Shrewsbury, Shropshire (0743 4523); open Monday to Saturday 9.00–17.30
General range of health foods, dairy products, free-range eggs, pure honeys, natural medicines, nutritional supplements. (NAHF)

Sidcup

Better Health 1 St John's Parade, High Street, Sidcup, Kent (01 300 1593); open Monday to Saturday 9.15–17.30, Thursday to 13.00
Sells, in addition to the normal range of wholefoods, fresh yeast, dairy goods (including yoghurt and goat's milk), herbs, spices, slimming and diabetic speciality foods. Also books and magazines on health topics. Deliveries locally. (HFTA)

Sidmouth

Southern Health Foods Frobisher House, Church Street, Sidmouth, Devon (039 55 4777); open Monday to Saturday 9.00–17.30
One of the well-established Southern Health Foods chain. For further information on this chain see entry on Southern Health Foods in Aldershot (page 000).

Skipton

Noble's Health Food Stores Ltd 15 Otley Street, Skipton, Yorkshire (0756 3396); open Monday to Friday 9.00–17.30, Saturday 9.00–15.30, closed Tuesday
Packaged health food groceries; dairy goods (including live yoghurt, vegetarian cheese); Allinson's bread; natural remedies; vitamin supplements. (NAHS)

Slough

Realfoods (Holland & Barrett) 269 High Street, Slough, Buckinghamshire (0753 23407); buses 401, 456, 457, 458; open Monday 9.30–17.30, Tuesday to Saturday 9.00–17.30
For further information on the Holland & Barrett chain see entry for Amersham (page 000).

Solihull

Prana Wholefoods Ltd (Holland & Barrett) 14 Drury Lane, Solihull, West Midlands (021 705 8646); bus 151 stops very near the shop; open Monday to Saturday 8.15–17.00
For further information on the Holland & Barrett chain see entry for Amersham (page 180).

Somerton

The Delicatessen Market Place, Somerton, Somerset (0458 72585); open Monday to Saturday 8.00–13.00, 14.00–17.30
The full title of this shop, which has been open nearly six years, is Delicatessen and Health Foods: the former is predominant. Stocks packaged wholefood groceries, dairy goods, speciality teas and coffees, herbs, spices, foreign foods.

South Shields

Health Food Centre 122 Fowler Street, South Shields, Tyne and Wear (0632 566390); open Monday to Saturday 9.00–17.00
Groceries, dried goods, full range of vitamin aids, diabetic products, gluten-free items. Wholemeal bread and cakes. Dairy products include fresh goat's milk. (NAHS)

Southampton

Holland & Barrett In Owen & Owen, 176 High Street, Southampton, Hampshire (0703 23891); open Monday to Saturday 9.00–17.30
For further information on the Holland & Barrett chain see entry for Amersham (page 180).

Radiant Health Centre Ltd (Holland & Barrett) 176 Portswood Road, Portswood, Southampton, Hampshire (0703 59802); buses 11, 11a, 13, 17; open Monday to Saturday 9.00–17.30, Tuesday 9.30–17.30
For further information on the Holland & Barrett chain see entry for Amersham (page 180).

Realfoods (Holland & Barrett) 4 Pound Tree Road, Southampton, Hampshire (0703 27493); open Monday to Saturday 9.00–17.30
For further information on the Holland & Barrett chain see entry for Amersham (page 180).

Southend-on-Sea

Natrafoods 17/19 Southchurch Road, Southend-on-Sea, Essex (0702 615415); adjacent to bus station; open Monday to Saturday 9.00–17.30
Modern self-service shop selling branded health foods, vegetarian speciliaties, honey and preserves, dairy goods, beverages, fruit juices, diet foods, confectionery, herbs, herbal remedies, books on health subjects. Wholemeal snacks prepared on the premises include bread pudding, oatcakes and carob cakes; the owner hopes to extend this to a full take-away service soon. (NAHS)

Southport

Holland & Barrett Station Arcade, Chapel Street, Southport, Merseyside (0704 31090); buses 3, 5, 7; open Monday to Saturday 9.00–17.30
For further information on the Holland & Barrett chain see entry for Amersham (page 180).

Southsea

Herbie's Spice Shop 241 Albert Road, Southsea, Hampshire (0705 753458); open Monday to Saturday 9.00–18.00
An excellent shop, opened in 1977. The owners are truly interested in good cooking. Huge range of herbs, spices and related essential oils which can be bought by mail order. Also home-made preserves and wide choice of honeys; flour, cereals, several kinds of muesli; pulses, nuts, dried fruits; vegetarian health foods; oils, juices, herbal and scented teas, grain coffee; natural cosmetics, vitamins, books on health. A juice bar is planned for the future. Deliveries in the area; vitamins sent by post, as well as herbs and spices.

Radiant Health Centre Ltd (Holland & Barrett) 63 Osborne Road, Southsea, Hampshire (0705 22662); open Monday to Saturday 9.00–17.30
For further information on the Holland & Barrett chain see entry for Amersham (page 180).

Sowerby Bridge

Health Food Centre 40 Wharf Street, Sowerby Bridge, Yorkshire (0422 32441); buses 24, 28, 60, 61, 62, X12 stop 50 yards away; open Monday to Saturday 8.30–13.00, 14.00–18.00, Monday and Saturday to 17.00, Wednesday to 13.00
This attractive and well-established shop, with natural wood fittings, is run at present by a yoga teacher but may change hands. A great number of dried wholefoods sold in bulk, vegetarian foods, vitamins, minerals, a small wine-making section, books on yoga. Fresh wholemeal bread delivered daily except Monday; fruity malt loaves fresh on Thursday. Local deliveries.

Spalding

Eggstasy Wholefoods 16 Victoria Street, Spalding, Lincolnshire (0775 61927); open Monday and Wednesday 09.30–17.00, Tuesday and Saturday 9.00–17.00, Thursday 9.30–13.00, Friday 9.00–18.00
Stocks its own packs of muesli, muesli base, oats, dried fruits, nuts, pulses, herbs, spices, prepacked proprietary health foods; vitamins, herbal remedies, biochemical and homoeopathic remedies; books (around 350 titles). Also a range of coffee beans; Chinese, Indian, Italian and other speciality foods. Hopes in near future to devote a section to home-brewing.

Stafford

Holland & Barrett 12 St Mary's Gate, Stafford (0785 52758); open Monday to Saturday 9.00–17.30, Wednesday to 13.00
For further information on the Holland & Barrett chain see entry for Amersham (page 180).

Staines

Holland & Barrett 4 Clarence Street, Staines, Middlesex (0784 58489); buses 116, 117, 203, 216, 218, 441, Green Line 718, 725; open Monday to Saturday 9.00–17.30
For further information on the Holland & Barrett chain see entry for Amersham (page 180).

Holland & Barrett 1 Elmsleigh, Staines, Middlesex; open Monday to Saturday 9.00–17.30
For further information on the Holland & Barrett chain see entry for Amersham (page 180).

Staveley

F. A. Clay, MPS 9 Church Street, Staveley, Chesterfield, Derbyshire (024687 2242); buses to Church Street; open Monday to Saturday 9.00–18.00, Wednesday to 13.00, Saturday to 17,00
A chemist now enjoying a revival in the shop's health section. Customers are turning from recognized chemist's vitamin and tonic preparations to health foods and natural vitamins. Stocks of health items increasing greatly each month as demand rises. (NPA)

Stevenage

Holland & Barrett 19 Town Square, Stevenage, Hertfordshire (0438 51417); buses 303, 384, 392, 716, 800, 801, SB1, SB2, SB3; open Monday to Saturday 9.00–17.30
For further information on the Holland & Barrett chain see entry for Amersham (page 180).

Stirling

Malcolm's Wholefood Cellar 18 Maxwell Place, Stirling, Central, Scotland (0786 62640); near post office; open Monday to Saturday 9.00–18.00, Wednesday to 13.00, Saturday to 15.30

An attractive basement shop which has been open now for three years. Most items sold in bulk, or in any smaller quantities from the sack: grains, cereals, flours, beans, pulses, pastas, brown sugars, dried fruits, shelled nuts, salts, over sixty herbs and spices. Up to 26 per cent discount for large orders. Also fruit juices, teas and herb teas, coffee and coffee substitutes, honeys, oils, meatless and gluten-free special foods; dairy produce, e.g. cheese and yoghurt; goat's milk; vitamin and mineral supplements; spreads (miso, tamari, tahini, peanut butter); natural cosmetics and shampoos; a good range of home-brewing requisites, including the hops and malt. Wholemeal bread, oatcakes, scones and biscuits baked daily on premises. The bakery is still in a fairly experimental phase: once it is working well a take-away service is planned followed by a wholefood restaurant if this proves feasible. Deliveries on request within ten-mile radius.

Stockport

Country Life Wholefoods 97/99 Princes Street, Stockport, Greater Manchester (061 480 2314); open Monday to Saturday 9.00–17.30
This is a branch of Hillstart Ltd's chain of health food shops in the Midlands and the North of England. For further information about this chain see entry on The Happy Nut House in Birkenhead (page 192).

The Health Shop 307 London Road, Hazel Grove, Stockport, Greater Manchester (061 483 1576); bus 192; open Monday to Friday 9.00–17.30, Wednesday half day
Helpful and informal service has helped to win a faithful clientele among younger residents. Children are welcome. A full stock of the usual wholefoods, natural vitamins, a vegetarian section, diabetic foods, fruit and vegetable juices, herbs and herbal remedies, slimming aids.

Stockton-on-Tees

Black's Health Food Centre 49 Bishopton Lane, Stockton-on-Tees, Cleveland (0642 69087); open Monday to Saturday 9.00–17.30
One of the three Black's shops owned by the athletic G. B. Black. (For further information see entry for Bishop Auckland, page 195).

Stoke-on-Trent

Armstrong's Herbal Stores 22 Town Road, Hanley, Stoke-on-Trent, Staffordshire (0782 25417); adjacent to bus station; open Monday to Saturday 9.00–18.00
A herbalist, established eighty years. Stocks dried wholefoods – grains, pulses, cereals – herbs and herbal remedies, natural medicines, herbal toiletries, sprouting seeds.

Stoke Poges

Wexham Bakery and Health Foods Wexham Street Village, Stoke Poges, Slough, Buckinghamshire (395 2014); buses to The Stag pub; open Monday to Saturday 8.00–17.00, Wednesday and Saturday to 13.00
Converted from an old bakery and granary, this shop has a charming atmosphere and high standards of service. Fresh dairy produce from local farms; free-range eggs; good 100 per cent wholemeal bread; fresh herbs; natural medicines; nutritional supplements; Beauty Without Cruelty and Hymosa cosmetics; the usual popular brands of health foods; home wine-making equipment. (NAHS)

Stourbridge

Holland & Barrett 59 High Street, Stourbridge, West Midlands (03843 2952); open Monday to Saturday 9.00–17.30
For further information on the Holland & Barrett chain see entry for Amersham (page 180).

Stowmarket

Truefoods 10a Bury Street, Stowmarket, Suffolk (04492 3723); open Monday to Saturday 9.00–17.00, closed Tuesday
Complete range of health foods, natural vitamins, mineral supplements. Wholemeal bread. (HSW)

Stratford-on-Avon

Holland & Barrett 19 Rother Street, Stratford-on-Avon, Warwickshire (0789 68580); buses 204, 207, 208, 212, 218 to Wood Street; open Monday to Saturday 9.00–17.30
For further information on the Holland & Barrett chain see entry for Amersham (page 180).

Stroud

Sunshine Health Shop
25 Church Street, Stroud, Gloucestershire (045 36 3923); open Monday, Tuesday, Wednesday, Saturday 9.00–13.00, 14.00–17.30, Thursday to 13.00, Friday 9.00–17.30
Complete range of wholefoods, sold packaged or in bulk. Dairy products include yoghurt, butter, cream; cheeses cut to requirements. Stocks the normal prepacked health foods; herbal remedies and supplements; fresh bread and cakes. Organic vegetables in season. Deliveries can be arranged. (NAHS)

Sudbury

Nutmeg 36 King Street, Sudbury, Suffolk (07873 73632); near bus station; open Monday to Saturday 9.00–17.00
An eclectic enterprise, with a health food shop (opened in 1974) extended to include a gifts and fancy goods section, and a restaurant catering for meat-eaters as well as vegetarians. Sells health foods such as wheat germ, bran, molasses and own mixed muesli at 65p for 2lb. Also preserves, honey, fancy fruitcake, teas, loose herbs, etc. Free-range eggs. Good choice of natural cosmetics. Wholefoods and herbal remedies for pets. Gifts include hop pillows and pot-pourri. The restaurant serves morning coffee and snacks. From 12.00 a self-service salad table is available, with a choice of hot meat or vegetarian dishes, e.g. quiches and home-made turkey pie. All food is prepared fresh twice daily. The restaurant is open in the evening from Thursday to Saturday: *à la carte* menu.

Sunderland

Fulwell Health Food Store Moran Street, off Sea Road, Fulwell, Sunderland, Tyne and Wear (0643 491506); buses 123, 124 to The Bluebell, Fulwell; open Monday to Friday 9.00–13.00, 14.30–17.30, Wednesday to 13.00, Saturday 9.15–13.00, 14.00–17.30

This popular and well-established health food store prides itself on personal service – customers are encouraged to make suggestions or, indeed, criticisms. The owner proudly claims that she would never stock anything unfresh or what she would not eat or take herself. Two local doctors are regular customers. Stocks goat's milk products, fresh yeast, vegetarian foods, herbs, spices, wholemeal bread and scones (fresh-baked daily); natural cosmetics; full range of packaged health foods; vitamin and mineral supplements. Deliveries locally.

Healthcare York Street, off High Street West, Sunderland, Tyne and Wear; open Monday to Saturday 9.30–17.30

Modern corner shop with a stock of packaged health foods, nutritional supplements slimming aids natural cosmetics. Deliveries within Sunderland.

Surbiton

Surbiton Wholefoods 20 Claremont Road, Surbiton, Surrey (01 399 2772); open Monday to Saturday 9.00–17.30

Opened in 1978, this is an offshoot of the Leatherhead Wholefood Shop (see page 271). Good stock of dried wholefood groceries; some fresh vegetables as available; dairy goods; free-range eggs; home-baked wholemeal cakes, scones, savouries; vitamins, minerals, herbal remedies; natural cosmetics; books on health subjects. (HSW, NAHS)

Sutton

Holland & Barrett 213a High Street, Sutton, Surrey (01 642 5435); bus 154 to Greenford Road, buses 80, 164, 280 to Nicholas Way; open Monday to Saturday 9.00–17.30

For further information on the Holland & Barrett chain see entry for Amersham (page 180).

Sutton Coldfield

A. P. Bennett Ltd 14 Chester Road, Sutton Coldfield, West Midlands (021 354 2140); most bus routes; open Monday, Tuesday and Friday 9.00–18.30, Wednesday 9.00–13.00, Thursday 9.00–18.00, Saturday 9.00–17.30
The health food side of this chemist's was opened in 1976. Good range of wholefoods; herbal and homoeopathic remedies. Advice given. The proprietor welcomes suggestions for additions to stock.

Holland & Barrett 20 Birmingham Road, Sutton Coldfield, West Midlands (021 354 3364); buses 102, 107, 110, 352, X99 stop 50 yards from shop; open Monday, Wednesday, Thursday, Friday 9.00–17.30, Tuesday 9.30–17.30, Saturday 8.30–17.00
For further information on the Holland & Barrett chain see entry for Amersham (page 180).

Sutton in Ashfield

Bacchus Home Brew and Health Store Outram Buildings, Outram Street, Sutton in Ashfield, Nottinghamshire (026 05 57448); buses to Idlewells bus station; open Monday to Saturday 9.00–17.00, Friday to 17.30, closed Wednesday
This fairly new shop specializes in home-brewing equipment and supplies, including malt, hops and grains. Stocks the general branded health foods, herbs, oils, muesli, soya products, honey, juices, dairy goods, e.g. cheese. Some natural cosmetics. The owners hope now to concentrate on expansion of the health food side of the business.

Swaffham

Swaffham Health Foods 27 Market Place, Swaffham, Norfolk (0760 21292); bus 434 (Norwich to Kings Lynn) stops outside shop; open Monday to Saturday 8.30–17.30
Wholemeal bread and flours; herbs and spices sold in bulk; soya protein foods; nuts and pulses. Also a large range of books. Deliveries in the immediate area.

Swansea

Ear To The Ground 37 Marlborough Road, Brynmill, Swansea, Glamorgan, Wales (0792 463505); open Tuesday to Saturday 10.30–18.00, Wednesday to 14.00; Stall 55b Swansea Market, Swansea, open every day except Thursday
Groceries and dried goods include wholemeal pastas, flours, flakes. Stocks Japanese products, e.g. miso. Large orders delivered, by arrangement. Policy here is 'If it's wholefood and you want it, we'll do our best to get it'. Plans to open a restaurant and prepare fresh food on premises.

Mumbles Health Fayre 622 Mumbles Road, Southend, Swansea, Glamorgan, Wales (0792 69624); Bus 1; open Monday to Saturday 9.30–13.00, 14.30–17.30
A pleasant sea-front shop with a view over Swansea Bay. Friendly service. Well-selected stock of dairy produce, free-range eggs, bread, the popular brands of health food groceries, natural medicines, nutritional supplements. Deliveries within two-mile radius. (NAHS)

Well's Health Food Store 5/7 Picton Arcade, Swansea, Glamorgan, Wales (0792 54053); buses 1, 2, 3; open Monday to Saturday 9.00–17.30
Complete range of health foods. Herbs, wholemeal bread, free-range eggs, goat's milk yoghurt, natural fruit juices, vitamins, biochemical remedies. A restaurant on premises serves fresh-cooked pizzas, flans, cauliflower cheese and other vegetarian dishes. (HSW)

Swindon

Joy-Bar (Healthfoods) 96 Victoria Road, Swindon, Wiltshire (0793 21831); buses 14, 15 stop outside shop; open Monday to Saturday 8.45–17.15
Good general range of packaged health and wholefoods; vitamins and herbal remedies; dairy goods (including yoghurt and cheese); teas and tisanes; herbs and spices; oils, honey and preserves; slimming aids; vegetarian foods; natural cosmetics; wholemeal bread; free-range eggs. Also books on health subjects.

Southern Health Foods 24 Canal Walk, Brunel Centre, Swindon, Wiltshire (0793 35460); open Monday to Saturday 9.00–17.30
One of the well-established Southern Health Foods chain. For further information on this chain see entry on Southern Health Foods in Aldershot (page 179).

Swinton

S. Hamblett Health Foods/Grocer 358 Worsley Road, Swinton, Greater Manchester (061 794 3362); buses 12, 26, 32, 39, 584 to Moorside Road; open Monday to Saturday 9.00–18.00
The general range of wholefoods includes breakfast cereals, dried fruits, honey, oils, fruit juices. Certain dairy products in stock, e.g. yoghurt and cheeses. Wholemeal bread. Vitamins. Deliveries locally.

Tamworth

Grape and Grain 25 Lower Gungate, Tamworth, Staffordshire (0827 62133); open Monday to Saturday 9.00–17.30, closed Wednesday
Health foods, vitamin supplements, herbal remedies, general groceries. Plans for a coffee and snack area. (NAHS)

Taunton

Southern Health Foods 5 Crown Walk, High Street, Taunton, Somerset (0823 75300); buses 274, 275, 276 to The Parade; open Monday to Saturday 9.00–17.30
One of the well-established Southern Health Foods chain. For further information on this chain see entry on Southern Health Foods in Aldershot (page 179).

Tavistock

Health Foods (Tavistock) 13 Brook Street, Tavistock, Devon (0822 3976); buses to Plymouth Road; open Monday to Saturday 9.00–17.30, Wednesday to 13.00
This modern self-service store in the town centre sells the usual wholefood groceries, wholemeal bread, dairy goods (including Bulgarian yoghurt), vitamins and herbal remedies, natural cosmetics, health food books and magazines. (NAHS)

Thurso

Thurso Health Food Centre 5 Princes Street, Thurso, Highland, Scotland (0847 3561 and 3156); open Monday to Friday 10.00–13.00, 14.00–17.00, Thursday to 13.00, and on all local and Scottish Bank Holidays
The northernmost health food shop on the British Isles mainland. Has been described by one of the Allinson family as among the best in the country. The huge stock is mainly weighed out from bulk and includes all the normal dried wholefoods. The management is very responsive to local needs; in the rare event that they do not carry what you want, it will be ordered for you. They mix their own fruit and nut muesli, and snacks are provided all year round for the tourist trade. In winter the shop runs yoga and slimming sessions and finds people to give talks and demonstrations on subjects related to health and wholefoods.

Tiverton

Greenslade's (Delicatessen and Health Food Centre) 38 Gold Street, Tiverton, Devon (088 42 2880); buses to Lowman Green; open Monday to Saturday 9.15–13.00, 14.15–17.00
Basically a delicatessen and grocery, with pies, gammons, etc. home-cooked and supplied by reputable local larders. Also a range of pre-packed health foods, vitamins, multi-minerals.

Tolworth

Health Food Centre 62 The Broadway, Tolworth, Surrey (01 299 3932); buses to The Red Lion; open Monday to Saturday 9.00–17.30, Wednesday to 14.00
A new shop in ultra-modern style. Sells the usual range of packaged wholefoods, herbal remedies, vitamin supplements, natural beauty products. Health foods and remedies for pets. A selection of books on all aspects of health and wholefoods. (HFW)

Tonbridge

Gastronomy 160 High Street, Tonbridge, Kent (0732 357696); open Monday to Saturday 9.30–17.30, Wednesday to 13.00
General range of packaged health foods and dried goods: pulses, maize, wholemeal flour, brown rice, wheat germ. Also raw cane sugar, herbal teas, vitamins.

Torquay

Grael 59 Abbey Road, Torquay, Devon (0803 211141); open Monday to Saturday 9.30–17.30
A wholefood shop and vegetarian restaurant run by the Grael Association, a community-based organization which prints and publishes books, has a book distribution service and teaches yoga and holistic psychology through its three outlets. The shop here sells both retail and in bulk, at discount prices including delivery. All the normal wholefoods and related products such as natural cosmetics. The restaurant prepares all food: pizzas, quiches, salads, American-style cakes, vegetarian hot dishes. Books – from their comprehensive stock – on health, wholefoods, nutrition, yoga and allied subjects.

Southern Health Foods 103 Union Street, Torquay, Devon (0803 23233); open Monday to Saturday 9.00–17.30
One of the well-established Southern Health Foods chain. For further information on this chain see entry on Southern Health Foods in Aldershot (page 179).

Totnes

Cranks Health Food Shop 35 High Street, Totnes, Devon (0803 862526);
open Monday to Saturday 9.00–17.30
A branch of the London Cranks. Entirely vegetarian, it sells an excellent range of packaged wholefoods, some under own label. Also Loseley dairy goods; own fresh compost-grown vegetables; free-range eggs; bread baked on premises; natural remedies and some homoeopathic supplies; nutritional supplements: natural cosmetics; books on related subjects. The associate restaurant is in Dartington, a nearby village; but here there is a take-away service with a delicious selection of bread, cakes and pastries, quiches, pizzas, nut and grain savouries, cheese baps, fruit flans and crumbles, fresh drinks. (HSW, NAHS)

Towcester

Goodness Foods 109 Watling Street, Towcester, Northamptonshire (0327 51653); open Monday to Saturday 9.00–17.30, Friday to 18.00
For further information on the Goodness Foods chain see entry for Braunston (page 203).

Truro

Health Foods (Truro) 1 Francis Street, Truro, Cornwall (0872 2991); buses to Francis Street; open Monday to Saturday 9.00–17.30, closed Thursday afternoon September to May
This shop has been called the best in Cornwall by satisfied customers. The owners are eager to help, either by explaining the uses of existing stock or by providing new items to fulfil local needs. Over 4000 lines already in stock; these include flours, sugars, whole grains, oats and barley flakes, muesli and granola breakfast cereals, beans and pulses, several kinds of rice, vegetable margarines, soya, goat's milk, locally produced cow and goat yoghurt, vegetarian cheeses, free-range chicken and duck eggs. Also home-baked wholemeal cakes and quiches, locally baked wholemeal bread, nuts from pine kernels to peanuts, dried fruit, cold-pressed oils, fruit and vegetable juices, diabetic and gluten-free foodstuffs, calorie-reduced diet foods, spreads, dried and fresh herbs,

spices, crystallized fruit, vitamin pills and dietary supplements, cosmetics, skin preparations, and shampoos. Locally produced compost-grown fruit, vegetables and herbs sold in season. Sells its own packed goods rather than branded groceries. Items prepared in the shop include muesli, bean soup mixtures, dried fruit salads, mixed nuts, curry powders, herbal mixtures. Honey, malt and molasses are put into jars from bulk containers. A good choice of take-away snacks freshly prepared, e.g. flapjacks, date slices, apricot cake, lemon curd tarts, vegetarian quiches. Deliveries all the way, from Penzance to Cardiff. (HSW, NAHS)

Tunbridge Wells

Holland & Barrett 5 Monson Road, Tunbridge Wells, Kent (0892 30553); buses 6, 97, 181, 186, 215, 281, to Five Ways or Opera House; open Monday to Saturday 9.00–17.30
For further information on the Holland & Barrett chain see entry for Amersham (page 180).

The Pilgrims Health Foods and Restaurant 37 Mount Ephraim, Tunbridge Wells, Kent (0892 20121 shop, 0892 20341 restaurant); open Monday to Saturday 8.30–20.00 (shop), 10.00–20.00 (restaurant)
A well-run and enterprising shop. Stocks all the usual health foods; a variety of locally baked wholemeal rolls and bread; fresh fruit and vegetables organically grown where possible; goat's milk products. Also delicatessen-type speciality foods. A take-away counter in addition to the excellent vegetarian restaurant (licensed), where all food is prepared on premises. Reopened in its present form in 1979, there are still further plans for its expansion.

Twickenham

Holland & Barrett 25 Heath Road, Twickenham, Middlesex (01 892 3835); buses 33, 90, 90b, 110, 202, 267, 270; open Monday to Saturday 9.00–17.30, Wednesday to 13.00
For further information on the Holland & Barrett chain see entry for Amersham (page 180).

Uckfield

Healthfoods 2a Church Street, Uckfield, Sussex (0825 3214); open Monday, Tuesday, Thursday, Friday 9.00–13.00, 14.15–17.00, Wednesday and Saturday 9.00–13.00
Stocks the general range of health foods, including wholemeal bread, fresh yeast, free-range eggs, herbs, spices, natural remedies and cosmetics.

Ulverston

Furness Health Food Store 59 Market Street, Ulverston, Cumbria (0229 53394); beside Victoria Road bus station; open Tuesday to Saturday 9.00–17.15
This shop was formerly Rytfoods in Fountain Street. The owners, who bought the business in 1972, moved it to the present premises in 1974 following expansion. Stocks all the popular wholefood grains and pulses, vegan and vegetarian foods, goat's milk, yoghurts, herbs and spices, flours, museli, vegetable and fruit juices, nuts, herbal remedies. The usual selection of vitamins and mineral supplements. Also a full range of home beer- and wine-making equipment. So named because the owners already have a large store under this title, in Barrow-in-Furness.

Upminster

Govani (Molend Ltd) 64 Station Road, Upminster, Essex (86 20044); bus to Station Road; tube to Upminster; open Monday to Saturday 9.00–18.00, Thursday to 13.00
Chemist with a large range of health foods, nutritional supplements, natural remedies.

Healthline 32 Corbets Tey Road, Upminster, Essex (86 20495); buses 248, 370; tube to Upminster; open Monday to Saturday 9.00–13.00, 14.00–17.30
The stock here is eclectic and includes rare grocery provisions and delicatessen-type foods as well as packaged health foods. Also Oriental pickles and spices; a good range of herbs and herbal remedies. Advice freely available on the use of any item. The owner, who only recently started here, plans to sell what he describes as 'healthy clothes' in the near future.

Uxbridge

Holland & Barrett 5 Pantile Walk, Uxbridge, Middlesex (0895 37841);
open Monday to Saturday 9.00–17.30
For further information on the Holland & Barrett chain see entry for
Amersham (page 180).

Wadebridge

Chapman's Dairy and Health Food Store The Platt, Wadebridge,
Cornwall (020881 2382); open Monday to Saturday 9.00–13.00, 14.00–
17.00, early closing Wednesday
A delicatessen and dairy, which also sells health and vegetarian foods:
cereals, both prepacked and from bulk (including a good range of
flours); live yoghurt; herbal and speciality teas; fruit juices and preserves; bread; free-range eggs; vitamin and mineral supplements;
herbal remedies; natural cosmetics; beer- and wine-making equipment.

Walkden

Walkden Health Food Centre 20 St Ouen Centre, Walkden, Worsley,
Greater Manchester (061 790 6406); buses to town centre; open Monday
to Saturday 9.00–17.00, Wednesday to 12.00
The usual range of packaged health foods, vitamin and mineral supplements, herbal remedies, natural cosmetics. Deliveries within five-mile
radius.

Wallasey

Holland & Barrett 9 Townfield Way, Wallasey, Merseyside (051 638
5839); open Monday to Saturday 9.00–17.30
For further information on the Holland & Barrett chain see entry for
Amersham (page 180).

Wallington

Noah's Health Food Stores 4 South Parade, Stafford Road, Wallington, Surrey (01 647 1724); buses 154, 157, 234, 234a to public hall; open Monday to Saturday 9.00–18.00
Renowned in the area for cheerful, friendly service and knowledgeable advice from Mr Noah and his staff. Bread made from 100 per cent stoneground wholemeal flour – and scones – delivered daily from local bakery; dairy produce (including goat's milk and vegetarian cheeses); fresh and dried yeast; cereals and grains; natural fruit juices; rennet; confectionery and essences; soya derivatives; oils; preserves and honeys; sauces, vinegars and sea salt; nearly 100 herbs and spices. When extension now taking place is complete, the range will be still greater, with such delicacies as natural ice-cream and sorbet. (HSW, NAHS)

Walsall

Holland & Barrett 4 Park Street, Arcade, Walsall, West Midlands (0922 23880); open Monday to Saturday 9.00–17.30
For further information on the Holland & Barrett chain see entry for Amersham (page 180).

Walsall Wholefoods 21 The Arcade, Digbeth, Walsall, West Midlands (0922 29871); open Tuesday, Wednesday, Friday, Saturday 9.00–13.30, 14.30–17.30
Open several years now, this collective keeps prices very low by making only sufficient profit to cover overheads. As much produce as possible is organically grown, and everything except Jordan's Original Crunchy is sugar-free. Stocks whole grains and flakes; muesli; several kinds of dried fruit; a huge variety of pulses; organically grown wholemeal flour; soya, olive and sunflower oils; free-range eggs; rennet-free cheese; spreads, including lots of kinds of honey; organically grown vegetables in season. Plans for expansion into natural healing methods: acupuncture and acupressure, yoga, massage, herbalism. (MWC)

Walton-on-Thames

Holland & Barrett 23 The Centre, Walton-on-Thames, Surrey (093 22 20774); buses 218, 461, 463, 716 (Green Line) to Hepworth Way; open Monday to Saturday 9.00–17.30
For further information on the Holland & Barrett chain see entry for Amersham (page 180).

Wareham

The Purbeck Wholefood Shop 17 West Street, Wareham, Dorset (092 95 2332); open Monday to Saturday 9.00–17.00, Wednesday to 13.00
Most wholefoods sold loose, in the interest of economy to customers. Nuts, beans, grains, dried fruit, pasta, flours, seeds, herbs, spices, honey, cooking oil. Also muesli (made to their own recipe), natural yoghurt, cheese, organically grown vegetables. A stall on market days at Dorchester, Weymouth, Wimborne and Lymington.

Warley

Cheese and Wine 475 Bearwood Road, Smethwick, Warley, West Midlands (021 429 1100); buses B82, 215, 220; open Monday to Friday 9.00–18.00
A well-established shop which, as its name implies, stocks primarily speciality and delicatessen-type foods; but it does sell the normal range of packaged healthfoods, free-range eggs, dietary supplements, wholewholemeal bread. A special feature is home-brew kits. Has an off-licence, so sells real tap-ale. Deliveries locally if amount worth while. (NAHFT)

Warrington

Country Life Wholefoods 69 Bridge Street, Warrington, Cheshire (0925 39758); open Monday to Saturday 9.00–17.30
This is a branch of Hillstart Ltd's chain of health food shops in the Midlands and the North of England. For further information on this chain see entry on The Happy Nut House in Birkenhead (page 192).

Warwick

Warwick Health Food Store 40a Brook Street, Warwick (0926 44311); buses to Warwick market place; open Monday to Friday 9.00–17.00, Thursday to 13.00
A general health food store selling the popular foods, herbal remedies, natural cosmetics. Also free-range eggs, fresh vegetables in season, Local deliveries.

Watford

Crystal Harvest 172 High Street, Watford, Hertfordshire (0923 20962); buses 142, 158, 306, 311/12, 347/8, 708, 719 stop 100 yards away; tube to Watford High Street; open Monday to Saturday 9.30–17.30, Wednesday to 13.30
This enterprising business hopes to extend in the near future into the realms of imported clothing and jewellery 'at competitive prices'. A fine stock of pulses, rice, dried fruit and nuts; some ninety herbs and spices sold loose; flour in bulk; fresh yeast; goat's milk and yoghurt; deep-litter eggs. Progressive discounts for bulk buying – ask for up-to-date price list. Deliveries in Watford area.

Holland & Barrett 2 Clarendon Road, Watford, Hertfordshire (0923 27393); buses 142, 306, 311, 719 to Beecham Grove, 321 347, W4 to Clarendon Road; open Monday to Saturday 9.00–17.30
For further information on the Holland & Barrett chain see entry for Amersham (page 180).

Weedon

Goodness Foods 40 High Street, Weedon, Northamptonshire (0327 41137); open Monday to Thursday 8.30–17.30, Friday 8.30–18.30, Saturday 8.30–16.30, Sunday 14.30–16.30
For further information on the Goodness Foods chain see entry for Braunston (page 203).

Wellington

Healthiway Stores Market Arcade, Wellington, Telford, Shropshire (0952 42526); open Monday to Saturday 9.00–17.00
High-protein foods, vegetarian foods, mineral supplements, herbal remedies, bread, free-range eggs, general groceries. (NAHS)

Sunseed 24 High Street, Wellington, Somerset (082 347 2313); bus 205; open Monday to Saturday 9.00–17.30
The new owner, who took over in autumn 1979, plans to retain the same basic stock and to build on it to include more dried wholefoods; a greater selection of wholemeal breads and take-away vegetarian snacks; beverages like iced herb teas. Wholefoods, packaged and canned vegetarian foods, cereals, herbs and spices, organically grown fruit and vegetables from local sources, vitamin and mineral supplements. Books and magazines on health; an enormous selection of home wine- and beer-making equipment. You can also get 'chat on anything we sell'! Deliveries for orders over £15 within radius of five miles. (HBWTA)

Wells

Cathedral Cuisine 23 Sadler Street, Wells, Somerset (0749 73667); open Monday to Friday 9.00–13.00, 14.00–15.30; Wednesday to 13.00, Saturday 14.00–17.00
Situated opposite the cathedral. Stocks the normal range of packaged health foods. (NAH)

The Good Earth Natural Food Store and Restaurant 4 Priory Road, Wells, Somerset (0749 78600); 100 yards from bus station; open Monday to Saturday 9.00–17.00, Wednesday to 13.00 (shop); Monday to Saturday 10.30–16.30, Wednesday 9.00–14.00 (restaurant)
This splendid enterprise, opened in 1974, comprises a wholefood shop, restaurant and marvellous store for self-sufficiency and small-holding supplies. Will send 'goods you thought you would never see again' all over the world. Housed in the old Palace Theatre building it caters in a totally unsentimental way for anyone who wishes to work the land and utilize every part of his produce. Everything from farriery tools to livestock requisites; equipment for bee-keeping, hunting and trapping,

fish-farming, dairy farming, butchery, alternative technology, and crafts like tanning and leatherwork. Books in profusion, telling you how to use the mounds of equipment in this Aladdin's Cave. For 60p you will be sent by return the complete catalogue to date of everything available by mail order. The wholefood shop sells a wonderful range of really fresh food; the locality has been combed for the best free-range eggs; organically grown vegetables; local cheeses. Wholemeal bread is baked daily on premises, and even the butter is fresh from local cows. Also a large range of loose herbs and spices; natural cosmetics and shampoos; loads of interesting books. In the restaurant next door, all the dishes – soups which are a whole meal, seasonal salads, desserts, savouries, flans, and delights like Somerset apple cake – are made daily by the staff. A selection of teas and freshly milled coffee; fresh fruit juice and wine if you wish. Real ale is served and scrumptious cider from their own farm; some of the wine is also made locally. In summer you can eat in the courtyard garden. No pressure to buy: the staff welcomes friendly visitors who simply want to look and admire – as well they may. There is also a Third World crafts shop. Deliveries within Wells. Long may their prospects flourish!

Welwyn Garden City

Holland & Barrett 36 Fretherne Road, Welwyn Garden City, Hertford-shire (07073 23565); buses G3, G4, G5, 369 to Welwyn Garden City station; open Monday to Saturday 9.00–17.30, Wednesday to 13.00
For further information on the Holland & Barrett chain see entry for Amersham (page 180).

Wembley

Bridge to Health 25c Bridge Road, Wembley, Middlesex (01 904 1370); tube to Wembley Park; open Monday to Saturday 8.30–18.00
Strictly vegetarian produce. A selection of honeys; fresh wholemeal bread daily; sugar-free live yoghurts; goat's milk. Vegetarian savouries prepared on premises. Deliveries in Wembley/Harrow area. (HSW, NAHS)

Holland & Barrett Unit 21, Central Square, High Road, Wembley, Middlesex (01 903 7446); buses 18, 79, 79a 83, 92, 182, 297; tube to Wembley Central and Wembley Park; open Monday 9.30–17.30, Tuesday to Saturday 9.00–17.30
For further information on the Holland & Barrett chain see entry for Amersham (page 180).

West Bromwich

Prana Wholefoods (Holland & Barrett) 43 Sandwell Shopping Centre, Queens Square, West Bromwich, West Midlands (021 553 5523); open Monday to Saturday 9.00–17.30
For further information on the Holland & Barrett chain see entry for Amersham (page 180).

Westbury

Grael Wholefoods 4 Church Street, Westbury, Wiltshire (0373 864802), Angel Mill Warehouse, Church Street, Westbury; both open Monday to Saturday 9.30–17.30
Threefold enterprise comprising warehouse, retail shop, restaurant. The shop sells from bulk: grains, beans, flours, sugars, teas, cereals, flakes, herbs, spices, oils and spreads (including pure nut butters and honeys). Also juices, preserves, chutneys; items such as cello bags; toiletries, e.g. natural soaps and shampoos; a good range of books on health and ecology. The warehouse sells all these in bulk quantities and will deliver anywhere in England within eight days; retail deliveries cover the whole of the West Country and north as far as Gloucester. The shop and small restaurant are shortly to be expanded. Meanwhile they provide tasty wholemeal dishes and snacks all day: everything is prepared fresh on premises daily. (SA, WT)

Westcliff-on-Sea

The Bryn Produce Company 146 Hamlet Court Road, Westcliff-on-Sea, Essex (0702 46277); open Monday to Saturday 9.00–17.30, closed Wednesday
A recently opened store with plans to extend the already comprehensive range of health foods to include books, cosmetics, etc. Also hoping to increase clientele by introducing promotional discounts each month on various herbal remedies. Muesli is mixed on premises. A range of hand-made biscuits and sweets. Deliveries locally when possible.

Gateway to Health 381 London Road, Westcliff-on-Sea, Essex (0702 40850); bus stop is at The Plough (most buses pass here); open Monday to Saturday 9.00–17.15, Wednesday to 16.15

Wholemeal rolls and scones, also carob cake, baked fresh. Normal range of health foods and confectionery. Good stock of vitamins and health aids like ginseng and Royal Jelly. Natural cosmetics. The shop is popular with showbusiness personalities playing at the Palace Theatre and Cliffs Pavilion. (NAH)

Mead's Health Food Stores Ltd 278/280 London Road, Westcliff-on-Sea, Essex (0702 45474); buses to Hamlet Court Road; open Monday, Tuesday, Thursday, Friday 8.30–13.00, 14.15–17.30, Saturday 8.30–17.30, closed Wednesday
Mr Mead proudly proclaims this shop (opened in 1912) to be the oldest health food store in Essex. Sells the popular brands of packaged wholefood groceries, honeys, teas and tisanes, coffee substitutes, vegetarian special foods, oils, mineral waters, herbs, spices, gluten-free foods and dairy goods (including eggs and yoghurt). (NAHS)

West Croydon

Friends Foods 2/4 St Michael's Road, West Croydon, Surrey (01 688 2899); adjacent to bus station; open Monday to Friday 10.00–17.00, Saturday 9.00–17.00 (shop); Monday to Thursday 12.00–15,00, Friday and Saturday 12.00–16.30 (restaurant)
The Rainbow Cooperative has now divided its activities into vegetarian restaurant at no. 2, take-away service at no. 3 retail and bulk wholefood

shop at no. 4. Excellent stock of dried wholefoods; whole grains, pulses, flours, cereals – including their own mixed muesli which is deservedly popular – seeds, nuts, dried fruit, herb teas and grain coffees. Huge range of herbs and spices. Also cold-pressed oils, sea vegetables, honey, spreads, cider vinegar, natural soap and shampoos, candles, cast-iron cookware. The restaurant has an attractive, peaceful garden and serves vegetarian dishes, curries, salads, quiches, soups, sweets, cakes. Next door is a full take-away service for all these dishes (prepared daily on premises).

West Kirby

Health Food Stores 111 Banks Road, West Kirby, Merseyside (051 625 8066); open Monday to Saturday 9.00-17.30, Wednesday to 13.00
Established over thirty years. Good range of packaged health foods, large selection of honeys, natural yoghurt and other dairy goods; free-range eggs, natural cosmetics, vitamin and mineral supplements, Denes pet foods. Plans to start a take-away service of wholemeal sandwiches (prepared to order) and fruit juices. Deliveries within ten-mile radius. (NAHS)

Weston-super-Mare

Sunflower Natural Whole Foods 67b Orchard Street, Weston-super-Mare, Avon (0934 414518); bus 105 to Boulevard; open Monday to Friday 9.30–13.30, 14.15–17.30, Thursday to 13.30
The owners would like to promote an interest in grain cooking for ecological reasons.: their shop is clearly run on idealistic rather than profit-making lines. The immediate aim is to increase the amount of food prepared on premises daily, using completely natural ingredients. At present only muesli is so prepared. Stock a complete range of organic grains, flakes, pulses, herbs; organically grown vegetables and fruit in season; macrobiotic and quality ordinary teas; vegan and natural cosmetics; macrobiotic specialities; cold-pressed oils; and other items like soya milk produce. Also books. Deliveries only within Weston.

West Wickham

Farrington's Health Foods 7/9 Beckenham Road, West Wickham, Kent (01 777 8721); bus 194 to swimming baths; open Monday, Tuesday, Thursday, Saturday 9.00–17.30, Wednesday 9.00–13.00, Friday 9.00–18.30
A well-established shop (under the same management as before its change of name) selling a complete range of packaged groceries; Loseley diary products; fresh wholewheat bread delivered daily; remedies and vitamin supplements; some fresh vegetables when available.

Weybridge

Weybridge Wholefoods 34 Baker Street, Weybridge, Surrey (0932 47649); open Monday to Saturday 9.00–17.30, Wednesday to 13.00
Large range of wholefoods, both in bulk and prepackaged. Also vitamin and mineral supplements, herbal remedies, natural cosmetics. Wholemeal bread, cakes, scones and savouries freshly baked each day on premises (BUFTA, BUMA, NAHS)

Weymouth

Weymouth Health Food Store 29 Maiden Street, Weymouth, Dorset (0305 782615); buses to King's Statue or Town Bridge; open Monday to Saturday 9.00–13.00, 14.00–17.30, closed Wednesday
This tiny shop, established in the 1940s, recently passed into new ownership. Staff are knowledgeable and willing to give advice on the stock: a good range of wholefoods, dietary specialities, vitamin and mineral supplements, herbs, spices, herbal remedies. Also fresh bread and compost-grown vegetables when available, depending on local supplies.

Whitley Bay

Health Food Centre 130 Park View, Whitley Bay, Tyne and Wear (0632 512412); open Monday to Saturday 9.00–17.00
Groceries and dried goods, full range of vitamin aids, diabetic and

gluten-free products. Wholemeal bread and cakes. Dairy products include fresh goat's milk. (NAHS)

Whitstable

Michael's 39 High Street, Whitstable, Kent (0227 274839); buses 4, 6 stop 200 yards away; open Monday to Saturday 9.00–18.00, Wednesday to 13.30

A friendly shop selling packaged wholefoods, herbs and spices, natural remedies, nutritional supplements. Selling space is undergoing enlargement to fit in more stock. Deliveries, for orders over £10, within ten miles.

Wigan

Platt's Herbal Store 6 Market Hall, Wigan, Greater Manchester (0942 30625); buses to town centre; open Monday to Saturday 9.00–17.30, Wednesday to 12.30

Harold Platt, grandfather of the present owners, began the business in the 1940s and it is still expanding. The founder was a colourful character renowned throughout Lancashire for his 'Platt's ointment' – I have myself seen this on sale as far afield as Slough – and the store has featured in a *Punch* cartoon by Bill Tidy. Huge range of loose herbs, particularly for medicinal use, and a good stock of herbal medicines; advice is still given on their use. Also grains, pulses, nuts, dried fruit and other wholefood groceries: the range is being extended. Vitamins and mineral supplements, diet and slimming foods, honey, fruit juices. A large selection of books on relevant topics. Goods by post anywhere. There is a new branch of the shop at St Helens. (BHMA)

Wilmslow

The Happy Nut House 4 Hawthorn Lane, Wilmslow, Cheshire (0625 526144); open Monday to Saturday 9.00–17.30

This is a branch of Hillstart Ltd's chain of health food shops in the Midlands and the North of England. For further information on this chain see entry on The Happy Nut House in Birkenhead (page 192).

Wimborne

Goodness Gracious 6a Leigh Road, Wimborne, Dorset (0202 882172); open Monday to Saturday 9.00–17.30, early closing Wednesday
A fairly new shop with the accent on vegetarian foods. Also fresh wholemeal bread, oils, juices, mineral waters; natural cosmetics, vitamins and mineral supplements; herbal remedies and treatments. Parking outside for customers. Deliveries locally.

Spill the Beans 7 West Street, Wimborne, Dorset (0202 888989); open Monday, Tuesday, Thursday, Saturday 9.00–13.00, 14.00–17.00, Wednesday 9.00–13.00, Friday 9.00–13.00, 14.00–17.30
Most wholefoods sold loose in the interest of economy to customers. Nuts, beans, grains, dried fruit, pasta, flours, seeds, herbs, spices, honey, cooking oil. Also muesli (made to shop's own recipe), natural yoghurt, cheese, organically grown vegetables. Four days each week you can find their market stall at Dorchester, Weymouth, Wimborne, Lymington.

Winchester

Culpeper 4 Market Street, Winchester, Hampshire (0962 2866); open Monday to Saturday 9.30–17.30
For further information on the Culpeper chain see entry for Bath (page 186).

Holland & Barrett 13 The Square, Winchester, Hampshire (0962 3242); open Monday to Saturday 9.00–17.30
For further information on the Holland & Barrett chain see entry for Amersham (page 180).

Windsor

Holland & Barrett 29 St Leonard's Road, Windsor, Berkshire (07535 64659); bus 48a stops just outside; open Monday to Saturday 9.00–17.30
For further information on the Holland & Barrett chain see entry for Amersham (page 180).

Witney

Beanbag Wholefoods 35 West End, Witney, Oxfordshire (0993 73922);
open Monday to Saturday 9.00–17.15
By the time this edition is published, the shop will probably have moved
from its tiny Edwardian-style premises, but the owner hopes to retain
its charm and intimacy (so unlike modern supermarket style). Goods,
sold from sacks, include an excellent range of wholefoods – pulses,
grains, dried fruit, nuts, cereals, specialist flours, raw sugars, dried fruit,
pasta, herbs, spices. Also free-range eggs; goat's milk; vegetarian
cheese; 100 per cent stoneground bread; organically grown vegetables.
Books on health foods and related subjects. Deliveries free locally; for
a small charge further afield.

Wivenhoe

The Wholefood Shop 5 Station Road, Wivenhoe, Colchester, Essex (020
622 3723); open Monday to Saturday 9.30–13.00, 14.00–17.30, Wednesday and Thursday to 13.00
Because of bulk buying and packaging carried out on the premises,
prices are competitive. Wide range of health foods; spices; herbs (including plants); cosmetics; books on various subjects. Cakes and biscuits baked on premises. Free car park near by.

Woking

Holland & Barrett 11 Church Path, Woking, Surrey (04862 61546);
open Monday to Saturday 9.00–17.30
For further information on the Holland & Barrett chain see entry for
Amersham (page 180).

Wokingham

Reading Whole Foods Wokingham Market, Wokingham, Berkshire; open Tuesday and Thursday only 10.00–17.30
Market stall supplied by Reading Wholefoods with economical goods sold from bulk: cereals, grains, pulses, nuts, seeds, dried fruit, sugars, pasta, yeast, herbs, spices. Also oils, honey, juices, preserves, yeast, a few food supplements. Fresh bread. Cookware, natural cosmetics, books on related subjects.

F. J. Searle & Sons Ltd 107 London Road, Wokingham, Berkshire (0734 785899); buses stop opposite shop; open Monday to Friday 8.30–13.00, 14.15–17.30, Saturday 8.30–13.00
Good health food range, together with pills, potions, cosmetics. Pâtés, pies and hams prepared on premises. Always willing to obtain any unstocked item for customers new and old. Deliveries locally.

Wolverhampton

Holland & Barrett 104 Mander Centre, Wolverhampton, West Midlands (0902 20937); open Monday to Saturday 9.00–17.30, Wednesday 9.30–17.30
For further information on the Holland & Barrett chain see entry for Amersham (page 180).

Roots 34 Pipers Row, Wolverhampton, West Midlands (0902 52147); open Monday to Saturday 10.00–17.00, closed Monday and Thursday afternoons
Opened in 1976, is looking for new premises to expand stock still further. Very good range of dried wholefoods sold loose; natural salt and sugars, yeast, molasses, spreads, e.g. tahini and miso, oils, organically grown grains, grain coffee, Dairy produce includes goat's milk and cheeses. Herbs and spices, Ecologically acceptable goods, such as toilet rolls from recycled papers. Natural toiletries and cosmetics. Large range of books and magazines on allied subjects, Wholewheat sandwiches freshly prepared on premises.

Woodbridge

Henry's 1 Thoroughfare, Woodbridge, Suffolk (039 43 3632/3140); bus to Thoroughfare; open Monday to Saturday 9.00–17.00
Health foods take their place in this delicatessen which also sells wines and spirits, cookbooks, kitchen equipment, wine- and beer-making equipment. Groceries, sandwiches, hot pies, fresh herbs in season, natural medicines, nutritional supplements. (NAHS)

Worcester

Beanfeast Natural Foods 21 Blackfriars Square, Worcester (0905 20622); buses to city centre; open Monday to Saturday 9.30–17.00
This totally vegetarian shop is determined to maintain its image with a strong accent on natural unprocessed foods, free from additives and preservatives. Stocks a large range of whole and vegetarian foods; flours, grains, beans, pulses, several varieties of honey, dozens of herbs and spices. Dairy products include goat's milk, yoghurt. The owners maintain that their super muesli is certainly the best value for miles around and they can supply a wide range of jams, spreads and pure juices. Bookshelf subjects are cookery. alternative medicine, astrology, etc. Retail bulk discounts arranged and wholesale list provided. Local deliveries. Future plans include opening a vegetarian restaurant and publishing a wholefood vegetarian cookbook.

Holland & Barrett 55 Broad Street, Worcester (0905 26710); open Monday to Saturday 9.00–17.30
For further information on the Holland & Barrett chain see entry for Amersham (page 180).

Holland & Barrett 22 Mealcheapen Street, Worcester (0905 28153); open Monday to Saturday 9.00–17.30
For further information on the Holland & Barrett chain see entry for Amersham (page 180).

Worksop

Wright's Herbal and Health Food Stores 1a Potter Street, Worksop, Nottinghamshire (0909 473349); buses to Bridge Street; open Monday to Friday 9.00–17.30, Thursday to 13.00
As the title implies, the emphasis is on herbs for medicinal and culinary use. Friendly advice given and the owner will arrange lectures for local societies on nutrition. Large supply of all herbal products. Packaged health foods. Twenty-five kinds of honey. Deliveries within ten-mile radius.

Worthing

Becket Health Store 4 Becket Buildings, Littlehampton Road, Worthing, Sussex (0903 66257); buses 204, 205, 206; open Monday to Saturday 9.00–17.30, Wednesday half day.
Tiny shop with personal service from experienced husband-and-wife team. Stock entirely vegetarian and almost all brand names are carried. Prewitt's products delivered daily. (NAHS)

The Good Health Food Stores 8/10 Gratwicke Road, Worthing, Sussex (0903 33596); buses to Queens Road or West Street; open Monday to Saturday 9.00–17.30, Thursday to 13.00
Packaged healthfoods, dairy produce, free-range eggs, vegetarian and diet foods, bread, herbal remedies, nutritional supplements. Deliveries locally.

Nature's Way Ltd 130 Montague Street, Worthing, Sussex; open Monday to Saturday 9.00–17.30
One of the chain of Nature's Way shops. For further information on this chain see entry for Beckenham (page 187).

Nature's Way Ltd 11 Warwick Street, Worthing, Sussex (0903 31274); open Monday to Saturday 9.00–17.30
For further information on this chain see entry for Beckenham (page 187).

Wrexham

Dutton's Health Food Shop 14 Central Arcade, Hope Street, Wrexham, Clwyd, Wales (0978 4447); open Monday to Saturday 9.00–17.15, early closing Wednesday
One of three shops owned by the Dutton family, who have many years' experience in the business. Don Dutton will give health food talks to any interested people in the district. Good range of dried wholefoods, muesli mixed on premises, decaffeinated wholeberry coffee ground to your choice. Also wholewheat bread delivered daily, dried fruit and nuts sold loose, herbs, vegetarian and diet special foods, free-range eggs, yoghurt, buttermilk and goat's milk, honey, juices, preserves, vitamin and mineral supplements, natural remedies, books on health. (NAHS)

The Granary 7 Church Street, Wrexham, Clwyd, Wales (0978 263302); open Monday to Saturday 9.30–13.30, 14.00–17.00, closed Wednesday
Opened in 1979, this is an offshoot of The Granary in Chester (see page 221). Stock is sold from bulk in any quantities required, and includes the usual wide range of grains and flakes, pulses, cereals, flours; dairy produce – yoghurts, organically produced butter, cottage and vegetarian cheese, goat's milk; good stock of herbs and spices (over 100 medicinal herbs), herbal oils, vitamins. Attractive range of general goods: bean sprouts, rock and sea salts, honeys, raw sugar preserves, oils, vinegars, pickles, spreads, beverages, e.g. herb teas. The book section stocks over 500 titles. Bulk supplies are the speciality; but no deliveries made.

Yeovil

Yeovil Health Food Store Glovers Walk, Yeovil, Somerset (0935 21785);
bus to bus station; open Monday to Saturday 9.00–17.30
Good range of speciality foods of the packaged variety, with emphasis
on vitamins; herbal remedies; natural cosmetics; requirements for
diabetic and gluten-free diets.

York

Alligator 104 Fishergate, York (0904 54525); open Monday to Saturday
10.00–18.00
Somewhat earnest but well-stocked shop run on cooperative lines.
Muesli made on the premises. Large selection of pulses, grains, nuts and
other dried goods sold loose, as are the herbs (excellent variety). Many
kinds of tea, Vegetables in season, organically grown: three-quarters of
the profits go to the growers. Oil sold from bulk. The partners encourage the recycling of bags and containers and the shop acts as a clearing
house for glass jars. Also toilet rolls made from recycled paper. Large
orders delivered within York if buyer unable to collect (list sent in
advance is appreciated). (FNWC)

York Health Food Store 1 Blake Street, York (0904 55948); buses to
Exhibition Square; open Monday to Saturday 9.00–19.30
Old-established shop frequented by 'the upper crust'. Tries to give
personal service. Great variety of honeys and nuts, dried fruits, herbs
and herb teas, flours, medicinal remedies, soya products. Deliveries
within ten miles of city centre.

York Wholefood 97 Micklegate, York (0904 56804); bus to city centre;
open Tuesday to Saturday 8.30–17.00 (shop); 10.00–16.30 (restaurant)
This recommended shop must be one of the first cooperative ownership
enterprises; it was opened in 1965. Has strong links with the Soil
Association and aims to provide as much really fresh, organically
grown food as possible. A network of local growers supplies fruit and
vegetables all year round; fresh herbs in season. There is a cooler with
goat's milk and yoghurts. The shop caters for macrobiotics, vegan and
plain wholefood aficionados, with a huge range of flours, mueslis,

pulses, grains, herbs and spices, soya products, free-range eggs, preserves and honey, local cheeses, yeast extracts, pure fruit juice, etc. Dried fruit, nuts, some cereals and sugars packed on premises can be bought in bulk. Wholemeal bread and other bakery items suppled daily by local wholefood bakery. Natural cosmetics stocked, but no vitamins or food supplements (on the grounds that if you shop here you don't need them). Dogs and smokers unwelcome. The restaurant, upstairs through the shop, specializes in salads, vegetarian dishes, wholefood snacks and quiches; the food is always freshly prepared. Deliveries by special arrangement only, within the locality. (HSW, NAHS)

3 Restaurants

In this edition, we offer simply a list of those restaurants in Britain which conform most to the definition of health food printed at the beginning of the guide. Their menus and prices change constantly; so, although the opening hours are given, it is suggested that you telephone in advance of your visit: you will then have the opportunity of discovering for yourself the degree to which the produce they use is organically grown. Availability of supplies is something not constant. If you do decide that a visit for a meal is worth while, you will at least be in an atmosphere of honest endeavour. The restaurants with an asterisk* are my special recommendations.

Apologies to those restaurants who are not included: we can't find you if you don't advertise or belong to any organization! But let us know about yourselves for the next edition.

Aberdeen

J. A. W.'s Wholefood Café St Katherine's Centre, 5 West North Street, Aberdeen, Grampian, Scotland (0224 25676); open Monday, Tuesday 9.00–15.00, Wednesday to Saturday 9.00–17.00

Altrincham

Nutcracker Café 43 Oxford Road, Altrincham, Greater Manchester (061 928 4399); open Monday to Saturday 9.45–17.00, Wednesday 9.45–14.00 Also health food shop.

Ambleside

Harvest Vegetarian Restaurant Compston Road, Ambleside, Cumbria (096 63 3151); open Monday to Sunday 12.00–14.00, 17.00–20.00 Also health food shop.

Aylesbury

Counterpoint 38 Buckingham Street, Aylesbury, Buckinghamshire (0296 85275); open Monday to Friday 10.00–15.00

Delikatesserie and Health Food Store 56 Kingsbury, Aylesbury, Buckinghamshire (0296 23487); open Monday to Saturday 8.30–17.30, Thursday 8.30–13.00   Also health food shop.

Aylsham

*Itteringham Mill Itteringham, Aylsham, Norfolk (026387 206); open Wednesday to Saturday evenings, bar opens 19.30 for dinner at 20.30; Sunday, bar opens 12.30 for lunch at 13.00. All meals have 6 courses.

Bath

White Rose Coffee and Tea House 34 Brock Street, Bath, Avon (0225 311955); open Monday to Saturday 10.30–22.00, Sunday 10.30–20.00

Bexhill-on-Sea

Brant's 66 Devonshire Road, Bexhill-on-Sea, Sussex (0424 220329); open Monday to Friday 10.00–17.00, Wednesday 10.00–14.00

Nature's Way 10 Devonshire Road, Bexhill-on-Sea, Sussex (0424 216860); open Tuesday to Saturday

Birmingham

Grapevine 207 Hagley Road, Edgbaston, Birmingham, West Midlands (021 454 0672); open Monday to Saturday 12.00–14.30, 17.00–22.30

Blackburn

Lovin' Spoonfull 42 Mincing Lane, Blackburn, Lancashire; open Tuesday to Saturday 9.00–17.30, Thursday to 15.00 Also health food shop.

Blackpool

Circle Health Food Vegetarian Restaurant 311 Dickson Road, Blackpool, Lancashire (0253 52865); open Tuesday to Saturday 12.00–14.00

Nibbles Vegetarian Restaurant 14a Milbourne Street, Blackpool, Lancashire (0253 25337); open Monday 12.00–15.00, Tuesday to Saturday 12.00–15.00, 17.15–20.00, Sunday 12.00–18.00

Bognor Regis

Harvest Health Food Restaurant 6/8 York Road, Bognor Regis, Sussex (024 33 28125); open Monday 10.00–17.00, Tuesday to Saturday 10.00–17.00, 19.00–23.00

Bournemouth

Flossie's Restaurant 73 Seamoor Road, Westbourne, Bournemouth, Dorset (0202 764459); open Monday to Saturday 9.00–18.00

The Salad Centre Post Office Road, Bournemouth, Dorset (0202 21720); open Monday to Friday 10.00–17.00, Saturday 10.00–14.30

Brighton

*Ceres 23 Market Street, Brighton, Sussex (0273 27187); open Monday to Saturday 10.00–17.00

Nature's Way Health Restaurant 35 Duke Street, Brighton, Sussex (0273 21574); open Monday to Saturday 9.30–17.00   Also health food shop.

Nature's Way Health Restaurant 89 Western Road, Brighton, Sussex (0273 26181); open Monday to Saturday 9.30–17.00   Also health food shop.

Slims 92 Churchill Square, Brighton, Sussex (0273 24582); open Monday to Saturday 9.30–17.30

Sunrise 16 North Road, Brighton, Sussex (0273 603188); open Monday to Saturday 12.00–17.00

Bristol

The Salad Kitchen 18 Park Row, Bristol, Avon (0272 24539); open Monday to Friday 9.00–19.00, Saturday 9.00–18.00   Also health food shop.

Bromley

Pure and Health Restaurant 5 Westmorland Place, Bromley, Kent (01 464 2913); open Monday to Saturday 10.00–19.30

Cambridge

*Arjuna Restaurant 12 Mill Road, Cambridge (0223 64845); open Monday to Friday 9.00–18.00, Saturday 9.30–17.00   Also health food shop.

Canterbury

Stanards 20 Love Lane, Canterbury, Kent (0227 52112); open Monday to Saturday 9.00–16.00   Also health food shop.

Cardiff

Field's 99 Wyeverne Road, Cathays, Cardiff, Wales (0222 23554); open Tuesday to Saturday 12.00–14.00, 18.00–23.00; closed most of August

Grain's Diner 5 Romilly Crescent, Canton, Cardiff, Wales (0222 21905); open Monday to Sunday 12.30–14.15, 18.30–21.00

Cheltenham

The Fruit and Nut Place 4 Henrietta Street, Cheltenham, Gloucestershire (0242 20577); open Monday to Saturday 10.00–16.30, Friday, Saturday 19.30–22.00

Clynderwen

Fedwen Stores Efailwen, Clynderwen, Dyfed, Wales (09947 339); open Monday to Saturday 9.00–18.00   Snack bar.

Colchester

H. Gunton: Health Foods 81/83 Crouch Street, Colchester, Essex (0206 72200); open Monday to Saturday 8.30–17.30   Also health food shop.

Cromer

The Singing Kettle Overstrand, near Cromer, Norfolk (026 378 204); open 7 days a week, Easter to October

Croydon

The Garden Café 2 St Michael's Road, Croydon, Surrey (01 688 2899); open Monday to Thursday 12.00–15.30, Friday, Saturday 12.00–16.30 Also health food shop.

Derby

The Lettuce Leaf 21 Friargate, Derby (0332 40307); open Monday to Saturday 10.00–19.30

Eastbourne

Ceres Health Foods 3/5 Bolton Road, Eastbourne, Sussex (0323 28482); open Monday to Saturday 10.00–17.30

Nature's Way Health Restaurant 196 Terminus Road, Eastbourne, Sussex (0323 26776); open Tuesday to Saturday 9.30–17.30 Also health food shop.

Edinburgh

*Henderson's Salad Table 92/94 Hanover Street, Edinburgh, Scotland (031 225 3400); open Monday to Saturday 8.00–22.40; to 23.30 during Festival.

Netherbow Arts Centre Restaurant 43 High Street, The Royal Mile, Edinburgh, Scotland (031 556 9579); open Monday to Saturday 10.00–16.00

Exeter

Grael Wholefoods 15 North Street, Exeter, Devon (0392 37782); open Monday, Tuesday, Thursday 10.00–14.30, 16.00–18.00, Wednesday 10.00–14.30, Friday, Saturday 10.00–19.00 Also health food shop.

Falmouth

Pepperpots 45 Killigrew Street, Falmouth, Cornwall (0326 315330); open Monday to Sunday (summer) 10.00–19.00, Sunday to 16.00; (winter) 11.00–16.00

Forest Row

*Seasons Kitchen Lewes Road, Forest Row, Sussex (034 282 3530); open Tuesday 10.00–17.00, Wednesday to Saturday 10.00–17.00, 18.00–21.00

Glasgow

Gemini 305 Sauchiehall Street, Glasgow, Scotland (041 332 8146); open Monday to Saturday 10.00–19.30

Guildford

Cranks Health Food Restaurant 35 Castle Street, Guildford, Surrey (0483 68258/71936); open Monday to Friday 10.00–17.00, Saturday 10.00–16.30

*Loseley Farm Shop Loseley Park, Guildford, Surrey (0483 71881); open Wednesday to Saturday 14.00–17.00 (28 May–27 September, including Bank Holidays) Also health food shop.

Hanley

Knotty's Vegetarian Evening Resource Centre 4, Mollart Street, Hanley, Stoke-on-Trent, Staffordshire (0782 266009); open Wednesday only, 19.30–22.30

Hastings

Brant's 45 High Street, Hastings, Sussex (0424 431896); open Monday to Saturday 10.00–17.00

Nature's Way Health Restaurant 23 Robertson Street, Hastings, Sussex (0424 431124); open Monday to Saturday 9.30–17.00   Also health food shop.

Uppercrust 34 Robertson Street, Hastings, Sussex (0424 432087); open Monday to Saturday 9.00–17.00

Hereford

The Marches 24/30 Union Street, Hereford (0432 55712); open Monday to Saturday 8.30–17.30   Also health food shop.

Holt

Larners of Holt 10 Market Place, Holt, Norfolk (026 371 2244); open Monday to Friday 8.30–17.00, Thursday to 13.00.   Buttery and delicatessen, specializing in lunchtime foods.

Hove

Granary 91 Church Street, Hove, Sussex (0273 734705); open Monday to Saturday 9.30–17.30, Sunday 10.00–15.00

Three Rooms 14 Blachington Road, Hove, Sussex (0273 779933); open Monday to Saturday 10.00–15.00

Hull

Nature's Larder 10/14 Paragon Square, Hull, Humberside (0482 27853); open Monday 8.30–17.30, Tuesday to Saturday 8.30–20.00   Also health food shop.

Ipswich

Marno's 14 St Nicholas Street, Ipswich, Suffolk (0473 53106); open Monday to Wednesday 10.00–14.00, Thursday to Saturday 10.00–14.00, 19.30–22.00

Leeds

Wharf Street Vegetarian Café 17/19 Wharf Street, Leeds, Yorkshire (0532 449588); open Monday to Friday 12.00–15.00, Friday 18.30–21.30, Saturday 12.00–21.30

Leicester

The Good Earth 4 Churchgate, Leicester (0533 538585); open Monday to Friday 12.00–14.30, Saturday 12.00–18.00

The Good Earth 19 Free Lane, Leicester (0533 26260); open Monday to Friday 12.00–14.30, Saturday 12.00–18.00

Llandrindod Wells

Van's Good Food Shop Clovelly, High Street, Llandrindod Wells, Powys, Wales (0597 3320); open Monday to Saturday 9.30–16.30   Also health food shop.

London

E2
Friends Foods 51 Roman Road, Bethnal Green, London E2 (01 981 1255); open Monday to Friday 10.00–18.00, Saturday 9.00–18.00 Also health food shop.

EC1
Crumbs 48 Holborn Viaduct, London EC1 (01 236 8970); open Monday to Friday 7.30–16.00

*****The Natural Snack** 188 Old Street, London EC1 (01 251 4076); open Monday to Friday 11.00–21.30, Saturday 11.00–16.00. Only wholefoods served here. It is part of The Community Health Foundation (East West Centre).

EC2
Harvest Vegetarian Restaurant 29 Copthall Avenue, London EC2 (01 628 6129); open Monday to Friday 9.00–15.30

EC4
Food For Health 15/17 Blackfriars Lane, London EC4 (01 236 7001); open Monday to Friday 8.00–15.00

Oodles 31 Cathedral Place, St Paul's, London EC4 (01 248 2550); open Monday to Friday 11.00–17.30

Oodles 3 Fetter Lane, London EC4 (01 353 1984); open Monday to Friday 11.00–15.00

Slenders 41 Cathedral Place, London EC4 (01 236 5974); open Monday to Friday 8.00–18.15

N6
Earth Exchange 213 Archway Road, London N6 (01 340 6407); open Monday, Tuesday 12.00–19.00, Friday to Sunday 12.00–22.00 Also health food shop.

NW1
Sunwheel Restaurant 3 Chalk Farm Road, London NW1 (01 267 8116); open Tuesday to Thursday, Sunday 12.00–21.00, Friday, Saturday 12.00–22.00 Macrobiotic.

NW3
*Manna 4 Erskine Road, London NW3 (01 722 8028); open Tuesday to Sunday 18.30–24.00

Pippin 83/84 Hampstead High Street, London NW3 (01 435 6434); open Monday to Sunday 11.00–24.00 Also health food shop.

SE10
The Source Wholefood Restaurant 106 Blackheath Road, Greenwich, London SE10 (01 691 1010); open Monday to Sunday 12.00–15.00, Tuesday to Saturday 19.30–23.30

SW1

Harrods Health Juice Bar (Ground floor) Harrods, Brompton Road, London SW1 (01 730 1234, ext. 3474); open Monday to Friday 9.00–17.00, Wednesday to 19.00, Saturday 9.00–18.00. Best juice bar in Britain.

*The ICA Restaurant (Justin de Blank) The Mall, London SW1 (01 839 6762); open Tuesday to Sunday 12.00–21.00

Wilkins Natural Foods 61 Marsham Street, Westminster, London SW1 (01 222 5727); open Monday to Friday 8.00–18.00

SW6

Windmill Wholefoods 486 Fulham Road, London SW6 (01 381 1281); open Monday to Saturday 11.00–23.00, Sunday 18.00–23.00 Also health food shop.

SW7

Holland & Barrett Health Food Bar 12 Gloucester Road, London SW7 (01 584 0372); open Monday to Saturday 12.00–15.00, closed Thursday Also health food shop.

Savours at The Meritor, Queensberry Place, London SW7 (01 589 9288); open Monday to Saturday 18.00–23.00

SW11

Di's Larder 62 Lavender Hill, London SW11 (01 223 4618); open Monday to Friday 10.00–19.00

SW16

The Wholemeal Café 1 Shrubbery Road, Streatham, London SW16 (01 769 2423); open Monday to Saturday 12.00–18.00, Friday, Saturday 19.00–22.00, Sunday 13.00–18.00

SW19

The Village Garden 28 Ridgway, Wimbledon, Village, London SW19 (01 946 4840); open Monday to Sunday 12.00–22.00

W1

Beverly Vegetarian Restaurant 24/25 Binney Street, London W1 (01 629 7123); open Monday to Friday 11.30–15.30

*Cranks Health Food Restaurant In Heal's Department Store, 196 Tottenham Court Road, London W1 (01 637 2230); open Monday to Friday 10.00–17.00, Saturday 10.00–16.00

Cranks Health Food Restaurant 8 Marshall Street, London W1 (01 437 9431); open Monday to Friday 10.00–20.30, Saturday 10.00–16.30 Also health food shop.

Cranks Health Food Restaurant In Peter Robinson Department Store, Oxford Circus, London W1 (01 580 6214); open Monday to Saturday 10.00–17.30, Thursday to 19.30

Healthy Wealthy and Wise 9 Soho Street, London W1 (01 437 1835); open Monday to Saturday 9.00–17.30

Highways Roxburgh House (Lower Ground Floor), 273/287 Upper Regent Street, London W1 (01 629 5389); open Monday to Friday 11.30–15.30

Jack Spratt 17 George Street, London W1 (01 486 5909); open Monday to Friday 10.00–21.00, Saturday 14.30–21.00

Justin De Blank 54 Duke Street, London W1 (01 629 3174); open Monday to Friday 9.00–15.30, 16.30–21.30; Saturday 9.00–15.30

Mandeer Restaurant 21 Hanway Place, Hanway Street, London W1 (01 323 0660); open Monday to Saturday 12.00–15.00, 18.00–22.30

Molton Brown 58 Molton Street, London W1 (01 629 1872); open Monday to Friday 12.00–15.00

Nuthouse Restaurant 26 Kingly Street, London W1 (01 437 9471); open Monday to Friday 9.30–17.00

Raw Deal 65 York Street, London W1 (01 262 4841); open Monday to Friday 10.00–22.00, Saturday 10.00–23.30

Wholefood Farm Bar Restaurant 110 Baker Street, London W1 (01 486 8444/ 01 935 3924); open Monday to Friday 8.00–20.00, Saturday 8.30–15.30, Sunday 9.30–16.30

W2

Oodles 128 Edgware Road, London W2 (01 723 7548); open Monday to Sunday 11.00–21.00

W11

Naturally Vegetarian Bistro 22 All Saints Road, London W11

A Taste of Honey 2 Kensington Park Road, London W11 (01 727 4146); open Monday to Friday 18.30–23.30, Saturday 12.00–15.00

WC1

Action Space Centre 16 Chenies Street, off Tottenham Court Road, London WC1 (01 631 1353); open Monday to Saturday 12.00–18.00

Oodles 113 High Holborn, London WC1 (01 405 3838); open Monday to Friday 11.00–21.00, Saturday 11.30–15.00

Oodles 42 New Oxford Street, London WC1 (01 580 9521); open Monday to Sunday 12.00–21.00

Sharuna Vegetarian Restaurant 107 Great Russell Street, London WC1 (01 636 5922/3/4); open Monday to Saturday 12.00–15.00, 18.00–22.00, Sunday 13.00–21.00   Hotel above restaurant.

WC2

*Food For Thought 31 Neal Street, London WC2 (01 836 0239); open Monday to Friday 12.00–22.00

*Neal's Yard Bakery and Tea Room 6 Neal's Yard, Covent Garden, London WC2 (01 836 5199); open Monday, Tuesday, Thursday, Friday 10.30–17.15, Wednesday 10.30–15.15, Saturday 10.30–16.15

Manchester

Nut'N'Meg 444 Palatine Road, Northenden, Manchester (061 998 4589); open Monday to Saturday 9.30–18.30 Also health food shop.

*On The Eighth Day Cooperative Ltd 111 Oxford Road, All Saints, Manchester (061 273 4878); open Monday to Saturday 12.00–15.00 Also health food shop.

Newcastle upon Tyne

Super-Natural Restaurant 2 Princess Square, Newcastle upon Tyne, Tyne and Wear (0632 612730); open Monday to Saturday 9.30–22.30

Nottingham

Maxine's Salad Table 56/58 Upper Parliament Street, Nottingham (0602 43622); open Monday to Saturday 9.45–16.30

Oldham

The Alternative 103 Union Street, Oldham, Greater Manchester; open Monday to Saturday 9.00–15.00, Wednesday, Thursday. Friday, Saturday 17.30–22.30 Also health food shop.

Oxford

Bevers 36 St Michael's Street, Oxford (0865 724241); open Monday to Friday 7.00–21.00, Saturday 9.00–22.30, Sunday 10.00–21.00

Health Foods Restaurant 3 King Edward Street, Oxford (0865 4307); open Monday to Saturday 9.00–15.30

Penzance

Olive Branch 3a The Terrace, Penzance, Cornwall (0736 2438); open Monday to Saturday 10.30–14.30, 16.00–19.00

Poole

The Inn A Nutshell Royal Clipper, Arndale Centre, Poole, Dorset (02013 3888); open Monday to Saturday 10.00–17.30

Reading

The High 8 High Street, Reading, Berkshire (0734 54621); open Monday to Friday 9.00–14.30   Also health food shop.

Richmond

All Manna Of Things 179 Sheen Road, Richmond, Surrey (01 948 3633); open in summer months only. Garden restaurant.

Ringwood

Savoury House 34 Christchurch Road, Ringwood, Hampshire (04254 3134); open Monday to Friday 9.00–17.00, Thursday to 14.00.   Salad bar.

St Peter Port (Guernsey)
Hansa Wholefood 16 Fountain Street, St Peter Port, Guernsey, Channel Isles (0481 23412); open Monday to Saturday 12.00–15.00, 18.00–22.30 Wholefood wine bar. Also health food shop.

Sheffield

Brick Rabbit Café 48 Langsett Road, Sheffield Yorkshire (0742 334049); open Wednesday to Saturday 18.00–22.00

Shrewsbury

The Good Life Barracks Passage, 73 Wyle Cop, Shrewsbury, Shropshire (0743 50455); open Monday to Saturday 9.30–16.00 (summer), 9.30–17.00 (winter)

Solihull

Salad Bowl 183 High Street, Solihull, West Midlands; open Monday to Wednesday 10.00–14.30, Thursday to Saturday 10.00–17.15

Southampton

Nutters 8 St Mary Street, Southampton, Hampshire (0703 26004); open Monday to Friday 12.00–14.30, Tuesday to Friday 18.00–22.00, Saturday 12.00–24.00

Street

Mad Hatter Crispin Hall, High Street, Street, Somerset; open Monday to Saturday 9.30–17.00

Stroud

Mother Nature 2 Bedford Street, Stroud, Gloucestershire (045 36 78202); open Monday to Saturday 10.00–16.15, Thursday to 13.00

Sudbury

Nutmeg 36 King Street, Sudbury, Suffolk (07873 73632); open Monday to Saturday 9.00–16.00, Wednesday to 14.00   Also health food shop.

Swansea

Well's Health Food Store and Salad Bowl 5/7 Picton Arcade, Swansea, Glamorgan, Wales (0792 54053); open Monday to Saturday 11.00–14.30 Also health food shop.

Torquay

Grael 59 Abbey Road, Torquay, Devon (0803 211141); open Monday to Saturday 9.30–17.30 Also health food shop.

Totnes

Cranks Health Food Restaurant Dartington Cider Press Centre, Shinners Bridge, Dartington, Totnes, Devon (0803 862388)

Truro

The Granary 36 St Austell Street, Truro, Cornwall (0872 77686); open Monday to Saturday 10.30–16.30.   Also health food shop.

Tunbridge Wells

The Pilgrims Health Food and Restaurant 37 Mount Ephraim, Tunbridge Wells, Kent (0892 20341); open Monday to Saturday 10.00–20.00 Also health food shop.

Wells

*The Good Earth Natural Food Restaurant 4 Priory Road, Wells, Somerset (0749 78600); open Monday to Saturday 10.30–16.30, Wednesday 9.00–14.00   Also health food shop.

Westbury

Grael Wholefoods 4 Church Street, Westbury, Wiltshire (0373 864802); open Monday to Saturday 9.30–17.30   Also health food shop.

Worthing

Nature's Way Health Restaurant 130 Montague Street, Worthing, Sussex; open Monday to Saturday 9.30–17.00 Also health food shop.

Vega 17 Warwick Street, Worthing, Sussex (0903 32920); open Monday to Saturday 9.45–17.00

York

Aquarius 108 Fishergate, York (0904 54750); open Monday to Thursday 17.00–22.00; Friday, Saturday 11.30–14.30, 17.00–22.00

York Wholefood 98 Micklegate, York (0904 56804); open Tuesday to Saturday 10.00–16.30   Also health food shop.

4 Health Farms

These establishments all serve health food – usually three times a day and in attractive surroundings. Because this book is supposed to direct you to health *food*, we have tried to keep the emphasis on eating and not to say too much about the other aspects of what people tend to call The Movement. But these other aspects are very closely inter-related and, just as natural medicines have to be mentioned because you are going to find them in health food shops, so the natural treatments that are given in health farms must be mentioned. However, we have deliberately omitted the various health farms all over the country which offer excellent facilities for exercise, sauna baths, massage and so on but do not serve health foods or are not in fact residential.

There are basically two kinds of establishment listed in the following pages (alphabetically under counties). There are hotels and guest houses where you will get food and can spend your holidays as you wish, and there are hydros and clinics where you will have an initial examination, probably for an extra fee, and where your programme of eating, rest and treatment will be designed for you by experts. The latter are considerably more expensive because they undertake to do more for you. Quite often a hydro will say that it cannot hope to make much impression on your health in less than a fortnight, whereas a hotel or guest house will fit you in for a weekend of good food if a room is free.

It is not residential but mention should be made of The Nature Cure Clinic, 15 Oldbury Place, London W1M 3AL (telephone 01 935 6213). This is a charitable institution established to provide natural healing for people of limited means. The value of nature cure treatment is confirmed in the patients' case histories, over some fifty years, which contain evidence of innumerable cures, often of so-called incurable diseases, after orthodox methods had failed. For those looking for a residential health centre, with yoga as its main feature, mention should be made of Swami Dev Murti Yoga Centre, Highfield, Lenham, near Maidstone, Kent (telephone 062 75 431). This establishment was founded in 1962 and has the only yoga teachers in Europe who have served a ten-year apprenticeship with their Indian Master.

Bedfordshire

Henlow
Beauty Farm Ltd, Henlow Grange, Henlow, Bedfordshire (0462 811111)
Open all year round. Railway station Hitchin (patients can be met by
prior arrangement). Accommodation 12 single, 21 double rooms, 7
suites
Leida Costigan opened first beauty farm in British Isles or Europe in
1958, and in 1961 it moved to this gracious Georgian house. Residential
courses for men and women, and fees include all normal treatments, diet
consultations, meals and recreation. Very extensive range of special
beauty treatments (drawn from all over Europe) also available on request
and at extra cost. Superb heated indoor swimming pool beside the
sprung-floored exercise room. For relaxation, an elegant bridge room,
library, TV lounge. Grange dining room, pillared terrace restaurant,
boutique, and, of course, hairdressing salon. For recreationand physical
exercise there is table tennis, billiards, tennis and badminton (grass
courts), archery, croquet, rowing and sailing. Wholefoods served whenever possible in restaurants, and special diets available.

Rates £210–£350 per week (one week minimum recommended).
No children or very young adults. Restaurants open to public for dinner.
Licensed.

Ickwell Green
Yoga For Health Foundation Residentia, Centre, Ickwell Bury, Ickwell
Green, near Biggleswade, Bedfordshire (0767 27 271). Open all year
round. Railway station Biggleswade (visitors can be met). Accommodation 16 double rooms
Headquarters of a national organization seeking to promote yoga.
Equally happy to welcome beginners and experienced. Possible to book
for any length of time, though a minimum of one week recommended to
get full benefit. However, special weekends are also organized – for
instance, for those with specific problems such as back troubles, migraine or even multiple sclerosis. Also training schemes for yoga
teachers. Attractive mansion, with lounge, television room, dining room
and yoga hall. Set in old gardens, Surrounding countryside offers
pleasant walking. Ickfield has fine village green, Bedford, the Shuttle-
worth Collection, RSPB HQ, Woburn Abbey and Luton Hoo all within
easy reach.

Accommodation includes full board, meals vegetarian wholefood, also non-vegetarian food on request.

Rates average £104.65 per week, including VAT. Restaurant open to public for dinner. (£3.50) if booking made in advance. Children welcome. Licensed.

Berkshire

Kintbury
Inglewood Health Hydro, Templeton Road, Kintbury, Berkshire (04886 2022). Open all year round, including Christmas. Railway station Kintbury (patients can be met.) Accommodation 67 rooms (all except 8 have own facilities)
Seventeenth-century house set in own grounds of fifty acres. Part of larger estate mentioned in Domesday Book. Hydro (founded 1975) fully equipped for many treatments: osteopathy, physiotherapy, mudpacks, massage, etc. In addition, yoga available on two of recommended minimum seven days. Golf, tennis, squash, riding, swimming, the gymnasium – and hot-air ballooning – available for more active guests. Special diets of course obtainable; wholefoods served where possible. Professionally qualified naturopaths, physiotherapists, doctors and nurses available for consultation.

Rates £118–£497 per week plus 10 per cent service (minimum stay of seven days recommended). No children. Licensed, but alcohol forbidden to patients.

Buckinghamshire

Newport Pagnell
Tyringham Naturopathic Clinic, Newport Pagnell, Buckinghamshire (0908 610450). Open all year round. Railway station Wolverton (taxis available). Accommodation for 69 patients
A naturopathic clinic, registered as charity and as nursing home. Because naturopathy not on NHS, charges made, but reductions considered in cases of hardship, and insurance schemes which cover private nursing can sometimes be invoked because this is a clinic, not a health farm. House is Georgian mansion designed by Sir John Soane; thirty acres of

landscaped gardens. Kitchen gardens provide organically grown food, an orchard, and wells from which pure water is drawn, free of fluoride. Lounges, Louis XIV library, colour TV, games room with billiards, table tennis, chess and cards. Surrounding countryside good for gentle walks. New swimming pool.

Food wholefood, vegetarian, and special diets prescribed on admission. New patients complete a form describing their condition and often arrive equipped with notes from their doctors. Trained staff assess their needs and may recommend acupuncture, balneotherapy, breathing exercises, diets, fasting, hydrotherapy, massage, herbal and homoeopathic medicines, osteopathy and chiropody, physiotherapy or skin hygiene treatments. Relaxation taught, linked with bio-feedback monitors, and yoga exercises held regularly. One-week bookings accepted, though two weeks the recommended minimum. Patients leave with full instructions on best diet and lifestyle for the future..

Rates £86–£175, no VAT. Children welcome, under supervision. Companions not receiving treatment welcome at reduced rates. Unlicensed.

Channel Isles

Guernsey
Hotel de Beauvoir, Rue Cohu, Castel, Guernsey, Channel Isles (0481 54750). Open all year round. Port St Peter Port (arrive by Sealink – guests can be met). Accommodation 29 twin-bedded rooms

Modern purpose-built hotel opened 1977, designed in style of old farmhouse originally on site. Has bar, sun lounge, television and attractive gardens with badminton or for just sitting and staring. Three miles from St Peter Port and within easy reach of beaches, museums and potteries.

A family hotel which caters mainly for carnivores but does offer special vegetarian facilities, and uses wholefoods when possible. The manager and his wife are vegetarians and take personal control of the vegetarian cooking – each year they cater for an increasing number of vegetarians. The hotel has its own water supply – important in Guernsey which can be subject to drought in the long hot summers.

Rates £56–£94 per week plus VAT, for bed, breakfast, evening meal. Bar and restaurant open to non-residents. Children welcome. Licensed.

Cornwall

Lelant
Woodcote Hotel, The Saltings, Lelant, near St Ives, Cornwall (0736 753147). Open March–November. Railway station Lelant (taxis available). Accommodation 2 single, 6 double bedrooms

Has been vegetarian hotel for over fifty years and has reputation for good food in peaceful atmosphere. Large house, with 2 chalets, in an acre of ground – partly growing compost-grown home produce – overlooking Hayle Estuary – a scheduled bird sanctuary. St Ives and Penzance within easy reach by bus or train, and area offers lovely walks, bathing, golf and riding. Television lounge, ironing and drying room, central heating, hot and cold water in all rooms, picnic lunches.

Vegetarian food, with emphasis on wholefood and a balanced diet. Vegan food readily available. Inspection of kitchens and vegetable garden welcomed. No longer run by the Blackallers, now under the management of John and Pamela Barrett.

Rates £60 per week, plus VAT. Restaurant open to public out of season, dinner about £3.50. Children welcome. Unlicensed.

Philleigh
The Glebe Country House, Philleigh, near Truro, Cornwall (958 205). Open all year round. Railway station Truro or St Austell (taxis can be arranged). Accommodation 2 single, 2 double rooms

A guest house specializing in wholefood meals. Building is Queen Anne, with large Georgian addition, and listed as historic monument, in village of Philleigh which has thirteenth-century church to which this house was once rectory. Four acres of grounds, including vegetable and flower gardens, and the space for bees, goats and poultry. Near King Harry ferry on the Fal, and within easy reach of Truro, St Austell and St Mawes. Beautiful surroundings offer sailing, boating, fishing and walking. Indoors are table tennis, mini-billiards, a television, washing and drying facilities, babysitting, special children's suppers by arrangement, and a welcome for well-behaved dogs with own food.

Food is totally wholefood, some supplied from the grounds, but vegetarian only if requested. All meals open to non-residents, speciality Cornish teas with home-baked scones.

Rates from £63 per week, plus VAT, excluding lunch. Restaurant open to public. Children welcome. Licensed.

Cumbria

Keswick
Orchard House, Borrowdale Road, Keswick, Cumbria. (0596 72830). Open March–October and alternate Christmases. Railway station Penrith or Windermere. Accommodation 4 single, 4 twin, 2 double rooms Small, intimate guest house in Lake District; central heating; hot and cold water plus tea-making facilities in all rooms; bed, breakfast, packed lunch and dinner. Food is lacto-vegetarian, though vegans catered for if notice given. Most baking done on premises and although food not strictly food reform effort made to provide good, fresh vegetables (some home-grown) and unadulterated foods. Both European and Eastern dishes served. One quiet lounge, one TV lounge, a fairly regular clientele and an effort to provide a calm atmosphere. Meditation is practised, smoking is not.

Rates £57.75 per week, plus VAT. Well-behaved children welcome. Unlicensed.

Devon

Okehampton
Kayden House Hotel, High Street, North Tawton, near Okehampton, Devon (083782 242). Open all year round except Christmas. Railway station Exeter (taxis; or visitors can be met). Accommodation 2 single, 4 double rooms, 3 suites
Small family-owned-and-run hotel opened in 1969 in a North Devon village near Dartmoor Forest. Exeter twenty miles, Okehampton seven miles, and good situation for inland holiday activities – golf, riding, walking and river fishing. Has lounge bar, colour TV room, and TV in each double room.

Food wholefood where possible, special diets and vegetarian meals on request.

Rates £64.50 per week, plus VAT. Restaurant open to public, average meal £4. Children welcome. Licensed.

Dorset

Eype

Eype's Mouth Hotel, Eype, Bridport, Dorset (0308 23300). Open all year round. Railway station Dorchester (taxis available). Accommodation 4 single, 23 double rooms

Country-house hotel on West Dorset coast with views over Lyme Bay. Beach – good fossil hunting and birdwatching – is five minutes' walk, boats for hire and golf courses and riding stables near by. TV lounge, sun lounge and large restaurant also used as ballroom. The Old Ship Bar serves snacks, coffee and drinks; Nick's Dive cellar bar has live music; dinner dance every Saturday night. Dogs welcome with own food.

Not strictly vegetarian or health food, but serves wholefoods where possible and fresh, organically grown vegetables when available. Regularly caters for vegetarians and people on special diets. Prices quoted are for two complete days including dinner, bed and breakfast and lunch both days. Special children's lounge, large car park and garden.

Rates £24.98–£28.89, per two days, plus VAT and 10 per cent service. Children welcome. Licensed. Dining room and bars open to public.

Lyme Regis

River House, Coombe Street, Lyme Regis, Dorset (029 74 3149). Open Easter–end September. Railway station Axminster (state time of arrival and taxi will be ordered for you). Accommodation 2 single, 9 double or family rooms

Vegetarian guest house, one minute's walk from sea front at Lyme Regis on Dorset coast. Good centre for walking, sailing, water-skiing, golf, bowls, tennis and, especially, birdwatching and fossil hunting. Indoors, TV room, a lounge and dining room. No lunches served – bed, breakfast and dinner only.

Ella and Arthur Rogers aim for relaxed and friendly atmosphere and are well-known for high-quality vegetarian cooking. The local baker produces coarse brown wholemeal loaves specially for them, and all vegetables locally grown and fresh daily. Special diets such as vegan and macrobiotic provided if arrangements made well beforehand.

Rates £41 per week, plus VAT. Children welcome. Unlicensed.

Weymouth
Weymouth Hydro, Greenhill, Weymouth, Dorset (0305 786893). Open mid-January–mid-December. Railway station Weymouth (taxis available). Accommodation 15 single, 7 twin-bedded, 1 double room

Small nature cure home which aims to provide a relaxed and informal atmosphere in which to assess patients and prescribe diets and treatments. House in residential part of Weymouth, overlooking sea. Swimming, bowling, golf, tennis and sailing all available near by. Three lounges, two are sun-lounges, dining room, TV room, coal fires and central heating. Guests not taking cures may accompany patients, space permitting, at 10 per cent off quoted price.

Vegetarian food reform; some grown in the gardens and all baking done on premises. Special diets available where necessary. Treatments include osteopathy, massage, sauna, steam baths, sitz, radiant heat, G5, slendertone, facials, beauty therapy and relaxation. Quoted rates cover all except beauty treatments. Length of stay depends on patient's condition, but minimum of ten days recommended. Outpatients accepted.

Rates £80.85–£93.45 per week, plus VAT. No children. Unlicensed.

Edinburgh

Edinburgh
Kingston Clinic, 291 Gilmerton Road, Edinburgh, Scotland (031 664 3435). Open all year round. Railway station Waverley (taxis available). Accommodation for 52 people in main house and annexe

Famous nature cure centre, based in an extraordinary house which one modern architect described as 'nightmare Gothic', set in ten-acre estate ten minutes from centre of Edinburgh. As well as Clinic, also home of Edinburgh School of Natural Therapeutics. House large, with comfortable modern furnishings, central heating and TV. Luxuriant grounds with tennis courts, summer houses, blossom trees, croquet lawn and secluded areas for quiet sunbathing or pondering. Within easy reach of shops, museums, Botanic Gardens and galleries of Edinburgh, and nearby Pentland Hills offer good opportunities for walks (packed lunches available).

Food wholefood, much grown in grounds, and all is vegetarian. New arrivals assessed; treatments and diets prescribed. Patients range from seriously ill to those simply overweight or out of condition. Treatments based on an understanding of the psyche as well as the body of the patient, and include diet, water treatment and manipulation – no drugs, fasting, enemas or other shocks. All treatments carefully explained so that patients can continue to eat and live well at home and remain in a good state.

Rates £77–£98 per week, according to situation of room, preliminary consultation fee £6. Children welcome, depending on age and requirements. Unlicensed.

Gloucestershire

Wotton-under-Edge
Coombe Lodge Yoga Centre, Wotton-under-Edge, Gloucestershire (045 385 3165) Open all the year round. Railway station Stroud (patients can be met). Accommodation 1 single, 1 double bedroom (both with bathrooms)
Coombe Lodge first opened in 1953 by Kathleen Keleny-Williams as vegetarian guest house, but is now just a yoga centre offering daily sessions. These are intended to bring new vision and purpose to pupils/guests who learn to accept the tranquillity of nature within them. Peaceful garden (the last edition of *The Health Food Guide* noted benefits of morning walks barefoot on dew-laden lawns here) and organic food play their restorative parts. Sometimes weekend sessions, when Hatha yoga, meditation, nutrition and philosophy of yoga are discussed. Very peaceful place.

Rates on application, Children over twelve years welcome. Restaurant open to public – if meal ordered beforehand, average price £3.

Gwent

Tintern
The Nurtons Field Centre and Guest House, Tintern, near Chepstow, Gwent, Wales (029 18 253). Open all year round. Railway station Chepstow (or take coach to Chepstow – local bus stops at bottom of drive). Accommodation 6 bedrooms, each for 1–4 people
Most people go to the Centre for weekend or week-long courses in sub-

jects like yoga, edible wild foods, spinning, weaving and natural dyeing, ornithology or organic gardening – but guest house also accommodates people on their own private holidays, space permitting. The Centre is in Wye Valley – designated as an area of 'outstanding natural beauty' – and within short drive of Forest of Dean and Brecon Beacons National Park. House mostly Victorian, partly Elizabethan, and recent redecoration revealed inglenook fireplace in sitting room. Has dining room, sitting room with TV, lounge–library and laboratory–studio–games room. Surrounded by twelve hectares of land given over to gardens, woodland, orchards and market garden in which vegetables organically grown for the kitchens (surplus sold to visitors).

Food wholefood and vegetarian, with macrobiotic bias. Much grown on premises (and visitors welcome to look round the market garden and discuss methods). All savouries, bread, etc. home-baked with 100 per cent wholewheat flour, and spring water always available. Rates cover bed, breakfast and dinner – packed lunches extra. Booking form advises suitable clothing and equipment for particular courses. Menus varied, and designed to tempt non-vegetarians as well as those used to vegetarian food. Family parties and well-behaved dogs welcome.

Rates £50 to £60 per week, plus VAT. Restaurant open to public for dinner, average price £3. Children welcome. Unlicensed (but may apply for licence soon).

Hampshire

Liphook
Forest Mere Hydro, Liphook, Hampshire (0428 722051). Open all year round except Christmas. Railway station Liphook (patients can be met). Accommodation 42 single, 21 double rooms, 1 suite
Now owned by Savoy Hotel group, Forest Mere first opened in 1962. A combination of hotel, retreat and spa, set in 250 acres of wooded grounds, where you can walk among lily ponds, gardens, pines and lawns. Eighteen-hole Liphook golf course close by. Elegantly furnished rooms all have colour TV, telephones, electric blankets. Fees cover all treatments prescribed by qualified consultants and also include physiotherapy, osteopathy, yoga and consultations. Amenities include sailing, tennis, badminton, croquet, swimming (with sheltered sunbathing terraces), golf, riding, billiards and cardplaying. Principal is a doctor.

State Registered Nurses available by arrangement for guests who have to be under medical supervision.

Rates £185–£274 per week (seven-day stay obligatory for first visit). No children. Unlicensed.

Hertfordshire

Tring
Champneys Health Resort, Tring, Hertfordshire (04427 3351). Open all year round. Railway station Berkhamsted (taxis available). Accommodation 63 rooms in the house and 20 garden rooms

Champneys was England's first health resort – opened 1925 and then reopened 1974 when modern medical techniques incorporated. Specializes in rest cures for stressed executives, suffering from tension, obesity and too much wine and work, but any jaded patient welcome, particularly with cardiac or weight problems. Some go just for a holiday. Large, elegantly furnished house in 170 acres of own parkland in Chilterns. Has drawing room, library, games room, light diet room and dining room, heated indoor swimming pool, tennis courts, outdoor gymnast trail, boutique, facilities for film shows, cards and chess, and Leisure Craft Centre offering evening classes in pottery, painting, etc. Evening entertainment includes dancing and musical evenings. Coach trips to local places of interest.

Food wholefood wherever possible, much grown in the grounds, and not vegetarian unless specially requested, or vegetarian diet deemed suitable for a particular condition. Diets vary from the very light (prescribed for the very heavy) to full meals with wine for those with no weight problems. Treatments include stress screening, physiotherapy, osteopathy, massage, hydrotherapy, heat treatments, facials, and there are also breathing sessions, yoga and exercise sessions, and lectures and discussions organized by the various members of the medical team. Outpatients accepted by appointment. Emphasis is on having a nice time while you get fit.

Rates £196–£483 per week, plus VAT (one week minimum recommended). Children over sixteen welcome. Licensed.

Radlett

Shenley Lodge Health Resort, Ridge Hill, Shenley, Radlett, Hertfordshire (0707 42424/5 and 43254). Open all year round except Public Holidays. Railway station St Albans or Potters Bar. Accommodation 1 single, 5 double rooms.

Large seventeenth-century house, set amidst quiet Hertfordshire countryside and surrounded by seven acres of woodland. Offers wide selection of facilities incuding gymnasiums, saunas, plunge pools, rest rooms and treatment rooms. All treatment and exercise formulated according to individual requirements as advised by the beauty therapists and gymnast. Most clientele at Shenley Lodge are daily although accommodation available for small number of residents. Programme based on a high-protein diet combined with exercise, sauna and massage. Beautiful walks around the countryside although Lodge not far from several towns.

Rates £30 per day plus 12$\frac{1}{2}$ per cent service. No children. Unlicensed.

Kent

Barham

Spinning Wheel Hotel, Pages Downs, Barham, near Canterbury, Kent (022 782 286). Open all year round. Railway station Canterbury or Aylesham (taxis; or visitors can be met). Accommodation 3 single, 22 double rooms, 1 suite

Tudor-style country house, which became hotel about fifty years ago, set in a Kentish village roughly half-way between Canterbury and Folkestone. Nine public rooms include TV room and licensed restaurant. Surrounding facilities include everything Canterbury and the North Downs have to offer, including cathedral, Pilgrims' Way, golf, riding, and tennis in Barham village.

Hotel specializes in wholefoods and much of baking is done on premises. Vegetarian, vegan and macrobiotic meals available, also non-vegetarian meals if preferred.

Rates £6–£8 per night, plus VAT and 10 per cent service. Restaurant open to public, average meal £4–£6. Well-behaved children welcome. Licensed.

Leicestershire

Ragdale
Ragdale Hall, Ragdale, near Melton Mowbray, Leicestershire (066 475 831). Open all year round, apart from one week at Christmas. Railway station Leicester (patients can be met). Accommodation 27 single, 21 double rooms, 3 suites, all with bathrooms
Georgian mansion in beautifully landscaped gardens in Quorn country. Owned by *Slimming* magazine and emphasis very much on weight loss. Has spacious lounge and foyer, a quiet room, TV, in all bedrooms, pleasant grounds to walk in and lovely countryside around. Tennis court, outdoor heated swimming pool, archery, boule, jokari; facilities for various indoor games, well-equipped gymnasium, and boutique for buying new clothes one size smaller.

Resident dietician advises on all meals – which are wholefood, vegetarian and generally deliberately small. Large qualified staff can provide massages, slendertone, hypnotherapy, panthermal, sontegra solarium, yoga, exercises and full range of facial treatments. There is a sauna, with plunge pool and impulse shower, and lessons are given on make-up and on diet régime for the future.

Rates £190–£300 per week (minimum stay seven days). No children. Licensed (but licence rarely taken advantage of).

Peeblesshire

Stobo
Stobo Castle Health and Beauty Spa, Stobo Castle, Peeblesshire, Scotland (07216 249). Open all year round. Railway station Waverley, Edinburgh (taxis; or patients can be met). Accommodation 6 single, 9 twin rooms, 1 suite
Georgian castle, complete with battlements, in Border country twenty-seven miles south of Edinburgh, six miles from Peebles. One of Britain's most expensive health farms. Run by Gaynor Winyard, once chairman of the Society of Health and Beauty Therapists, emphasis is on relaxation, weight loss and beauty treatments. So some treatments geared more to women than to men, and to redress the balance there is an executive health course especially for men. Grounds include terraces, trout lake, and Japanese water garden. Surroundings – Tweed valley

and the Eildon Hills – are beautiful. Impressive views from all windows. Facilities in the area for golf, shooting, fishing, horse-riding, tennis and squash. There are two large reception rooms, oak-panelled dining room, telephones and TV in every room. They try to provide a relaxing atmosphere and promise to 'pamper you into good health'. Quite a few famous regular clients.

Resident dietician tells you what you may eat from the choice of food – mostly wholefood but not vegetarian. Organically grown vegetables and fruit, River Tweed salmon, Border beef, lamb and game. Private water supply piped into the castle from nearby mountain spring. Treatments include sauna, steam baths, massages, Faradism and Galvanism, a solarium, exercises in gym, yoga, reflexology, aroma-therapy, chiropody and full range of facials – all included in the overall price, even if you have your hair set daily. Day visitors offered a top-to-toe service. No treatments on Sundays.

Rates about £325 per week. Children over sixteen welcome. Unlicensed.

Perthshire

Callander
Brook Linn Vegetarian Guest House, Callander, Perthshire, Scotland (0877 30103). Open May–September. Railway station Stirling (taxi may be ordered to meet train). Accommodation 2 single, 4 double rooms
Vegetarian guest house on hillside overlooking valley in which Callander lies. Good centre for exploring mountain and loch scenery of the Trossachs and Scottish Highlands. Pony-trekking and water ski-ing facilities available nearby, also tennis, bowls, golf and boating. Public Highland games take place on certain dates. Storage heaters and coal fires in lounge and dining room, and TV in dining room for when meals finished.

Food is vegetarian wholefood, fresh fruit and salads served every day. As many vegetables as possible grown in the terraced garden of 1½ acres. Vegans, and those on special diets, can be catered for if notice given, and special children's teas; again, give warning first. Take your dog if it's good, but feed it yourself. Guests asked not to smoke in house.

Rates £66.50 per week, including VAT, plus 5 per cent service. Children welcome. Unlicensed.

Crieff

Roundelwood, Drummond Terrace, Crieff, Perthshire, Scotland (0764 3806). Open all year round. Railway station Gleneagles (patients can be met). Accommodation 21 single rooms, 9 double rooms with bathroom or shower, 2 suites with bathroom/shower.

Founded in 1977, Roundelwood nestles in side of hill overlooking picturesque town of Crieff. Beautiful stone building in Scottish baronial style, to which a modern wing has been added. Original house has gracious public rooms and most bedrooms have magnificent views over Highland foothills. Health Centre offers various programmes based on individual requirements and facilities include gymnasium, sauna bath, hydrotherapy pool, whirlpool tank and impulse shower. Treatments available include physiotherapy, hydrotherapy, massage, weight reduction, smoking cessation and stress control. Trained nursing staff and a qualified physiotherapist on premises. Near by are several golf courses, tennis courts and bowling greens with water ski-ing and sailing only a few miles away on Loch Earn. Facilities also available locally for hack-riding.

Rates £150 per week plus treatment charges (minimum of three-day stay recommended). Children over five years welcome. Restaurant open to public, average price £4. Unlicensed.

Powys

Llanymawddwy

The Rectory Guesthouse, Llanymawddwy, near Machynlleth, Powys, Wales (065 04 375). Open Easter–November. Railway station Machynlleth (visitors can be met). Accommodation 1 single, 5 twin-bedded rooms Vegetarian guesthouse in area designated by Snowdonia National Trust as of outstanding natural beauty. Good centre for walking, visiting mountains and waterfalls, birdwatching, sailing on Bala Lake or investigating the Centre of Alternative Technology at Machynlleth. Indoors, emphasis on homely accommodation, with guests and owner, Mrs Etheridge, sitting around coal fire after dinner chatting. No television.

Wholefood served at all times, with salads and home-made bread a speciality, and pure spring water always available. Tariff covers bed, breakfast, packed lunch and three-course evening meal with coffee or herb tea.

Rates £55 per week, including VAT. Restaurant open to public if space – rare. Children welcome, but only two at a time. Unlicensed.

Strathclyde

Arran

Kincardine Lodge, Lochranza, Arran, Strathclyde, Scotland (077 083 267). Open April–October. Railway station Ardrossan (steamer to Brodick, taxis available). Accommodation 1 single, 6 double rooms

Small hotel, set in coastal village on Isle of Arran, which specializes in 'activity' holidays, each organized by qualified leader. Available activities are painting, photography, spinning and dyeing, weaving and yoga. These 'holidays with a purpose' are particularly popular with those going away alone. You can stay without taking part in organized events and make use of nearby facilities for sea and river fishing, sailing, cycling, tennis, golf, pony trekking and hill walking. For evenings there is lounge but no TV (on an activity holiday, evenings taken up with discussion groups).

Mountains, glens and sea coves of Arran are beautiful – sub-tropical plants flourish in predominantly mild weather – and you can observe wild red deer, many varieties of sea birds, hawks and owls, and also grey seals.

Charges cover bed, breakfast and evening meal, with packed lunch optional extra. Wholefood served where possible and vegetarian meals readily available.

Rates ordinary, £55 per week, plus VAT; activity, £65 per week, plus VAT. Children welcome. Restaurant open to public. Unlicensed.

Iona

Argyll Hotel, Isle of Iona, Argyll, Strathclyde, Scotland (06817 334). Open Easter–mid-October. Railway station Oban (ferry to Iona). Accommodation 10 single, 3 double, 5 twin-bedded rooms

Small, vegetarian hotel on beautiful island of Iona. Main building built as an Inn in 1850, with small rooms overlooking sea, but more recent extensions provided large dining room and extra single rooms with basins. Island offers sandy beaches, golf, fishing and historic Iona Abbey.

Isle of Mull easily reached by ferry, and also mainland Scotland by longer trip. Two lounges, sun lounge and large dining room – no TV. Aim to provide informal atmosphere in which guests can be sociable or in retreat as they choose.

Wholefoods served wherever possible, vegetables cultivated organic-. ally in garden, tomatoes in greenhouse, all baking done on premises. Vegan, macrobiotic and other diets available on request.

Rates £47–£73 per week, plus VAT and 10 per cent service. Restaurant open to public if space, average price £2.33 lunch, £4.40 dinner. Children welcome. Unlicensed.

Suffolk

Coddenham
Shrubland Health Clinic, Shrubland Park, Coddenham, near Ipswich, Suffolk (0473 830404). Open all year round except last two weeks December. Railway station Ipswich (taxis available). Accommodation 22 single, 11 double rooms, 1 Russian lodge
Improve your health in the stately home of Lord and Lady de Saumarez. Large Georgian house in 200 acres wooded parkland, and classical gardens laid out by Sir Charles Barry. Elegantly furnished and completely modernized, with fully equipped treatment rooms. Has dining room, light diet room, library, orangery, billiard room and gymnasium, heated swimming pool covered in winter, and hairdressing and beauty salon. TV and telephone in each bedroom. Nearest town Ipswich but Lavenham, Kersey and Constable country all within easy reach. Nearby facilities for riding and playing golf.

Vegetarian food reform, and specific diets and treatments prescribed individually. No cooked meals – salads, raw vegetables, soups, cheese, yoghurt and wholewheat bread. Garden produce organically grown in grounds, bread and yoghurt freshly made on premises. Available treatments include manipulation, sauna, steam cabinets, massage, underwater massage, Kneipp water therapy, herbal baths, peat baths, colonic irrigation and specialized physiotherapy – all prescribed by resident doctor. Exercise classes and beauty treatments provided for those who want them.

Rates £105–£176 per week, plus VAT, plus £8 initial consultation fee. No children. Unlicensed.

Surrey

Camberley
Tekels Park Guest House, Camberley, Surrey (0276 23159). Open all year round. Railway station Camberley (from which guests can be met). Accommodation 15 single, 8 double rooms

Secluded estate of fifty acres in Surrey, owned by Theosophical Society. London thirty-five miles and a fifteen-minute walk from Camberley. Estate is a wildlife sanctuary with badgers, foxes, squirrels and many kinds of woodland bird. Good setting for leisurely walks and bird-watching. Has drawing room with TV, large sun room, silent room, and a small lecture room (seating 50–60 people) makes it suitable for small conferences.

Entirely vegetarian catering using wholefoods as much as possible – vegan and other special diets can be served by prior arrangement. Discussion, meditation and healing sessions sometimes held and intention to provide a harmonious atmosphere where such activities flourish. Day visitors welcome but asked to book meals in advance if possible.

Rates £60 per week, including VAT and 10 per cent service. Children welcome. Dining room open to visitors by arrangement. Unlicensed.

Godalming
Enton Hall Health Clinic, Enton Hall, near Godalming, Surrey (042 879 2233). Open all year round, except Christmas. Railway station Milford (from which patients can be met). Accommodation 51 single, 17 double rooms, 1 suite in main building, annexe and chalets

One of the natural health centres which aims not only to cure patients but to teach programmes of diet and relaxation to keep them healthy. Treatment, public rooms and some bedrooms are in main house, other accommodation in attractive chalets in grounds. Recently they have leased off their home farm, but still have fifty acres lawns, parkland and woodland, and extensive kitchen gardens where vegetables organically grown for the table. Pleasant area for walking, in and beyond grounds, and West Surrey Golf Course adjoins. Chauffeur-driven cars available for further exploration – you might visit Petworth House, Loseley House (which supplies Enton Hall with dairy produce), the arboretum at Godalming, Wisley Gardens or the Roman villa at Fishbourne. Large dining room, lounge, drawing room, sun lounge and room for indoor games where smoking allowed. TV and telephone in each bedroom.

Individually prescribed diets entirely wholefood, vegetarian if wanted or deemed suitable. As in all the clinics, charges cover food and all necessary treatments – differences in price relate only to size or location of bedroom. Charge of £6 for initial consultation. Outpatients accepted – inquire for rates. Treatments include osteopathy, physiotherapy, massage, irrigation, sauna, blanket baths, slendertone and steam cabinet. Yoga and exercise classes arranged. For inpatients, a stay of at least ten days recommended.

Rates £191.26–£323.20, including 10 per cent service and VAT. Children over fourteen welcome. Meals can be served to guests of patients, at £1.50 for salad lunch, £2.00 for dinner. Unlicensed.

Hindhead
Grayshott Hall Health Centre, Grayshott, Hindhead, Surrey (042 873 4331). Open all year round. Railway station Haslemere (taxis; or guests can be met). Accommodation 51 single, 31 double rooms, 1 suite

Late nineteenth-century main building with modern wings, in forty-seven acres landscaped grounds backing on to 700 acres National Trust land. Grounds provide all-weather swimming pool, croquet lawn, golf course, driving range and hard tennis courts. Indoors are billiard room, private cinema, gymnasium, hairdressing salon, beauty parlour and boutique. Colour TVs can be rented for rooms. This is a luxury health centre, popular with sporting and showbusiness people. Offers a holiday as well as good food and treatments necessary for cures and reconditioning.

Diet varies according to needs of guests – usually begins with a fast and goes on to more varied food, always wholefood, often vegetarian, much grown in grounds. All baking and yoghurt-making done on premises. Alcohol frowned on, though guests have been known to sneak out to local pubs. Treatments include sauna, massage, underwater massage, electrotherapy, physiotherapy, hydrotherapy, relaxation therapy, meditation, yoga and a solarium.

Rates from £154 per week, plus VAT and 10 per cent service. Meals available to public by arrangement, for £3.50. No children. Unlicensed.

Sussex

Brighton

Metropole Health Club, Brighton Metropole Hotel, Kings Road, Brighton, Sussex (0273 775432). Open all year round. Railway station Brighton (taxis; or guests can be met). Accommodation 65 single, 192 double rooms, 16 suites

Brighton Metropole is an enormous Victorian hotel, popular for holidays but also known as a conference and exhibition centre – with all the cocktail lounges, dining rooms and butteries that suggests. Also within Hotel is Health Club where guests or day visitors can take advantage of massages, underwater massage, steam baths, sauna, a solarium, a gymnasium and wide range of beauty treatments. Within Health Club, special diets can be arranged – slimming, vegetarian, vegan, etc. General diet is food reform, and stay of at least three days recommended to get full benefit of treatments.

Rates full week at Metropole costs about £175, but rates for brief stay or day visit at Health Club vary, so inquire. Children welcome. Licensed.

Yorkshire

Ilkley

Craig End Lodge Vegetarian Guest House, Cowpastures Road, Ilkley, Yorkshire (0943 4876). Open Easter–November, and at Christmas. Railway station Ilkley (taxis available). Accommodation 3 single, 4 double, 2 twin-bedded rooms

Craig End Lodge is 100-years-old stone house, standing in $\frac{3}{4}$ acre own grounds, surrounded by trees. It is on Ilkley Moor, near famous 'cow and calf' rocks, and moor offers woods, water, rocks to climb and easy moorland paths. Garden arranged for croquet and putting. Golf, tennis, riding and swimming available near by. TV in dining room, for between meals.

Food entirely vegetarian (vegan if necessary) wholefood – mostly raw – and proprietor, Mrs Enid Hunter, does own baking and is known for her good vegetarian cuisine.

Rates full board from £70.32 per week, including VAT. Bed and breakfast only, from £40. Restaurant open to non-residents, at £1.95 lunch, £3.95 dinner, but book in advance. Children welcome. Unlicensed.

Otley

Chevin Hall Health and Beauty Hotel, West Chevin Road, Otley, Yorkshire (0943 462526). Open all year round except 2 – 3 weeks at Christmas. Railway station Leeds, then Ilkley train to Guiseley Station, where you can be met. Accommodation 6 single, 5 double rooms, 6 suites

Large, stone hotel set in five acres of private grounds, with views over Wharfe valley. Specializes in residential and non-residential courses on health and beauty – usually running from Monday to Friday of any week. More popular with those temporarily jaded by civilization, and with showbusiness people, than with people with actual ailments. TV in all bedrooms and two lounges. Outside sauna, swimming pool, tennis court, gardens and its own kitchen garden. Good centre for exploring Yorkshire Dales – if you feel you can spare the time from treatments.

Food mainly vegetarian wholefood, and all guests are on reducing diets – mostly of fresh fruit and vegetables. Available treatments include Roman spa bath, sauna, Turkish baths, Swedish massage, traxator, G5, slendertone, vibratory massage couches, pedicure, manicure, sunray, facial massages and instruction in make-up. Cathiodermies, depilation and skin peeling are extra. Aim is to reduce weight, improve posture soften skin, brighten eyes and eradicate tension lines from face – and to provide a relaxed and friendly atmosphere in which all this can happen.

Rates £165–£180, including VAT. No children. Unlicensed.

Contributors

Dana Balfour is a cookery writer, with a special interest in wholefood and natural remedies.

Michael Balfour has been the editor of *The Health Food Guide* since it was first published in 1970. He is a publisher (Garnstone Press/Geoffrey Bles) as well as a writer on diverse subjects.

Joanne Bower, Honorary Secretary of the Farm and Food Society (see Part Three) and freelance journalist, has carried out research into the effects of factory farming and written articles on antibiotics in animal husbandry for the *Ecologist*.

Margaret Yourdi Brady, MSc is a lecturer on food and health, a member of the council of the Soil Association, author of articles in *Health for All* and *Here's Health*, and author of two books on babies and children, and a film, *Our Daily Bread*.

Harvey Day is a well-known writer on health subjects with a long list of published works to his credit, including several of the books in Thorsons' *About* series – *About Molasses*, *About Honey*, etc.

Lilian Donat, member of the Institute of Journalists and student of nutrition, is founder and director of the London School of Yoga, and contributor of articles to various magazines.

Brian Furner, horticultural journalist and author of four books on organic gardening, is a member of the Organic Gardening Society of Great Britain, Fellow of the National Vegetable Society, Fellow of the Royal Horticultural Society, and gardening correspondent to the Soil Association's journal.

Doris Grant is the author of *Your Daily Bread, Your Bread and Your Life*, and other books, a contributor to *Here's Health*, *Prevention* and *Grace*, and author of articles.

Ruth Harrison is an environmental journalist, author of *Animal Machines* and *Animal Farming*, and contributor to *Can Britain Survive?*, *Animals, Men and Morals*, the *Ecologist*, *Your Environment*, *Punch*, etc.

Lawrence D. Hills is director of the Henry Doubleday Research Association, author of six Comfrey reports, *Russian Comfrey* published by Faber & Faber, and contributor to *Here's Health* and *Health for All*; he has done all the research on Comfrey carried out in the last thirty years.

Arthur Hollins runs Fordhall Organic Farm in Shropshire and has been producing the well-known Fordhall cheeses and live yoghurts for more than thirty years.

Olivia Holt NDD was production supervisor of the dairy products for Loseley Park Farms, which are available in many health stores.

John D. Hyde MNIMH is a qualified consultant medical herbalist and a director of the National Institute of Medical Herbalists.

Leslie Kenton was health and beauty editor of *Harper's & Queen* for many years.

Jay Landesman, author, freelance journalist and publisher, has contributed articles on macrobiotic eating to the *Sunday Times*, *Oz*, *East West Journal* and *Seed, Journal of Organic Living*.

Dr Barbara Latto is secretary of the McCarrison Society, and has contributed articles to the Soil Association journal, vegetarian magazines and continental health publications.

Joan Beatric Lay ND MBNOA MCSP is an osteopath and naturopath, and has contributed articles on cookery and treatments to *Health for All*.

Leah Leneman, a long-time vegan, is a postgraduate student in history at Edinburgh University.

Claire Loewenfeld holds the Dr Bircher Benner Diploma and is an expert on the growth and use of herbs, contributor to many health magazines and author of books on herbs, nuts and fungi.

Muriel Mackay FIAP FSTA is a naturopathic consultant and director of the Health and Beauty Clinics in the Midlands and Edinburgh, whose lectures have been published by the Soil Association and the British Naturopathic Association.

Sam Mayall is an organic farmer and flour miller whose products are available in many health food shops; contributor to the Soil Association's journal.

Keith Michell is an internationally renowned actor and director.

Dr Reginald F. Milton BSc PhD FRIC FI Biol. is a consulting biochemist, scientific editor of *Health for All* and consultant to the Soil Association.

Miriam Polunin is editor of *Here's Health* and author of several books, the most recent of which is *Minerals – What They Are And Why We Need Them*, and many articles.

James Pym was a regular contributor to the *British Vegetarian* and *Science of Thought Review*.

Ruby Rae has worked on various health food magazines, and is an assistant editor of *The Health Food Guide*.

Gregory Sams runs Harmony Foods (see Part Three) and has been a contributor to *Seed, Journal of Organic Living* as well as to *IT, Oz, Galdalf's Garden* and others.

Jack Sanderson is former editor, and now editorial adviser, of the *Vegan*, the journal of the Vegan Society, and deputy president of the Vegan Society.

John Robert Philpott Soper CBE is retired from the Colonial Agricultural Service, and was honorary secretary of the Bio-Dynamic Agricultural Association (see Part Three) and contributor to its journal, *Star and Furrow*.